CAPTURED BY ALIENS

The Search for Life and Truth

in a Very Large Universe

Joel Achenbach

Simon & Schuster

SIMON & SCHUSTER
Rockefeller Center
1230 Avenue of the Americas
New York, NY 10020

Designed by Ruth Lee
Manufactured in the United States of America

10 9 8 7 6 5 4 3 2 1

Library of Congress Cataloging-in-Publication Data
Achenbach, Joel.
Captured by aliens: the search for life and truth in a very large universe / Joel Achenbach.
p.   cm.
Includes bibliographical references and index.
1. Life on other planets.   I. Title.
QB54.A23   1999
576.8'39—dc21   99-37592   CIP
ISBN 0-684-84856-2

Portions of this work previously appeared, in different form, in *The Washington Post*.

To a great teacher, Chris Morris

(P. K. Yonge Laboratory School, Gainesville, Fla.)

CONTENTS

CAPTURED
BY ALIENS

PART ONE

ADVENTURES
IN THE
UNKNOWN

The astronomers took rooms on the beach at Guajataca. They had enough time for a swim before dinner. The two men were in the prime of their lives, popular, accomplished, and as the waters of the Atlantic enveloped them they would have been forgiven a moment of unrepressed joy. In a matter of hours they were going to conduct an experiment unlike any other in the history of science: They would point a gigantic radio telescope at the Andromeda Galaxy, at the Great Nebula in Triangulum, and at two other nearby galaxies, and listen for signals from alien civilizations.

Frank Drake and Carl Sagan were colleagues at Cornell University. Drake had achieved fame among the select group of astronomers who studied the universe in the radio portion of the spectrum. Sagan was rapidly becoming known on a grander scale, his reputation spreading as a visionary, a poet of science, a quote machine for reporters on deadline. He had been a frequent guest on *The Tonight Show*, serving as Johnny Carson's house astronomer. He'd recently authored a groovy book called *The Cosmic Connection*, a meditation on Mars and Venus, the distant stars, the galaxies, the intelligent civilizations yet to be discovered. He was flush with ambitions and dreams and plans, his mind awhirl with amazing ideas and books that he must someday write.

I

Dialing Up Andromeda

Puerto Rico, 1975

Though roughly the same age, they were dissimilar in style and speech, Drake being matter-of-fact and Sagan loquacious. Drake guessed that in our own galaxy, the Milky Way, there are ten thousand civilizations that are potentially detectable through radio astronomy. Sagan made Drake look like a pessimist. He calculated that there are a million civilizations awaiting discovery. A million planets with intelligent, communicative organisms—a galaxy rioting with consciousness! It was a dazzling thing to contemplate.

The night before the experiment, they grabbed what sleep they could. A grumbling employee woke them, as instructed, at 4 a.m., and without breakfast or coffee they began the long drive on dark, curving roads into the interior of the island of Puerto Rico. Drake was the wheelman, alert for stray animals in the road. This could be a paradigm-shattering day, and Drake didn't want to blow it by crashing into a goat or veering off a cliff. The villages were still asleep, even the chickens. Bullfrogs sat on the roadside like fallen boulders. The road went up and down over rugged terrain known as karst. The land is uplifted limestone, an ancient seabed turned into haystack hills by the collision of the Caribbean and North American tectonic plates. There are no valleys to speak of. The geology was ultimately the reason the astronomers had to come here. In 1959, the Air Force and Cornell had looked for a place to build a radio telescope, a reflector dish a thousand feet across. Officially it would study the moon and the planets, and would need to be near the equator, so that these objects would be almost directly overhead. But it would also serve a military purpose, to scan the upper atmosphere for disturbances caused by nuclear tests. To build something so large, the engineers needed to use a natural bowl, a depression, and in this part of northern Puerto Rico there were sinkholes amid the limestone hills. A hole near the town of Arecibo offered the perfect site. But getting there, even on a paved road, was not easy in 1975. It was only ten miles as the crow flies from the hotel on the beach, but it was twenty miles by road and took more than an hour of white-knuckle driving.

In the passenger seat, Sagan ate scraps from the previous night's dinner. No one needed to tell him how important this day might be. He had spent his life thinking and lecturing about extraterrestrial life. So much had been speculation and guesswork. Sagan was desperate for data. Maybe this would be the day.

Theirs would not be the typical scientific experiment. A little secret of the research process is that rarely does anyone really "discover" something. A discovery in the classical sense is a bit like a triple play in baseball, a chance event that has little to do with the way the game is normally played. Science is usually carried out in such a way that discoveries are collaborative and gradual; it's an incremental process. The idea that new research proves that old research is wrong is a popular fiction. Most scientists conduct research just beyond the margin of what is already known, asking narrowly defined, prosciutto-thin questions that only a handful of colleagues—a tiny fraternity of hunched, squinting specialists—would consider interesting. These are people who spend a lot of time in front of computer screens. They scribble equations on chalkboards. They gather data. They parse and tweeze and calibrate. They make things precipitate in solutions. Their parents and siblings and children have the vaguest notion of what they do and why they do it, since the pay is not great and the only glamor comes from seeing one's name in publications like *Geophysical Research Letters*. Few journal articles address large questions, because the Great Unknowns are composed of countless smaller ones, each of which, in turn, branches into unknowns that are tinier still. At the terminus of the branching lie infinitesimal mysteries, the capillaries of the unknown, and through those passages squeeze the lives of most of the men and women who are loyal to the scientific method.

Sagan and Drake were not incrementalists. They weren't trying to wriggle down some microtubule of the unknown. They were going for the big kill, the eye-popping discovery, the globe-shattering revelation. Good Lord, they were going to find aliens! A signal from a totally unknown race of advanced biological entities! This was hardly crossing the "t" on some other fellow's work! And the most amazing thing was, they might make their discovery in a matter of hours. They might point, listen, find. It was quite possible that they might discover an extragalactic civilization before lunchtime.

The Arecibo experiment was surrounded by theoretical uncertainty. No one on Earth knew if there were intelligent beings, or any life at all, beyond the confines of our planet. This was true in 1975, and it remains true today. All assertions about extraterrestrial life continue to be based on thin evidence, or, as is more often the case, mere "plausibility arguments," as Sagan put it. Nonetheless, it is widely presumed by people in the scientific community that what has happened on Earth could happen elsewhere in the cosmos. Even if the rise of intelligent life on Earth was extremely improbable, the result of a series of flukes, there still might be many intelligent civilizations in outer space. The universe is ridiculously huge. There's a lot of turf out there where strange biological events can take place.

But there's a nagging doubt in all the theorizing. It's something called the Fermi Paradox. The Fermi Paradox is an important tool in thinking about extraterrestrial intelligence. The story goes that the scientist Enrico Fermi was having lunch one day in 1950 at the atomic laboratory at Los Alamos, and he and his colleagues began talking about extraterrestrial civilizations. What was said at the table isn't precisely known. The gist of the story is that Fermi ventured that, in less than a million years, human beings would figure out how to travel in space faster than the speed of light. This would make it possible, presumably, for humans to colonize the galaxy. But if humans could do it, then other civilizations, if they exist, could also do it. And so the galaxy ought to be already thoroughly colonized. The galaxy has been around for something like ten billion years, and the ancient civilizations, the Old Ones, should have finished the job by now.

Fermi posed a question to his colleagues: "Where are they?"

It was such a simple question, and yet it resonated. People repeated the question (or some variant—one version is, "Where is everybody?") as though it were a piece of poetry, a mantra, a piercing insight into the nature of reality so brilliant that only an atomic scientist could possibly have thought it up. Where are they? A cosmological conundrum lay folded up in ten-dimensional space inside those three little words. All that space out there and not one civilization has managed to develop interstellar travel? It made no sense! They should be here! Where in tarnation are those damn aliens?

In the popular literature and in movies the aliens know how to make an entrance. They apparently have fabulous maps to show them the location of the White House. In the 1953 Arthur C. Clarke novel *Childhood's End*, the aliens manage to arrive by page 6. A guy looks up at the sky and there they are.

For a moment that seemed to last forever, Reinhold watched, as all the world was watching, while the great ships descended in their overwhelming majesty—until at last he could hear the faint scream of their passage through the thin air of the stratosphere.

Boom, they're here. The aliens take over. That is simply how it's done, theoretically. So why doesn't it happen in real life? This is the Fermi Paradox. There are reports of flying saucers and alien abductions, but these ETs stubbornly elude scientific verification. The hard evidence is missing. We hear no alien radio signals, see no alien rocket-ship exhaust, detect no alien picnic trash.

This absence of extraterrestrials has dogged the dreamers. It has challenged them to think about how they could reconcile the absence of galactic starships with their continued belief that these smart creatures are out there somewhere. Perhaps the extraterrestrial civilizations don't survive long enough to become star trekkers. Perhaps interstellar travel is too physically and energetically and biologically difficult for even the most advanced civilizations. Perhaps the aliens are reticent, or withdrawn, or simply don't like to go out much. Philosophers grapple with the question of whether expansionism, exploration, and colonization are universal impulses or some kind of human quirk. Would aliens tend to be colonizers, or is that some kind of anomalous European/Polynesian personality flaw?

Drake had tried to find a way to quantify the probable abundance of alien civilizations. He was the perfect partner for Sagan. Whereas Sagan was voluble and theatrical, Drake had a straight-ahead, low-key quality that seemed to ground the enterprise in common sense. Drake had a round face, eyeglasses, and white hair combed straight back. His hair had turned white when he was twenty-five, a family trait, and one that didn't bother him in the slightest. It meant that the graying process had no significance, that it did not carry with it, as it did for most people, the intimation of mortality. He looked like a tinkerer, someone who had strange gadgets in his garage and couldn't wait for the next issue of *Popular Mechanics*. His name would be forever linked to the search for extraterrestrial life, because he was the Drake of the increasingly influential Drake Equation.

The Drake Equation, like the Fermi Paradox, is one of the conceptual tools that everyone uses to pry open the mystery of aliens. The idea originated as an outline for a meeting in Green Bank, West Virginia, in 1961. A bunch of astronomers and visionaries, including the young Sagan, wanted to have a serious, rigorous conference about extraterrestrial intelligence for the first time in human history, so far as anyone knew. Drake needed a way to organize the discussion. He decided that the number of extraterrestrial civilizations who are beaming radio signals at us could be expressed as the product of a series of fractions. He jotted down some symbols. What he wrote gradually evolved into the famous equation:

$$N = R^* f_p n_e f_l f_i f_c L$$

This meant that the number of radio civilizations in a galaxy at any given moment depends on, in order, the number of stars (often expressed as the rate of star formation), the fraction of stars with planets, the fraction of planets with habitable environments, the fraction of those planets that actually have life, the fraction with intelligent life, the fraction of cultures engaged in interstellar communication, and, finally, the average lifetime of communicative civilizations. Only the first factor was well understood in 1961, and by 1975, when Drake and Sagan went to Puerto Rico, there had been little progress on the rest of the equation. (It would be another two decades before astronomers had solid evidence of any planets outside our solar system.) Different people could plausibly plug in different numbers, which is why Drake came up with the figure of ten thousand civilizations and Sagan managed a million. In Sagan's universe there was simply nothing that unusual about an advanced civilization. It was what was to be expected. A planet is a potential abode of life; life becomes complex; complex organisms become astronomers. This was one of the essential ideas of what might be called Saganism. He estimated that a new civilization comes into existence in the galaxy every ten years. He was an optimist.

Listening for signals from the Andromeda Galaxy was a big-picture experiment even within the context of the already ambitious science of SETI (the Search for Extraterrestrial Intelligence). Sagan and Drake decided that the most easily detected civilizations might not be the closest but the brightest. There might be a few supercivilizations that were capable of exploiting all the energy of a star, not just the tiny fraction that happened to be intercepted by some solar cells on the surface of an orbiting planet. These civilizations would be so flush with power they could easily afford to send radio beacons across intergalactic space. Therefore, Sagan and Drake decided to look at entire galaxies—conglomerations of billions of stars—and see if they could pick up evidence of one of these powerful, gregarious societies. If the astronomers did find such a signal, one thing was certain: The source would be a civilization far more advanced than our own. We would be humbled as never before. Any lingering sense of being special in the universe, of being uniquely blessed, would be eradicated in a single irreversible moment.

Sagan and Drake arrived at Arecibo at 6 a.m. The radio telescope was straight out of a science-fiction novel, a twenty-fourth-century contraption plunked down in the Puerto Rican jungle by the bold visionaries of the Space Age. How could they not be thrilled? This was the farthest edge of science, the knife blade slicing into the fabric of space-time. The dish at Arecibo, Drake had calculated, could hold 357 million boxes of cornflakes. It reputedly could hold the equivalent of the world's annual beer consumption. The transmitting and receiving equipment dangled hundreds of feet overhead, suspended on cables slung from three concrete towers on the periphery of the dish. Each tower was taller than the Washington Monument. (What it really looked like was a James Bond set—as it would become years later in the otherwise forgettable *Goldeneye*.)

Sagan was beside himself with anticipation. He was too good a scientist to get ahead of the facts, but he had thought about extraterrestrial life for so many years, run through all the logic of the Drake Equation, and felt that it simply ought to work. They ought to be out there. It made perfect sense.

They decided not to search Andromeda first, because it was the biggest of the targets and would have required a great deal of maneuvering of the telescope. It stretches three degrees across the sky, six times the width of a full moon. (In the desert or on a mountaintop the galaxy is visible to the naked eye, though it usually requires a slightly averted gaze—faint objects are easier to see when you don't look directly at them. Humans had seen it since the dawn of time and hadn't known until the early twentieth century what they were looking at.) Instead of Andromeda, the astronomers picked M33, the Great Nebula in Triangulum, which is another nearby spiral galaxy but not as large, and more easily examined with the telescope at Arecibo. The telescope had new equipment that enabled it to scan 1,008 frequencies simultaneously. There was not yet a computerized method for determining what was noise and what was an intelligent signal, so Drake and Sagan would have to decide that themselves as they stared at a computer screen. The screen would show dots.

The data came streaming in.

Nothing at first, just random dots.

Still nothing.

They looked at the screen. Minutes passed. All they saw was noise. In the world of alien-hunting there is always much discussion of the signal-to-noise ratio, and in this case it was infinitesimal. No signal, all noise.

About an hour into the experiment, Drake noticed that Sagan was slumping in his chair.

After two hours, he slumped lower.

Drake began to feel sympathy for his partner. Sagan hadn't done much of this SETI work before. He had written about it and talked about it on television. Sagan didn't know that the general rule of any SETI experiment—and indeed a rule that dominates the study of extraterrestrial life—is that at the end of the day you have to deal with the fact that you haven't found any actual aliens.

In a couple of days Sagan was staring out the windows, unable to keep his eyes on the screen that stubbornly remained an alien-free zone. Finally, after five days, they had no choice but to give up. The experiment was over. They'd found nothing. Sagan went into something of a funk. For a week he was depressed and confused. He couldn't understand what had happened. Did they have the wrong frequencies? How could it be?

Seattle, 1996

"Okay, it's very far away," Sagan said. "So you have to have a very fancy civilization. But in a hundred billion stars there's not one civilization? I can't imagine."

Twenty-one years after he had gone to Arecibo, he was sitting in the living room of a rented house near Lake Washington. He'd come to Seattle for another round of chemotherapy in his long battle with myelodysplasia, a rare blood disease. He no longer resembled the famous Carl Sagan, the Carson-show guest, the star of *Cosmos*. He was bald, bony, washed out from chemotherapy. Like everyone else, I had seen him countless times on TV, and he'd always had a certain rock-star/professor look, a quirky handsomeness, with flyaway hair and usually some kind of leather jacket, or (if he was strolling a seaside cliff) a windbreaker. It was a shock to see him so ravaged by disease, but the disorientation lasted only a few minutes as his voice took command of the situation. He still enunciated with gusto, still put italics on his favorite words. And his mind, I quickly realized, had not been diminished by the poisons in his blood.

He had the disease whipped, he said. The most recent tests showed him clean of any cancerlike cells.

"No myelodysplasia. No anomalous cells. Nothing," he said.

Sagan talked about his body in dispassionate, clinical terms. Of myelodysplasia he said, "There is some faint evidence that it is due to benzene and other aromatic hydrocarbons, but that's merely faint."

We talked about space, alien abductions, superstition, the short-comings of science education, and the failure of SETI programs to detect intelligent life in the universe. Only later would I realize how patient Sagan was, answering questions he'd been asked by reporters and students and ordinary citizens countless times. He offered some possible explanations for the silence of the stars.

"One possibility is they're as little interested in communicating with us as we are with them. Another possibility is that no one has the adequate technology to broadcast over intergalactic distances. Especially if you're broadcasting in all directions to all the nearby galaxies. It does take huge amounts of power. . . . Another possibility is that a different technology is used to communicate, because they don't want to get us at this early stage. They only want to communicate with a civilization that has managed to avoid self-destruction by being around long enough to invent Zeta waves. I don't know what Zeta waves are, but they're much better than radio. . . . Then there's the Zoo Hypothesis, and that is a variant of the prime directive. [On *Star Trek: The Next Generation* there is a rule that the Federation starships and their crews not interfere with young, developing civilizations.] That is, that they're here. And they're studying us. But they have such enormous technological capabilities that we don't know that they're here."

He had discussed all these ideas before, and they all still sounded vaguely desperate.

"How could they be studying us if we don't see them?" I asked. "I'm not an advanced enough life form to tell you."

I wrote up a profile of Sagan for *The Washington Post*, but couldn't shake the topic. I kept thinking about Andromeda, that immense whirlpool of silent worlds. I kept thinking about Drake and his lovely little equation full of unknowns. I kept thinking of Fermi's three-word question, and about the tales of UFOs that would serve as a potential answer. I kept thinking about the Space Age, this theme that dominated popular culture from the time I was a kid. I kept thinking about aliens—friendly, evil, slimy, humanoid, reptilian, insectile, all manner of creatures spawned by the gifted brains of writers and filmmakers and this new crowd of experiencers, channelers, and abductees. I kept thinking about space—how strange it is that we know so much about the structure of the universe, about the galaxies and stars and planets, about interstellar dust clouds, comets, asteroids, about neutron stars and pulsars and quasars, and yet know nothing about what lives out there. The situation is virtually intolerable.

That day with Sagan, I started on a long journey through alien country. What follows is my report, my travelogue, and the reader is warned that the alien question takes the investigator down some strange avenues. I have notebooks about the flying-saucer crash in Roswell, New Mexico, stacked on top of notebooks about the origin of life and the rise of photosynthesis. I have been hypnotized in an attempt to recover any possible repressed memories of an alien abduction; I watched the founding of the Mars Society, which is bent on colonizing the Red Planet; I met the people who would design spaceships to take us to Alpha Centauri; I tromped through the jungle in Central America looking for debris from the mountain-sized object that came from space and wiped out the dinosaurs. A book about aliens became a book about life, intelligence, humanity, spirituality, the scientific method, the Enlightenment, postmodernism, truth, fiction, love and death, the fate of the Earth, and the destiny of our species. It expanded a bit.

This is what happens with all alien books, particularly the ones about direct alien encounters. They suffer from a runaway broadening effect, until finally they become the Most Important Book of All Time. In some spectacular, hyperventilating fashion, the writer grows as a person. I have made a yeoman attempt to avoid any such personal growth. There are no last-minute conversions, no one is saved. The only thing that happens—not to give away the plot—is that in the end everyone feels lucky to be alive.

Throughout this enterprise Sagan has served as my spiritual guide, because I knew that wherever I went, and whatever I investigated, he had been there already. He tamped down a lot of brush over the years, hacked through many a thicket. I kept seeing his boot prints, fibers from his clothing. I heard echoes of his polysyllabicized exhortations. When you're lost in a topic as vast and nutty as this one, it's nice to know that someone has been lost in the same place before. Sagan knew that to understand the alien question you had to understand a million different things, from the nature of life itself to the contingencies of evolution to the prevalence of planets capable of maintaining water in its liquid state. If nothing else, I can say I came to a new appreciation of water. It's the damnedest stuff, when you think about it.

What I didn't know as I spoke with Sagan was that there were big things about to happen in the field of extraterrestrial life. A hot topic was about to become scorching. There were discoveries and revelations only a few months away. There were paradigms poised for a seismic shift. Saganism was on the rise even as the man himself was fading.

Sagan felt certain that we had brethren among the stars. It stood to reason. It simply made perfect sense. And yet the aliens had not yet revealed themselves. At the age of sixty-one, sick, dying, Sagan still couldn't answer Fermi's question.

Where are they?

2

The Goldin Age

A few weeks after I talked to Sagan, I managed to get an appointment to see Dan Goldin, the administrator of the National Aeronautics and Space Administration. Sagan might offer guidance to the scientific questions about extraterrestrials, but I needed someone who would talk about our own future as a space-faring species, someone who knew programmatically what we are doing to meet the aliens halfway, as it were. Unfortunately, shockingly, I could not find the NASA headquarters. The NASA headquarters spans a city block a short distance southwest of the Capitol, and yet somehow it eluded me. I mention this only to make the point that I approached NASA with an open mind and few preconceptions, even in the matter of location.

Eventually the interview was rescheduled and I found the place and soon was on the ninth floor, in a large corner office with a view of the freeway. The thing you notice in Dan Goldin's office is Dan Goldin—he is an occupying force, a man who fills a room, an executive with expensive cowboy boots peeking from under his trouser cuffs. He is a self-made man from the Bronx, probably the only guy in the building who went to City College of New York, surely the only senior official with a mere bachelor's degree. What he lacks in lofty academic credentials he makes up for with chutzpah, volume, raw smarts, and pure animal drive. ("As you can see, I'm very intense," he told me one day after his guard had been lowered.)

The office is filled with models of spaceships and space planes and lots of framed pictures of stars and planets. He gave me a fifty-cent tour, and pointed to a little rocket-ship model on his coffee table, the X-30. "I canceled it," he said. "That was a dream, and was un-achievable. They wanted to go from Mach 0 to Mach 25."

He had two sides when it came to fantasies and visions. On the one hand, he relentlessly slashed budgets and declared that missions must be "faster, better, cheaper." Of those, cheaper may well have been the ultimate criterion. But even as he played the pragmatist, Goldin wanted desperately to get people thinking, to inspire them, to force them to imagine a different sort of space agency, a gee-whiz place where incredibly cool stuff could happen—like in the old days. Goldin had never given up the dream of the Space Age, the ultra-futuristic scenario in which every American citizen has a jumpsuit, a rocket car in the driveway, and two tickets on a starship to Alpha Cen-tauri. I got the impression of a man who is impatient with our progress in escaping the bonds of Earth. I heard him say things like, "In the twenty-first century we will be citizens of the solar system." And: "We're in the business of making dreams come true." If he were to lose his job at NASA, he could always go to Disney and become the head greeter at Tomorrowland.

"The future will remember us for reaching out with the human mind and spirit to touch the universe," Goldin told me in a hushed voice. After a while with Goldin, you simply get used to the fact that he sounds like Captain Kirk, minus the clenched fist and pajama top. It's easy to imagine him navigating his way through the galaxy, liberat-ing noble races from the iron grip of antidemocratic madmen.

Staring at a picture taken by the Hubble Space Telescope, he said, "The human life is more than survival. You need food, you need shelter, but you also need intellectual nourishment."

Goldin is someone who doesn't speak so much as expound, de-clare, and intone. He has a way of inspiring himself. He'll start talking and then, listening to his own words, get even more excited. A biolo-gist might describe him as "autocatalytic." Big ideas will launch him out of his chair. He'll begin pacing, gesturing—the listener gets the sense that Goldin is repressing an urge to shout. At one point during our conversation he leaped into the air and touched the ceiling, which was his way of demonstrating the concept of reaching for big answers. Even sitting down, he translates everything into a frenetic sign language. He splays his fingers, rolls his fists like someone doing the dough-rolling portion of the patty-cake rhyme, and palms the air like a mime trapped behind glass.

My goal that first day was to ask Goldin about the great unanswered questions in science. He couldn't have been more thrilled at the topic. The Unknown puts him into a rapture. He is something of a would-be Sagan. The world was going to need someone to take up Sagan's slack someday (tragically soon, as it turned out), and Goldin seemed to be a candidate. He's a big-picture guy.

"I stand humbled about what we don't know," he said.

He mentioned galaxies. "We make observations and we have no idea how they form. We're still having a major debate about how old is the universe. We still don't know! I mean, these are questions of *unbelievable magnitude*. . . . Is there a terrestrial-sized planet around another star? Around any star?"

He started prowling his office, as though somewhere in the room we might find the answers.

"So—another question, antimatter. We have detected antimatter up to the small-particle level. But does helium antimatter exist? Does hydrogen antimatter exist? Are there antimatter galaxies? I don't know!"

These were interesting things to think about. Could there be intelligent beings made of antimatter? Imagine if a human being tried to shake hands with an antimatter creature. Blammo! Total annihilation.

"We ought to be humble. We ought to be *humble*. I mean, as I grow older, and as I learn more, I become more overwhelmed by what we don't know."

Sagan might not have said it quite that way. Sagan might have lingered a bit on what we do know, on the tremendous achievements of science in a sometimes unappreciative and hostile world run by stupid politicians and doctrinaire theologians. He would make the point that the human species has amassed a fantastic and honorable store of knowledge about the nature of reality both large and small, that we know the structure of the cosmos and the organs of a living cell and the constituents of an atom. No longer do we need to be ruled by superstition and shamanism, Sagan would say. There is always this tension in talking about the enterprise of knowledge: Do we feel smart yet? Do we know a lot or a little? Is our species notable for its grand scientific achievements or for its continued backwardness compared with what might be the cosmic norm for intelligent civilizations?

"I'll give you another one," Goldin said. "Is life unique to Planet Earth?"

Life had become Goldin's obsession. Just months earlier, he had told a group of scientists that he wanted to search for life on Mars. He had said, "Maybe we will learn that the same building blocks of life washed over both planets simultaneously. We could find fossils of cells with elements of proteins similar to those here on Earth. We might find the one fossilized cell that is the missing link between the planets. We might find actual life—imagine that!"

Anything would do, really, as long as it was an example of an extraterrestrial organism. The one thing missing in the science of extraterrestrial life is a sample of the stuff. A scientifically verified sample of an alien organism would make the study of extraterrestrial life seem more like a real science, and less like a flight of fancy. It could be anything. A slug. A mite. Some green, viscous slime. Replicating goop. Anything. We need a sample! Until then, we have a problem.

Because of the sample problem, no one at the end of the century can honestly say whether Earthlife is a freak event—a miracle, even—or merely a naturally occurring product of ordinary chemistry. A comparison sample would suddenly show us the natural order (or strangeness) within our terrestrial biochemistry. No one knows if the history of life on Earth, the patterns of evolutionary change, follow some general process that is uniform across the cosmos. Is photosynthesis a common adaptation on inhabited worlds throughout the universe? Are alien plants also green? Does it invariably take roughly 3.5 billion years for life to emerge from its single-celled rut and turn into a swimming creature with eyeballs? How inevitable is the emergence of large animals with complex central nervous systems and a tendency to build telescopes and nuclear warheads?

A few decades ago, the study of life beyond Earth was usually called "exobiology." But there weren't many full-time exobiologists.

Critics derided the field as the "science without a subject." The optimists, such as Sagan, felt that experiment and observation would boost exobiology to a new level of credibility, and indeed there were moments when the field seemed to be strengthening—sudden spasms of interest culminating in international conferences on life in the universe. But then nothing would happen. That is the only constant in the field: Nothing happens. Or something positive will happen, only to be followed by a crushing negative. Sometimes people think they have finally made the breakthrough, that they've detected alien biology, but always the doubters have muscled their way into the debate, raised alternate theories, debunked key assertions, and when everyone has had his or her turn to talk, the very thing that had seemed for a moment so wonderful and clear and obvious has become just another fuzzed truth—another revolutionary discovery mired, submerged, and permanently bogged in fuzz.

You could imagine the impatience of the old-time exobiologists. How much longer would it be? What would it take? Occasionally the only recourse was to change the name of the field and see if that helped. "Exobiology" started to feel stale in the 1980s, and gave way to "bioastronomy," the term favored at the triannual meeting of experts on the subject. A decade later, with "bioastronomy" having failed to deliver any extraterrestrial life, NASA began to favor a new word, "astrobiology." There are some subtle differences in what the words mean ("astrobiology" implies a greater emphasis on the structures of the universe, such as galaxies and stars and planets, upon which life might evolve), but basically this was the same product in a new package. This was changing dice in a game of craps after throwing too many snake-eyes.

Goldin's astrobiological interests had become turbo-charged a few months earlier, in the fall of 1995, when scientists in Switzerland and the United States announced the discovery of seven planets outside our own solar system. This was a bombshell in the astronomical community. There had been, over the years, a few claims to the discovery of "extrasolar" planets, but the planets had a way of evaporating under closer scrutiny. This batch seemed different. The discovery had the beautiful twin virtues of dogged work and sudden inspiration. For years, the Swiss and American astronomers had been looking at stars for hints of wobbling caused by the gravitational tug of orbiting planets. The painstaking search hadn't produced anything. These astronomers were risking their careers. Searching for planets, something unseen and fundamentally speculative, was right at the borderline of credible science, a bit like the radio search for intelligent civilizations. We were searching for ourselves once again: Our knowledge of planets was based on a single example of a planetary system, one in which planets were divided fairly neatly into two categories, the inner rocky planets and the outer gas giants. The astronomers found no sign of anything, year after year. Then the Swiss team had a fabulous thought: What if a gas giant were in an exceedingly tight orbit, whipping around the star not in a matter of years but in just a few days? With this more liberal vision of how a planetary system might be constructed, the Swiss team quickly found a planet orbiting the star 51 Pegasus. The American team of Geoff Marcy and Paul Butler confirmed the discovery and then examined their data for similar planets. They found six more. The planets had been right there all along, hidden in the numbers.

The discovery of planets did not equal a finding of extraterrestrial life. But this was progress. A major factor in the Drake Equation might be solved in the foreseeable future. We were getting somewhere, finally.

Soon after that came what might have been an even more important finding. Pictures of the surface of Europa, one of Jupiter's moons, bolstered the theory that Europa has a subsurface ocean. The icy surface was smooth, with hardly any craters. One crater was surrounded by a smooth plain, as though liquid water had gushed forth from the impact. There were what appeared to be icebergs, great rafts of ice groaning through a frozen sea. Long cracks in the ice fit with the theory that there was liquid below. And life? Yes, there is probably life there, one enthusiastic researcher said at a NASA press conference. His unguarded quote raced across the news wires.

And something even bigger was in the offing. As I met with Goldin, scientists at NASA's Johnson Space Center in Houston were looking at a meteorite from Mars—a rock that contained, they thought, tiny fossils of ancient Martian life. A sample at last! Their discovery, when announced a few months later, would be sensational, and controversial. It would put the issue of extraterrestrial life right into the scientific mainstream. People would be running around saying they knew it all along—that life is a "cosmic imperative."

No wonder Goldin seemed so stoked about life beyond Earth

and all those other imponderables. In his fervor to get NASA in the mystery-busting game, he had made a master list of Big Questions. He showed me his early draft. His questions were something of a mission statement for NASA's efforts in space science. He prefaced the questions with an introductory paragraph, written in Goldinspeak.

> We look as far as the aided eye can see and reach as far as the translated hand can extend, to open the air and space frontiers. We boldly explore the unknown to make discoveries that inspire the soul, provide intellectual nourishment for the mind and enrich our lives. We seek an understanding of our origins and the laws of nature. We develop tools and knowledge to help preserve our freedoms and provide hope and opportunity to our children.

This was *Star Trek* talk. Captain Kirk was vowing to boldly go where no man has gone before.

The list was printed out in big, bold letters, some of them uppercase, clearly a work in progress:

1. How did galaxies, stars, and solar systems, and planetary bodies of all kinds form and evolve?
2. Is life of any form, however lowly or complex, carbon-based or other, unique to planet Earth?
3. By looking out at other planetary bodies and at Earth as a planet, can we develop predictive environmental, climate, natural-disaster, resource-identification, and resource-management models to help insure sustainable development and a high quality of life?

Needless to say, there would have to be some editing for style. But the gist was there: NASA was going to explore the fundamental mystery of life in space. The approach would be the opposite of that used by Sagan and Drake at Arecibo in 1975. Rather than start at the top, with a search for the most advanced civilizations in our part of the universe, Goldin wanted to start at the bottom, with the search for a possible habitable environment. He wanted to do what you might call "bottom-up astrobiology." NASA, he decided, would look for evidence of life in a steady, incremental, potentially boring fashion, starting with the search for planets, water, organic chemistry, and various "biomarkers" that might suggest the presence of living things. The NASA scientists wouldn't search for aliens, just life itself—bacteria, viruses, the simplest and least prepossessing organisms.

Goldin wasn't against the SETI idea, exactly. He would have supported the program if it were politically feasible (a certain Nevada senator, Richard Bryan, had killed the agency's most recent attempt to find alien radio signals). But he also thought SETI had become a cult of sorts. It was too focused on radio. The people who believed in radio as the medium for finding aliens tended to be such true-believers they didn't even like the idea of searching in the infrared. SETI assumed that intelligent beings would transmit in the radio portion of the spectrum, because radio waves penetrate the atmosphere of planets and can pass through regions in space of thick dust and gas. But what if the aliens preferred to use strobe lights? What if they had invented some new form of electromagnetism (Sagan's Zeta waves, maybe)? What if they had cable? A SETI search was too restrictive, Goldin felt. A radio search would never find life of an advanced, intelligent, but non-radio-transmitting nature—meaning a SETI search would never find dolphins or elephants or even Victorian novelists. Jane Austen didn't transmit interstellar radio signals and hence would not meet the SETI standard for "intelligence." Picasso? Didn't send radio signals into space, hence was not an intelligent being. Shakespeare? He didn't even know how to make a phone call! Not intelligent, the Bard.

Goldin believed it made more sense to keep one's options open, to entertain every possibility, to continue to imagine a universe full of surprises. Indeed, one day some months after that initial interview, Goldin told me he thought that intelligent civilizations in our galaxy might choose to explore our world with tiny robotic craft that we do not even notice.

"You could have nanoprobes," Goldin said.

We looked at each other silently.

Then I confessed that I didn't understand.

Goldin explained that it would be efficient for aliens to send miniaturized probes across interstellar space, probes that might be no larger than a single cell. This would solve a lot of the problems with interstellar travel. If the craft is tiny enough, you don't need much energy to send it across trillions of miles of space. The alien probes could be the size of proteins, Goldin said.

I reflexively looked around the room, at the tabletop, and scanned my clothing. We could be covered with the stuff—the invisible spunk of distant aliens.

It was a crazy idea, but at least it had the sanction of the NASA administrator, the most powerful person in the human quest to understand and explore the universe. So it couldn't be too crazy.

Einstein said the most incomprehensible thing about the universe is that it is comprehensible. This is the boast of a scientist. We can know things. We can apply our big primate brains to complex problems and gradually construct a plausible and satisfying model of reality. Perhaps in prehistoric times we built models of necessity, as tools for guessing the location of game to be hunted, fruits to be gathered. Today we dare to model the entirety of nature, from galaxies to atoms.

And yet there is one model that has been difficult to construct—a good, solid model of the biology of the cosmos. We don't know what lives in the universe beyond our own planet. It's not merely that we have an incomplete grasp of the situation. We don't know anything at all. You can take everything conclusively known about extraterrestrial life and fit it on the period at the end of this sentence (with room left over for about seventeen angels).

Nonetheless, the model-building spirit is resilient, and so people press ahead with scenarios and stories about extraterrestrial life. They invent myths. In the last half-century, many inquiring, curious, searching individuals have put together a powerful narrative of alien invasion. Millions of people are now immersed in the UFO "enigma," to use a favorite word.

3

The Biggest Unknown

Even if they're not sure what's happening, they're convinced that . . . well, that *something* is going on. (Cue the spooky music.) The belief system is constructed of countless individual pieces of evidence, most of it flimsy, but some of it not readily explained away (emphasis on the *readily*). There are hundreds of thousands of documented sightings. There are trained pilots who have seen what looked like metallic craft come to a dead stop in midair and then speed away at impossible velocities. Mysterious blips showed up on radar screens and appeared to buzz the White House and the Capitol one night in 1952. Three silver-skinned creatures with pointed ears and crab-claw hands abducted two men from a fishing pier in Pascagoula, Mississippi, in 1973. The Space Shuttle astronauts saw a UFO that narrowly avoided being blasted by a laser beam from a secret U.S. government Star Wars facility in Australia. The aliens had the nerve to abduct the United Nations secretary general in 1989—an incident that, unsurprisingly, he claims never happened.

The Book of Ezekiel in the Bible has a dead-on account of a flying saucer landing and an alien stepping out. Jimmy Carter saw a UFO in 1969. He insisted, when he became president, that his aides open up the files on UFOs to see what was there. Bill Clinton, soon after winning the 1992 election, told his Arkansas crony Webster Hubbell to find out two things when he went to work at the Justice Department: "One, Who killed JFK? And two, Are there UFOs?"

These spaceships, or whatever they are, have lurked on the cultural fringe for the entire second half of the twentieth century. Surveys have generally shown that a sizable fraction of the public believes that aliens have come to Earth in flying saucers (it seems to depend on how the question is framed—one survey may say 22 percent believe in UFOs, another will say 47 percent and another 57 percent). For many years the UFO phenomenon went in waves, spurred by flurries of sightings (for example, in 1947, 1952, and 1973), but by the 1990s it had stabilized somewhat, becoming less susceptible to spikes in public attention. Other millennial fears loomed, like asteroid impacts and Y2K computer meltdowns, but the aliens persevered in a competitive market. On February 17, 1999, NBC televised a two-hour "special-event" documentary titled *Confirmation: The Hard Evidence of Aliens Among Us?* The question mark laboring at the end of the title could not alter the fact that the program was a ufological orgy, narrated in

Gothic style by the Dracula-like actor Robert Davi, and coproduced by Whitley Strieber, America's most popular alien-abduction victim. As ever, the Hollywood-produced footage of flying saucers (not always clearly labeled) was far more impressive than any "real" footage. A thoughtful viewer would surely come away with the conclusion that, although some UFO cases are clearly hoaxes, others are more convincing, and so alien visitors, the abductions, and the massive government cover-up are all quite possible. These are respectable beliefs. This wasn't some cable program, this was on the very mainstream NBC. Thus a person could entertain the idea of an alien invasion and still be a member of polite, rational society.

Even someone desperate to be skeptical may find it hard to imagine that all these cases are confabulations and misinterpretations and fantasies. Can a cultural belief system this powerful and durable be constructed entirely from vapor? People may say to themselves: Isn't it possible that the Earth has been just a *little bit invaded?*

The finest minds in the UFO world are coming to the conclusion that the aliens are involved in some kind of elaborate breeding program. They want our genetic material. This is what might be called the Central Irony of the UFO world. The belief in aliens is, at first glance, a firm embrace of the Copernican Principle. Humans are not the center of the universe. There are other intelligences. They are, indeed, smarter and more advanced. And yet these other intelligences are obsessed with us. They come across mind-boggling reaches of space to meet us, experiment with us, mate with us. We have such enchanting DNA, they just can't stay away. Ufology, for all its generosity in filling the universe with life, nonetheless has a distinctly anthropocentric flavor.

To the dismay of UFO researchers, the mainstream scientific community has remained completely and adamantly opposed to the notion that aliens are visiting our planet. Even worse, the capitalist system has turned the whole thing into a commercial gimmick. The ufologists want to prove that there's a cosmic Watergate going on even as we speak—the Ur-conspiracy—and yet there are all these aliens on backpacks, lunch boxes, and T-shirts. There are aliens bobbing from rearview mirrors and staring off bumper stickers. There are alien keychains, alien snow domes, alien tattoos. Human civilization is apparently more adept at cashing in on conspiracies than at solving them. For an easy-to-fix lunch you can try Chef Boyardee Flying Saucers & Aliens beef ravioli and pasta, in which three different aliens will zoom through the intergalactic tomato sauce aboard their disk-shaped noodles. (I dare say ordinary SpaghettiO's were once considered sufficiently thrilling.) Aliens have become an all-purpose multimedia revenue generator.

In the wee hours of the morning, a man living in a trailer 60 miles from Las Vegas, Art Bell, speaks into a microphone and through the magic of radio brings the disparate elements of the alien presence and the government cover-up to an estimated eight million worldwide listeners. Meanwhile, Hollywood has made the alien story a staple narrative in the same way that the Western used to be. In fact, the aliens are so overexposed that the movie studios have resorted to parodies and send-ups, like *Mars Attacks!* and *Men in Black*. Just in case there were any children left who hadn't been exposed to aliens, the Public Broadcasting Service recently started airing the British-imported program *Teletubbies*, featuring cuddly aliens who coo and gurgle and have televisions in their bellies. Aliens now inhabit every environmental niche on Earth; perhaps their true biological significance is not their intelligence but their hyperadaptability. In the pre-teen movie *Spice World*, the Spice Girls get off their bus to go pee in the woods, only to meet four aliens (with, as one Spice Girl describes it later, "really squidgy faces") who ask for autographs and tickets to their next concert. It is hard to think of another commercial gimmick that has been so rapidly disseminated through the American culture. The aliens can be marketed to all audiences, perhaps because they lack any specific sex or race or age. They're one step beyond Michael Jackson.

So deeply has The Alien penetrated popular culture that we need only the slightest visual clue to identify a character as being from a different world. In the 1960s, it took antennae, as in *My Favorite Martian*. Now it's the bald oversized head with big almond-shaped black eyes, skinny limbs, and hands with four, or sometimes three, fingers. The alien in the notorious Roswell "autopsy" documentary had six fingers, which many ufologists believed was proof that it was a hoax—because their research showed that the real Roswell aliens had only four.

• • •

Most observers make a sharp distinction between legitimate exobiology and "kooky" ufology, but I think of them both as part of a fabric of curiosity. Everyone has different methods; some methods will hold up better over time. The common thread is a sincere desire to understand the universe, to find truth and meaning in a time when we are overwhelmed with astronomical data. The UFO story is really what you'd call a heresy, a heresy of modern astronomy.

The official story of the night sky is that it is full of stars that may have planets, and planets that may have life. People nudge that story a bit further. They tweak it. They populate the officially sanctioned structures of astronomy with creatures of their own devising. So, too, do they assume that some subset of mysterious aerial phenomena is the result of an alien presence. They see things they cannot explain, that seem unfamiliar, "not of this Earth." They see so many things that the Air Force spent twenty-two years investigating the phenomenon, notably through Project Blue Book, before finally giving up in 1969.

Some of us are crazier than others, but we're all searchers. We're all scientists of a sort. Even if you don't buy into the invasion scenario, it's hard to go through life without pausing to wonder what's out there. On a dark, moonless night, the stars explode upon our consciousness, challenging us to figure out the significance of all that brilliance. We find ourselves asking big questions. Why does the universe exist? What's the point of it? Why are we here at all? It's almost as though we don't really know who we are in the immensity of space.

The Copernican Principle holds that there is nothing special about our place in the universe, that we are an ordinary world in orbit around an ordinary star in an ordinary galaxy that is but one of perhaps fifty billion galaxies. As the "known universe" has grown in size during the twentieth century—its expanse recalibrated as new and better telescopes reveal ever-more-distant structures—we have fully absorbed the notion that the Earth is but a speck, a smudge, an insignificant granule in the brain-hammering enormousness of space.

We're nothing.

We're the crud in the grouting of the bathroom tiles in the cosmic palace.

This is clearly an existential crisis.

To deal with this crisis we develop stories. We are storytellers by nature. It may be a side effect of some other evolutionary adaptation. Perhaps in the long years of perfecting our ability to stalk prey, or find the ripest fruit, or communicate to our brethren the dangers of the nearby pack of wild beasts, we developed more narrative skill than we know what to do with. The stories we tell the most passionately are those that deal with our relationship to the gods, to the bringers of life, to the entities that have powers beyond the mortal sphere. Those stories have been forever changed by modern science. The gods still live, but they must exist in a larger and more crowded universe. Science impinges upon our mythmaking. Astronomy in particular has altered the narrative, or at least its scale. No longer must we account merely for the beings that live in the sky or under the Earth or in a heavenly sphere that surrounds our planet. We have to account for all the gods and spirits and entities that exist in a universe vast beyond imagination. We are told that there may be other universes outside our own. We have to wrestle with cosmologies that blow our minds.

There are people who are at peace with our demotion from the center of the universe. They feel that it will be character-building, that an awareness of our true place in the universe will counterattack some of the worst vanities of our kind. Anthropocentrism has had nasty applications in everyday life, from the destruction of the environment to the exploitation of animals to the repression of peoples who had not yet accepted the religious doctrines of those who felt their centrality most keenly. Anthropocentrism is a kind of cosmic bigotry. It is now far more fashionable to embrace an extreme counter-anthropocentrism, in which humans are not terribly special in any way at all—neither in the cosmos nor on Earth.

If people lack much specific knowledge about life beyond Earth, they can at least claim to have the correct attitude about it. It is expected of intelligent people that they accept the abundance of alien life, at least in theory. There has arisen what might be called the water-cooler paradigm of life in the universe. People standing by the water cooler will suddenly find themselves yapping about aliens in space—maybe they just saw the "alien-autopsy" video on TV again—and they'll invariably offer up some version of the following words: "You know, it would be incredibly arrogant of us to think that we're the only intelligent beings in the universe."

And they are right when they say this. They have astronomy on

small, specific, and virtually incomprehensible. In fact, each of the first four questions involves something small. The universe, the subject of the first question, started out small. It started out so small it had no physical dimensions whatsoever. At the earliest moment in its history that can be described in physical dimensions, it was still far smaller than the head of a pin. How it sparked into existence is a matter limited to theory rather than observation. One idea is that it began as a quantum fluctuation of energy in a vacuum of nonzero density. (Yes, *one of those.*)

The fundament of matter and energy? Could be "strings," little trembling loops that vibrate in ten dimensions of space-time. Space itself is created by these strings. Unfortunately, string theory is baffling to all but a few geniuses. For the moment the world of physics relies on the so-called Standard Model for an explanation of the subatomic world. There are flaws with the model. No one understands gravity, at least not at the quantum level, the territory of quarks and leptons and all those exotic critters produced in collisions in particle accelerators. The Standard Model lacks simplicity. Physicists would like to junk it, and come up with something more aesthetically pleasing. Not long ago, Japanese researchers working in a zinc mine three thousand feet below the surface determined that the enigmatic particles called neutrinos have mass—a devastating blow to the solar plexus of the Standard Model. Neutrinos weren't supposed to have mass.

Consciousness? Possibly the most elusive question of them all. There is a general agreement that consciousness isn't a single operating element of the brain but, rather, an "emergent" property, the product of multiple mental processes. Some scientists would say that the brain is fundamentally a complex machine, and that in principle there is no reason a computer could not someday be designed to think for itself. Others, the "mysterians," believe that consciousness cannot be reduced to a wiring diagram. Again, the riddle plays out on a microscopic level, requiring an understanding of tiny cells working in strange and perplexing patterns.

Now you see the magic of question five. Aliens are large. Aliens (as we imagine them) exist at the same scale as humans, roughly speaking. They are dynamic. Aliens do things. Aliens pilot starships and cruise across the galaxy and invade other planets. Aliens don't dither around, they have an agenda. No one has ever met a lazy alien.

They're so enterprising! (This is why they're so otherworldly—they seem never to need to take a nap, drink a beer, or go fishing.) Aliens can whip one of their tentacles around your neck and hurl you across the room and into a vat of acid faster than you can say "chronosynclastic infundibulum." Aliens lay fat, oozy, pulsing, glow-in-the-dark eggs containing their repulsive larval offspring. What I'm trying to say is that, even though aliens are completely strange, we can still relate to them.

And aliens might have answers. If the aliens were to send us a message, or land on the White House lawn (crushing Sam Donaldson in the middle of his nightly stand-up), they might well pony up the answers to some of the other big unknowns. They might simply announce that life on Earth began with the arrival of a radiation-resistant virus sent by the Vegans.

But even aliens wouldn't know everything. Presumably they wouldn't be able to see the future (let's assume that even advanced interstellar travelers can't indulge in time travel), and the future, as it turns out, is the most intriguing unknown of them all. One reason the concept of making contact with aliens is so dazzling to the imagination is that it is a narrative that shows where humans are going. It's reassuring in its eradication of uncertainties. Contact with our space brothers would affirm our sense of being involved in a progressive development, an emergence from terrestrial barbarism, on our way toward cosmic citizenship. We want something like that because we know there are more awful alternatives. We know that as we use the tools of science to discover the secrets of nature, we are also misusing that knowledge to design weapons of mass destruction and other technologies that ravage the Earth. There has been, in this era of thrilling science, a steady drumbeat of doom. I suppose I should add another question to my list:

6. What is going to happen to us?

Despite their reputation, aliens are not new, contemporary, or "futuristic." They are not a Space Age production. They are a dusty old idea, moldy, desiccated—a doodad from the attic of civilization. As prodigiously documented in such books as Steven J. Dick's *The Biological Universe* and Karl Guthke's *The Last Frontier*, the extraterrestrial-life debate goes back to antiquity, which means that it probably goes back even further than that. The terms were different in ancient times, but it's clear that as soon as philosophers began debating the place of man in the universe they seized upon the possibility of life elsewhere. How could an intelligent person avoid the concept? You merely had to look at the sun and the moon and the planets and the stars to sense that our world was not the totality of the Creation.

For the better part of two thousand years the alien question was phrased in terms of the "plurality of worlds." No one talked of "aliens," per se. It was assumed that a "world" would be one with a diversity of life, including intelligent life, but penciling in the details of those intelligent beings didn't seem so great a compulsion. The ETs were forever separated from us by great gulfs of space, even if the true extent of those distances was not yet known. Ancient people had no ex-

4

Aliens in the Archives

ample of artificial flight, and they did not have the same predilection for seeing spaceships and flying saucers that would arise centuries later. Only after we had invented the technology of flight would people start seeing what they initially called Mystery Airships.

The early thinkers on extraterrestrial life did the very same thing that the most recent, current, technology-wielding astrobiologists do: They guessed. Epicurus (341–270 B.C.) wrote a letter to Herodotus asserting that there were "infinite worlds both like and unlike this world of ours" and that "we must believe that in all worlds there are living creatures and plants and other things we see in this world. . . ." Others followed the Epicurean logic. The philosopher Metrodorus of Chios produced a book, *On Nature*, that asserted, "It is unnatural in a large field to have only one shaft of wheat, and in the infinite Universe only one living world."

Aristotle didn't believe in aliens or the plurality of worlds. The concept violated his complex doctrine of "natural place," in which water and earth had to move downward—only fire and air could go up. If there were other worlds, water and earth moving downward upon them would be moving upward *relative to our own world*, and to Aristotle this made no sense. We are all trying to make sense of this thing, including those of us who, like Aristotle, don't know what we're talking about.

Because Aristotle was the most influential by far of ancient philosophers, his beliefs put a damper on the alien idea for centuries to come. Eventually the medieval Christian scholars who resurrected the writings of antiquity began freshly debating the notion of extraterrestrial life. Nicholas of Cusa (1401–64) suggested that life "in a higher form" existed in the heavens. "It may be conjectured that in the area of the sun there exist solar beings, bright and enlightened denizens, and by nature more spiritual than such as may inhabit the moon—who are possibly lunatics—whilst those on earth are more gross and material." The new model of the solar system advanced by Copernicus in 1543 not only destroyed the geocentric universe but implicitly called into question man's centrality in the Creation. Christian theologians began a long encounter with the doctrinal complexities implicit in a plurality of worlds. Would Christ's martyrdom on Earth redeem the souls of inhabitants on other planets? Or would He have to do it all over again in all those other places? Wouldn't that be a hell of a lot harder than just saving all the decent souls on Earth? (A contemporary answer, if one might summarize, is that salvation on a cosmic scale is not so terribly intimidating when you happen to be the Supreme Being of the Universe.)

Into this fray stepped Giordano Bruno. He was not, as is commonly supposed, burned at the stake for endorsing the plurality of worlds. Bruno was burned because he was a heretic who denied Christ's divinity. One is tempted to say that, if the church elders were going to burn someone, Bruno wasn't a bad choice. Bruno does, however, get credit for establishing the water-cooler paradigm fully four centuries before the debut of *The X-Files*. His book *On the Infinite Universe and Worlds* appeared in 1584 and declared what many people believe today: God's vast Creation makes sense only if there are countless planets in the universe that have intelligent life. Bruno prefigured the modern debate, too, in his ability to develop a passionate belief system without any actual proof. For Bruno, inhabited worlds were aesthetically pleasing, correctly reflecting God's omnipotence. His greatest achievement may simply have been his willingness to die for his beliefs—he supposedly refused to recant, and thrust away a crucifix even as flames were licking up from his feet. He was no Einstein, but he was all man.

The data about other worlds did begin to trickle in shortly after Bruno's death. In 1609, Galileo, with his telescope, discovered the mountains of the moon, the four largest satellites of Jupiter (including Europa, the darling of 1990s exobiology), and untold numbers of stars previously unseen by human eyes. He was the first to realize that the light of the Milky Way is radiated by individual faint stars. Johannes Kepler, the astronomer who put together the first accurate picture of how the planets orbit the sun, realized that the plurality of worlds called into question the centrality of human beings in the grand order of the universe. If there were other Earths, he said, "How can we be masters of God's handiwork?"

Bernard le Bovier de Fontenelle's *Conversations on a Plurality of Worlds*, published in 1686, was an enormous best-seller, going through thirty-three editions in France. He said there were surely aliens out there, on the moon, the other planets, everywhere among the stars. At about the same time, the astronomer Christiaan Huygens argued that the Copernican model of the universe forced us to admit that the Earth was not special. Huygens noted that he and his fellow astronomers talked often about the alien question when looking together through a telescope. "But we were always apt to conclude, that 'twas in vain to enquire after what Nature had been pleased to do there, seeing there was no likelihood of ever coming to an end of the Enquiry." More than three hundred years later, people still haven't come to the end of the inquiry, and what we know about aliens today is essentially what Christiaan Huygens knew. All we have are what Huygens called "probable Conjectures," which, for him, included the belief that aliens must have hands.

By the eighteenth century, many credible thinkers believed that the other planets in the solar system were inhabited. Immanuel Kant wrote in 1755 that the material that makes up life is lighter and finer the farther one goes from the Sun. "The excellence of the intelligent creatures, the speed of their thought . . . become more perfect and complete the farther their habitation is from the sun." In 1836, Thomas Dick, an English writer, calculated the number of intelligent beings on the various worlds of the solar system, using as his standard the 280 people per square mile in England. Thus, he estimated that Jupiter had 6,967,520,000,000 inhabitants. The total for the solar system came to more than twenty-one trillion sentient beings. But that didn't include the solarians, the inhabitants of the Sun. Dick wrote, "It would be presumptuous in man to affirm that the creator has not placed innumerable orders of sentient and intelligent beings . . . throughout the expansive regions of the sun."

Presumptuousness—bad then, bad now.

Improvements in the telescope made it possible in the 1800s to see the moon and Mars much more clearly. That opened the way for Percival Lowell. Lowell is a role model for all those who wish to misapprehend nature in the service of wishful thinking. Long before people misunderstood what was going on in Roswell, New Mexico, Lowell towered over the landscape, misunderstanding what was happening on Mars. He had a science-and-mathematics education from Harvard, but as an astronomer he was an amateur, professing professionalism. He had a preternatural gift for discerning through a telescope the existence of canals on the surface of Mars. The Italian astronomer Giovanni Schiaparelli in 1877 had found some lines on Mars that he called *canali*, "channels," which turned into "canals" by the time it reached America. A few British and French astronomers managed to see canals as well, and then William Pickering of Harvard found some in 1892, as well as forty lakes and many high clouds. Other astronomers couldn't see any of this cool stuff—dooming their names to obscurity for lack of vision. (Decades later, during the height of the "flying-saucer" era, there would again be a division between those who saw mere fuzzy blotches and globular lights and those who could discern, amid these shapes, the clear evidence of extraterrestrial spacecraft.)

Schiaparelli continued to see the channels and did not rule out the possibility that they were the product of intelligence. Conceivably, he said, they were natural features, akin to the English Channel. The great leap forward came when Lowell teamed up with Pickering and, in 1894, built a fabulous observatory near Flagstaff, Arizona. Over the years, Lowell saw four hundred canals with an average length of fifteen hundred miles. He announced that they were clearly artificial. He drew detailed maps, named the canals, and deduced an entire narrative of what was going on. He said an ancient Martian race was fighting to survive on a dying planet with a dwindling water supply. When Lowell opened his observatory he said his goal was to investigate life on other worlds. "There is strong reason to believe that we are on the eve of a pretty definitive discovery in the matter," he said. The great searchers are always right on the verge.

Lowell understood that the citizens of the 1890s viewed themselves as scientifically literate, and he wanted to assure them that his theories were not tainted by speculation or some other mush-brained trait. In *Mars as the Abode of Life*, he wrote, "In our exposition of what we have gleaned about Mars, we have been careful to indulge in no speculation. The laws of physics and the present knowledge of geology and biology, affected by what astronomy has to say of the former subject, have conducted us, starting from the observations, to the recognition of other intelligent life."

The aliens, in other words, weren't some flight of fancy. This was real science. Lowell claimed that four threads of evidence indicated the existence of Martians. First, there was nothing antagonistic to life in the broad physical conditions on the planet. Second, there's a shortage of water on the surface, so the inhabitants would have to use irrigation if they were to survive. Third, there are markings that fit the model of a global irrigation system. Fourth, there are spots that look as if they could be irrigated oases. He had it nailed four ways. He also threw in some other amazing observations. The surface of Mars has one-third the gravity of the surface of Earth, and Lowell felt this might have led to the evolution of Martians three times as large and twenty-seven times as strong as Earthlings. Moreover, they would have to be far more advanced intellectually, since the irrigation system is so elaborate and the planet itself is, he believed, much older. "A mind of no mean order would seem to have presided over the system we see—a mind certainly of considerably more comprehensiveness than that which presides over the various departments of our own public works."

Compared with the Martians, we were morons.

Soon after Lowell began announcing his wondrous discoveries, H. G. Wells introduced the world, in 1898, to an invading horde of Martians who used a death ray to scorch the English countryside before succumbing to common bacterial infection. The Wells novel was accompanied by a rash of sightings of mysterious airships, including one that supposedly crashed in Texas, leaving bodies on the ground, a nifty hoax that prefigured the Roswell case half a century later. The 1890s initiated a pattern of intellectual energy that still exists in the world of extraterrestrial life—science mixes with fringe science, fiction mixes with fact, and the public takes it from there.

As aliens became fashionable, there remained a few skeptics, notably Alfred Russel Wallace, the great evolutionary biologist who developed, with Darwin, the theory of evolution. Wallace wrote an influential book, *Man's Place in the Universe* (1903), that argued that the Earth was almost certainly the only inhabited world in the universe. He knew this because, he said, the Sun is providentially placed at the approximate center of the universe. This placement causes a constant and steady infall of matter, a gift of gravitational centrality. That matter, in turn, allows the Sun to burn for eons, and provide a steady source of heat and light to our planet. Only a few hundred stars near the center of the universe could also have such long lifetimes, he said. He argued that the Earth had numerous features that were amenable to life but would be unlikely to appear on other worlds near the cosmic center. Thus man was alone in the universe, as he should be. Wallace had preserved the central Biblical premise that the emergence of the human species is a central event—indeed, the purpose—of the Creation. If there were many other inhabited worlds it wouldn't seem quite right: "It would imply that man is an animal and nothing more, is of no importance in the universe, needed no great preparations for his advent, only, perhaps, a second-rate demon, and a third or fourth-rate earth."

Unfortunately for Wallace, he had misplaced the solar system. The Sun is not at the center of the Milky Way galaxy (or the universe—the distinction at that time remained unclear). New research soon after Wallace's book appeared showed that our solar system is on a spiral arm of the galaxy, tens of thousands of light-years from the center. Nor did the Sun rely on the infall of matter to maintain its fires. Rather, the Sun relied on nuclear fusion. The Sun wasn't on fire at all! Fusion is a completely different and greatly more efficient method of generating heat than an ordinary fire. Wallace couldn't have known this, and remained, for his work on evolution, a great figure of science, but on the question of the plurality of worlds and the existence of extraterrestrial life he ensured he would go down in history as a prime example of Getting It All Wrong.

The Martian canals, meanwhile, became increasingly hard to detect, and eventually vanished. In 1910, for example, George Ellery Hale, using the new sixty-inch telescope on Mount Wilson in California, took a closer look at Mars and found no canals. A bigger, better tool had erased them from the Martian surface altogether.

Although the scientific orthodoxy in the first years of the twentieth century tended to be conservative, there were always a few prominent and well-credentialed mavericks able to promote more exciting scenarios that captured the public imagination. "Professor Pickering of Harvard Observatory has lately shown that the moon is probably not quite so lifeless as has been supposed," the young Robert Goddard said at his high-school graduation in 1904. "It is easily supposed that carbon dioxide gas, which supports plant life, occurs abundantly, as has already been suggested, near volcanic crevices. This being the case, we know that there is some possibility of plant life, for the temperature of the moon's surface during the lunar day is above zero, and below the boiling point of water. Moreover, Professor Pickering has found certain so-called 'variable spots.' Near sunset and sunrise, the lunar day being about as long as twenty-eight terrestrial days, the spots are almost invisible, but near the lunar noon they vary from gray to intense black. At that time, since shadows are geometrically impossible, there must be a real change in the nature of the reflecting surface. Organic life resembling vegetation seems to be an adequate explanation if we conceive the vegetation to flourish during the lunar day, and to wither and die as night approaches."

Pickering had it all wrong. But Goddard, possessed of this misinformation, was not deterred from his destiny, which was to become the singular pioneer of rocketry in the United States, a man who envisioned trips to the moon and actually helped make them happen. The NASA center in Greenbelt, Maryland, is now named after him. The point being that perhaps in the thrall of undocumented belief a person might be inspired to great intellectual achievement. Perhaps (and one just blurts this out as a thought experiment) it is good for people to believe in things that are not technically true. Some ideas are wrong but *useful*. Great civilizations are not constructed by the hopeless and the skeptical and the permanently unamused.

As the twentieth century progressed, it became increasingly apparent that the Earth was a world utterly unlike the other planets of the solar system. We were blue, wet, warm; our planet was an anomaly. On no other planet could a creature stand on the beach and stare at the lapping waves of the ocean. Liquid water on the surface of a planet is a marvel, a magic trick on a rock hurtling through a cold, pressureless void. And yet the aliens did not vanish from our minds. The aliens proliferated. Their salvation came through fiction. In 1926, Hugo Gernsback began publishing a magazine of fiction called *Amazing Stories*. So was born American science fiction, and in it the Alien would be a staple character. No longer an astronomical obsession, extraterrestrials became a literary device. Robert Heinlein, Ray Bradbury, Arthur C. Clarke, Frank Herbert, Philip K. Dick, Ursula Le Guin—they all investigated at great length the nature of alien life forms (Isaac Asimov spent more time on robots). Lowell's canals may have disappeared with the improvement of telescopes, but the Martians of literature were indestructible.

For many young boys there was a series of novels about Mars, written by Edgar Rice Burroughs, that were utterly addictive. The first, *A Princess of Mars*, told the tale of a man named John Carter, an American who simply wished himself to the Red Planet. He found himself among a race of green-skinned, four-armed Martians. Over the course of eleven books he fell in love, fought in wars, and buddied around with the aliens in the dying days of their great civilization. Testosterone spurted from the pages. This was Lowell's Mars, fictionalized by a master storyteller. One young lad who read the John Carter books, all eleven of them, and who would mention them the rest of his life, was a New York City science whiz named Carl Sagan.

5

The Gatekeeper

When Carl Sagan became a credentialed astronomer in the 1950s, he chose to study planets, a subject that had become rather tame and quaint. Planets, after all, were old news. They had been discovered by ordinary people, by shepherds, by people who noticed that some of the "stars" wandered strangely among the others as though following a private agenda. The ancients knew of five planets in the night sky, and astronomers added three more after the invention of the telescope, culminating in the 1930 discovery of the eccentric planet Pluto (which some astronomers now say is not a proper planet at all but, rather, a stray object from a distant zone of planetoids in the outer reaches of the solar system). By 1930, planetary science had already become something of a fringe area of study, in good measure because the serious scientists had been repulsed by the goofy notions of Lowell.

In science there must never be a question of the seriousness of one's work. Respectable research is grounded in the sober parsing of empirical evidence. For a planetary scientist there was always the danger that naïve acquaintances would ask questions about Martians, or perhaps about those very hyper creatures who live on the Sun. Planets were for pagans and wackos and astrologers. (Astrologers—the great embarrassment of the astronomical community. They had all been astrologers once, in ancient times, and they would never go back.) The no-nonsense academics preferred to study things like galaxies, quasars, the chemistry of stars, the interstellar dust, the electromagnetic radiation that permeates the cosmos, the origin of the universe, and other matters involving objects and structures that have the redeeming feature of being far away. As Sagan put it, "There was a kind of view that the seriousness of astronomy was proportional to the distance of the object."

Sagan also came of age at a time when extraterrestrial life was in some disrepute. In the 1950s, life seemed a much chancier phenomenon than it would in the late 1990s. No one yet knew that life had originated on Earth extremely early in the planet's history—which suggested that it wasn't a particularly miraculous event. No one knew that life could survive in boiling springs or in the pores of rocks many miles below the surface. No one knew that microbes could go into suspended animation for millions of years, and that some could actually survive a trip through empty space, possibly spreading the blessing of life from one planet to another. No one knew how resilient and pervasive and precocious life could be. As for intelligent life, that was simply beyond the pale. That was flying-saucer stuff. The preeminent biologists felt that intelligence on Earth was itself a fluke. Intelligence, they argued, was the unlikely result of numerous evolutionary accidents. Leading this line of argument was the biologist George Gaylord Simpson, who declared that the odds of intelligence existing elsewhere in the universe were minuscule.

Sagan never wavered. He was a believer in ETI—extraterrestrial intelligence—from the beginning. The physicist Freeman Dyson once asked him about his conviction that intelligent civilizations are common.

"Why are you so sure?" Dyson asked.

"I feel it in my bones," said Sagan.

Dyson would say later in an interview, "He was quite aware that it was a matter of faith and not of knowledge."

In 1957, Sagan published one of his first popular articles, in the *University of Chicago Magazine*, titled "Life on Other Planets?" He contended that the seasonal color variations on the surface of Mars coincided with the availability of water. Thus, he wrote, the variations probably signaled the presence of vegetation. "If the reported color changes are real, there seems to be no other reasonable interpretation," he wrote.

Sagan did not want to be the spiritual heir to Lowell. He wanted to be the guy who found real examples of extraterrestrial life, not phantom civilizations at the limit of telescopic observation. He labored on almost every NASA program to explore the planets with unmanned spacecraft. He had his fingers in real experiments, offered his insights to anyone with a chance of digging up an extraterrestrial microbe. He did his own experiments on the origin of life, cooking up strange molecules in his lab. Sagan knew a little bit about everything from particle physics to paleontology. He was like Lowell in that he had a fertile imagination, a gift for communication, and a willingness to engage the public at large. With his booming baritone voice, his explosive consonants—he could make the word "billions" sound like ordnance from a mortar—and his obvious delight in all things cosmic and intergalactic and fantastic, Sagan became a translator of science for the masses. He could make the equations for the fusion of hydrogen in the stellar interior sound like poetry. Students jammed his lectures, first at Harvard and then at Cornell, and the hall could never hold all those who wanted to hear the man who talked so excitedly of the vast . . . the imponderable . . . the unimaginable cosmos. Everyone knew the Sagan voice and tried to imitate those oral italics. One day the school paper at Cornell ran a parody of Sagan in which every other word was italicized or boldfaced.

Sagan was certain that intelligence is a cosmic phenomenon of which we are but the local example. This was not some casual, offhand, water-cooler paradigm. It was a carefully considered philosophy, what Sagan referred to as the Assumption (or "Principle") of Mediocrity. From the late 1950s to the mid-1990s, Sagan endeavored to convince his students, his colleagues, and his television viewers that they should abandon any sense that the universe was created for their glorification. We aren't superior beings, Sagan said. We're not even smart, compared with most technological civilizations. We might be the dumbest technological civilization in the entire galaxy. We're the "dummies," he wrote. He reasoned that since we had only recently achieved advanced technology, such as radio astronomy, it would be highly unlikely for us to make contact with a civilization that by pure chance had only recently mastered radio astronomy. On a cosmic scale, there would be no way to escape the truth of our dumbness. At a conference in Boston in 1972, Sagan said, "There is almost certainly no civilization in the galaxy dumber than us that we can talk to. We are the dumbest communicative civilization in the galaxy."

It was a kind of self-denial, perhaps. The truth was that Sagan himself was gifted with a startling intelligence, the kind of candle-power rarely seen. Many people would say he was the smartest person they'd ever met.

Over the course of forty years, Sagan became the gatekeeper of any serious discussion of extraterrestrial life. He became the go-to guy for anyone with a new idea. Sagan managed to think through virtually every angle on the topic, entertain every possibility, even the ones that evaporated in the harsh light of day. Only he could decide, as gate-keeper, if a creative idea should be allowed into the lecture hall or in-stead left outside, panting on the sidewalk.

Being gatekeeper required that he keep the hippies and dream-ers in line. In 1970 Sagan and Drake paid a visit to Timothy Leary, the Harvard professor turned LSD guru. Leary had taken up residence involuntarily in a California prison for the criminally insane. Leary summoned the astronomers for some help in a matter of relocation. He said the Earth was going to be destroyed in a nuclear conflagration and that he was going to build a spaceship and launch three hundred hand-picked people to another solar system. "Which of the stars is my best bet?" Leary asked. In such moments, it is necessary for cooler heads to explain a bit of the geometry of the universe and the funda-mentals of space travel. Sagan and Drake told him that interstellar travel was still impossible. They said that no one had yet discovered a single planet beyond our own solar system. Leary took the news hard. Sagan and Drake had done their job. Nonetheless, years later, Leary would manage to get his ashes launched into space on a rocket—not his original idea, but close enough.

(Sagan's resonance with the drug culture was no fluke. Like many college professors of his day, he found he could liberate his imagination through the use of cannabis. One of Sagan's biographers, William Poundstone, has discovered that Sagan documented his men-tal journeys in an anonymous account printed in Lester Grinspoon's 1971 book *Marihuana Reconsidered.* Sagan ("Mr. X") argued for the le-gitimacy of drug-induced insights, saying they shouldn't be dismissed the next morning. He revealed that one time, while high, he pounded out eleven short essays in a single very intense hour: "I have used them in university commencement addresses, public lectures, and in my books.")

The ufologists needed Sagan. He could offer them entry into the realm of credibility. He certainly was one of the few astronomers who would entertain sensational ideas. But Sagan over the years grew increasingly irritated with the UFO community. He'd been a teenager when the flying-saucer era began in 1947, and he certainly found the idea of alien visitors believable. In his book *The Demon-Haunted World*, he recalled being unable to find a good counterargument to the theory that UFOs were alien spaceships. It seemed the logical conclusion: "Why shouldn't other, older, wiser beings be able to travel from their star to ours?" But then he became a scientist, and he saw how human biases lead to mistaken conclusions. He read *Extraordinary Popular Delusions and the Madness of Crowds*, by Charles Mackay (1841). Over the years he would continue to follow the UFO reports, but he found the alien explanation increasingly less persuasive. He was drawn ever closer to the conviction that this represented a mass psychological phenomenon, that it had nothing to do with life beyond Earth, astronomy, advanced technology, propulsion, or interspecies contact. That conviction ran head-on into his natural instinct to be open, to entertain the fantastic and the just-barely-possible. Even after he became a professional astronomer he was reluctant to say that the extraterrestrial explanation for UFOs amounted to nonsense.

In 1968, the Committee on Science and Aeronautics of the House of Representatives, led by a congressman highly sympathetic to the claims of ufologists, held a symposium on flying saucers. Every speaker, except Sagan, declared that UFOs were of extraterrestrial origin. Sagan was completely outgunned. He took a neutral position. Though he gave a brilliant summary of the scientific reasons to be skeptical of flying-saucer stories, he stressed that it remained an open question. "I do not think the evidence is at all persuasive, that UFOs are of intelligent extraterrestrial origin, nor do I think the evidence is convincing that no UFOs are of intelligent extraterrestrial origin." In enemy country he was floating the little-bit-invaded theory. Sagan also left open the possibility of intelligent aliens elsewhere in the solar system: "Moderately unlikely" was how he characterized that scenario.

The ufologists didn't like Sagan's attitude, but they still cared about what he said. Sagan stood sentinel at the entrance to the halls of legitimacy. If he could get inside the Ivory Tower despite talking about all this alien business, then surely so could the flying-saucer people. Or so they hoped.

One long, cold, eerie night in San Diego, I heard a Sagan story from Richard Hoagland, a leading member of the small scientific field of exoarcheology. Exoarcheology is a niche science because by definition it is the study of archeological sites on other worlds, all of which, as far as we know, are uninhabited and thus have no archeology whatsoever. Hoagland, a former CBS news science consultant with an undergraduate degree in astronomy, has done more than anybody else to popularize the "Face on Mars," which he believes is a monument built by an extinct race of Martians. NASA says the Face is just a mesa that in blurry images looks like a face. The skeptics say there is no plausible reason why Martians would build a monument that looks like a human face, given that (as George Gaylord Simpson would be the first to say) the avenues of evolutionary development are so diverse that no alien race would look anything like *Homo sapiens*. But Hoagland and his allies would say this simply shows a lack of supple thinking. Humans could be descendants of the Martians, for one thing. (There is much interest in UFO circles in finding evidence for a non-Darwinian rise of the human species. Like some religious fundamentalists, many ufologists are concerned that this evolved-from-monkeys stuff is being shoved down our throats by the scientific elite.) Hoagland has also found, in photographs taken by the Apollo astronauts, evidence of artificial structures on the moon. That he has endured ostracism from the scientific community merely confirms that science is resistant to new ideas. That's why Sagan became so crucial to Hoagland—Sagan showed a willingness to entertain unorthodox notions.

"He was Mr. ET," Hoagland told me as we stood in the parking lot of a suburban hotel. "The standard was, if you can't get Carl interested, there must be nothing there."

One day in 1985—this is the Hoagland story—the National Commission on Space held hearings in Palo Alto, California, to discuss future exploratory missions in the solar system. Sagan was among those invited to speak. Hoagland, a fringe figure at best, was not invited and had to listen to the presentations as a somewhat unwelcome guest. At one point Gerard O'Neill, a Princeton professor who

wanted to build giant space stations and populate them with thousands of people, asked Sagan what would sustain the public interest in Mars explorations over a long period. O'Neill noted that the public had quickly lost interest in Apollo. Sagan turned and looked directly at *Hoagland*, even though he was about to address O'Neill.

"Because, Jerry, Mars has surprises," Sagan said—looking, Hoagland recalls, right into his eyes.

Hoagland told me the story with great drama. He believes that Sagan was, in code, telling him to keep pressing forward with his research. This was Sagan, the skeptic, endorsing Hoagland's investigation of the Martian monuments. But there was more! In the crowded hallway outside the commission hearings, Hoagland found himself in the midst of an intense conversation with somebody or other. Suddenly he heard a voice in his ear. It was just a whisper.

The voice said, "Keep up the good work, Rich."

A trip to Mars isn't complete without a stop at Phobos, the moon rumored to be artificial. NASA

Hoagland whirled around. Whoever had whispered to him had vanished. But he knew that it must have been Sagan. He couldn't be absolutely sure. But it had to be. It made sense. "Keep up the good work, Rich." That was Sagan's message.

A whispered endorsement from an unseen person who might have been Sagan—in the field of exoarcheology that's like having *official validation*. Sagan later wrote a devastating article in *Parade* that debunked the Face on Mars, but Hoagland still suspected that Sagan was not being straight about his true feelings. This is something that many people in the UFO world felt: that Sagan was a closet believer, that he knew things he couldn't afford to say. He'd write one thing, but think another.

There was that look in the meeting . . . and that whisper.

• • •

The reality was that Sagan wrote exactly what he felt, and this some-times caused him problems. He was an open book, and many of the pages were a bit fantastic by academic standards. Early in his career, Sagan would speculate that Phobos, one of the two moons of Mars, was a hollow object, an artificial satellite left by some ancient civiliza-tion. (For a time in the 1970s the license plate of his sports car said PHOBOS.) He could be maddening in his oscillations between imagi-nation and skepticism. Even in the case of the Face on Mars, he sur-prised his colleagues with a late reversal, saying that although it was almost surely a geological formation it ought to be re-examined by the Mars Global Surveyor. Over the course of forty years, as he wan-dered here and there through the field of exobiology, there were mo-ments when it appeared he might get lost. There were times when his ambitions, his wildly grasping mind, and his craving to find life be-yond Earth threatened to carry him into the shadowy realm of undis-ciplined speculation and pure nonsense. He would step right to the edge of the much-dreaded bog of mush, fuzz, and intellectual spew, and then, at the last second, when it appeared he would pitch forward and vanish beneath the scum, he would miraculously lurch backward onto solid ground.

He was intent upon solving the mystery of intelligent life in the universe, but he also wanted tenure. His colleagues didn't always join his fan club. The Harvard professors were flummoxed by Sagan and, in a moment that would rank among his few professional failures, the university declined to offer him tenure. The oldest university in America simply could not absorb the imagination of the man who would become the most famous scientist of his generation. The peo-ple who knew him best understood that Sagan, though somewhat op-portunistic, fundamentally believed in his mission of educating the public. He was a populist. He felt he could get anyone—a welfare mom, a hillbilly, a bored elementary-school student—interested in as-tronomy, biology, the history of philosophy, Aristotle, the Pythagore-ans, the Egyptians . . . all that cool stuff. He knew he could convert the masses. He would gently correct their errant ways, one of which was a credulousness for outmoded religious doctrines.

Sagan wouldn't go so far as to confess himself an atheist, but in his cosmology there was no obvious sign of God. There were lots of living things, but no one absolutely in charge. A god didn't seem necessary or even terribly useful. "Maybe there is one hiding, maddeningly unwilling to be revealed. Sometimes it seems a very slender hope," he wrote in *Pale Blue Dot*.

In his early days Sagan struggled with the Fermi Paradox. He agreed with the presumption of the Fermi Paradox that interstellar travel is possible in theory. He also agreed that an intelligent civilization would at some point colonize the galaxy. And, finally, he agreed that there was no good evidence that the flying saucers were truly spacecraft from another world. But this idea of a universe bereft of intelligent species other than our own—he couldn't believe it. So he came up with a different solution to Fermi. He speculated that the aliens aren't here now but have been here in the past. This was something he simply reasoned out, based on a paper calculation. He concluded that the aliens might have been to Earth during the historical era, sometime in the past five or six thousand years. In one of his most famous early papers, "Direct Contact Among Galactic Civilizations by Relativistic Interstellar Spacecraft," published in 1962, Sagan—a freshly minted Ph.D. from the University of Chicago—argued that a statistical analysis of the probable abundance of advanced galactic civilizations, and the rate at which such civilizations would likely launch starships for other solar systems, led to the conclusion that Earth would be visited by aliens every few thousand years. He assumed that interstellar spaceflight would be simply a matter of finding the right equipment and technique. The aliens might use enormous "Bussard ramjets," spaceships named after the scientist who first imagined them, which would scream through the depths of space with a scoop about seventy-five miles wide. The scoop would gather random hydrogen atoms into a central fusion reactor.

Archeologists, he wrote, should be attuned to the possibility that they could discover extraterrestrial artifacts. Sagan had put the concept of "ancient astronauts" in play.

"We expect that there is today in the Galaxy an integrated community of diverse civilizations, cooperating in the exploration and sampling of astronomical objects and their inhabitants," he wrote. They would "sample" planets with intelligent life every ten thousand years, and those with advanced civilizations every one thousand years. "Consequently, there is the statistical likelihood that Earth was visited by an advanced extraterrestrial civilization at least once during historical times."

About this time, Sagan sent one of his writings to a Russian scientist named Iosef Schmuelovich Shklovskii. Shklovskii was a lion of Soviet astronomy, with a wide-ranging interest in the evolution of stars, the uses of radio astronomy, supernovas, and, as it turned out, extraterrestrial intelligence. When he got Sagan's missive he quickly wrote back, using a Russian proverb: "The prey runs to the hunter." They were of like minds, and Shklovskii had already started writing a book on the subject of alien intelligence. Sagan arranged for the English translation of the book, and then, in an extraordinary bit of hubris, wrote an equal amount of material on the topic, mixed it with what Shklovskii had written, and produced a single volume with two authors, *Intelligent Life in the Universe*. At no point in the process did the authors ever meet.

In their book they raised the possibility that legends of ancient cultures should be carefully examined for clues that they had been inspired by the landing of extraterrestrial beings.

> With the numbers we have discussed, it seems possible that the Earth has been visited by various Galactic civilizations many times (possibly [about ten thousand] during geological times). It is not out of the question that artifacts of these visits still exist—although none have been found to date—or even that some kind of base is maintained within the solar system to provide continuity for successive expeditions. Because of weathering and the possibility of detection and interference by the inhabitants of the Earth, it might have appeared preferable not to erect such a base on the Earth's surface. The Moon seems one reasonable alternative site for a base. Forthcoming high-resolution photographic reconnaissance of the Moon from space vehicles—particularly, of the back side—might bear these possibilities in mind.

If you just looked in the right place, you might find the most amazing thing ever discovered.

Sagan's brand of science is full of things not yet ruled out, marvels still conceivable. A 1976 *New Yorker* profile, not terribly flattering, quoted him as saying, "Someone has to propose ideas at the boundaries of the plausible, in order to so annoy the experimentalists or observationalists that they'll be motivated to disprove the idea." Saganism as a philosophy allowed that any particular scenario might be unlikely, but collectively they added up to a probability. Take enough far-out gee-whiz ideas and mix them together in a vat and something alive and conscious and technological will burble from the slime.

Shklovskii and Sagan hooked up in Armenia in 1971, at the Byurakan Astrophysical Observatory, for an epochal Russian-American conference on extraterrestrial intelligence. The Russians, it seemed, were even more gung-ho about contacting aliens than were the Americans. Something about socialism opened up the minds of scientists, and phenomena such as extrasensory perception became a matter of obsessive investigation. The Soviets, rumor had it, would do anything to get an advantage over the richer and more technologically proficient Americans. The Americans might make it to the moon first, but it wouldn't mean anything if the Soviets could strike a deal with the rulers of the galactic empire.

Galactic empires had been a common theme in science fiction—in Isaac Asimov's *Foundation* trilogy of the early 1950s, there is an empire stretching across twenty-five million inhabited planets—and this kind of astropolitical structure was easily imagined by farsighted astronomers. Some of the optimists thought that the first sign of an alien civilization would not be a message but rather a visibly obvious artifact, perhaps a massive construction project. Freeman Dyson talked about what came to be known as Dyson Spheres, artificial shells constructed around stars in order to capture a large percentage of their radiation. Sagan had wondered aloud if the aliens might have constructed buoys around black holes to warn interstellar travelers of major navigational hazards. These buoys could perhaps be detected from Earth. A Russian named V. L. Ginzburg argued that entire civilizations might exist inside atoms. What might look like a fundamental particle could be an entire universe!

Sagan wouldn't go quite that far, but he did think that perhaps there were undiscovered laws of physics right under our noses, such as tachyons, particles that moved faster than the speed of light, so fast they could appear to travel back in time. Maybe that's what the aliens used for their interstellar signals, rather than slow-as-molasses radio waves.

These were heady times. The imagination ran unconstrained through a sparkling universe. Even as they spoke at Byurakan, human beings walked on another world, the Earth's moon. Dreams were coming true—a temporary phenomenon, as it turned out.

6

Space Age Blues

It didn't really mean much—cosmically—that Sagan and Drake had gotten diddly-squat in their search of Andromeda and the other nearby galaxies. Their tool of choice, Arecibo, was a magnificent thing to look at but was still a primitive and crude device for registering alien signals on an intergalactic scale. The aliens would have had to be awfully voluble—desperate to be heard. Sagan and Drake could assure themselves that absence of evidence is not the same thing as evidence of absence.

The bigger problem for the visionary wing of the astronomy community was that the intellectual climate had started to change. There were defections from the camp. There were scientists, like SETI researcher Ben Zuckerman, who began to suspect that the radio search would never work, because the aliens might not be sufficiently abundant, and might not be out there at all. The 1970s had seen the rise of a new set of doubts, a sense of constraints. Intellectuals fell in love with the word "limits," and all that it implied, including a rejection of the avarice and gluttony and warmongering and ecological destructiveness of Western civilization. The Luddites were on the move. It was a terrible time to be a person like Dan Goldin, a technophile. Vietnam had been a technological horror. Then, in the span of a couple of years in the mid-1970s, SETI lost its luster, Martian life virtually became an extinct concept, and millions of people gave up the dream of rocketing through space. NASA became so lost and confused that it briefly flirted with the idea of becoming an energy agency, putting up windmills and solar collectors.

The Space Age all but died.

Normally the people who didn't like SETI hadn't really spent much time thinking about it, or failed to understand that this wasn't a search for little green men in flying saucers. When Senator William Proxmire gave SETI a "Golden Fleece Award" in 1978 for being a waste of government money, he argued that even if we detected aliens out there they'd probably be thousands of light-years away, and would already be dead by the time we got the signal. What would be the point? That seemed to be the gist of Proxmire's argument. If you can't hold a galactic kaffeeklatsch, who cares if there are civilizations on other worlds? (Sagan eventually managed to quell the Proxmire rebellion with a visit to the senator on Capitol Hill.)

Less easily mollified were the astronomers who found the SETI scheme too great a long shot. The astronomer Michael Hart made the loudest objection. He revisited the Fermi Paradox and found it compelling. He wrote a paper arguing that there probably weren't many aliens out there, and submitted it to *Icarus*, the scientific journal of which Sagan was editor. *Icarus* rejected the paper. ("It might have frightened Carl. He was riding high at that point," suggests Zuckerman.) Hart managed to get his paper published in the *Quarterly Journal of the Royal Astronomical Society* in June 1975. It was called "An Explanation for the Absence of Extraterrestrials on Earth." Hart asserted that, in principle, interstellar flight is possible, and that it ought to lead to the colonization of the entire Milky Way galaxy in something like 650,000 years. Maybe it would take twice that long, he wrote, but in galactic terms it was readily accomplished. Yes, the distances and travel times are daunting, but there are numerous ways some civilizations might get around that. "For a being with a lifespan of 3000 years a voyage of 200 years might seem not a dreary waste of most of one's life, but rather a diverting interlude."

Hart dealt with the theory that aliens might simply decline to visit the Earth and other worlds, preferring to hang out at home and study the universe remotely. This is the Contemplation Hypothesis. Hart said it didn't stand up to scrutiny. This hypothesis, he said, would make sense only if it could be applied to every race of intelligent beings on every planet through every stage of its social and political history. That was a stretch. So, too, did he reject the idea that civilizations destroy themselves. Could it be true of them all? Another possibility was the UFO Hypothesis, that aliens *are* visiting Earth. He couldn't abide it. "Since very few astronomers believe the UFO Hypothesis, it seems unnecessary to discuss my own reason for rejecting it," he wrote. He concluded that there might be a few civilizations here and there, but the number would have to be small and might be zero. We were the first civilization in the galaxy, most likely, "even though the cause of our priority is not yet known."

Naturally, Hart's argument annoyed the visionaries. We don't want to be alone, particularly in a place that's one hundred thousand light-years in diameter. The Hart scenario would have us drifting in a rubber lifeboat in an ocean that is never interrupted by land. There's not a single island with a palm tree. We drift aimlessly, going nowhere, drinking rainwater, and catching turtles—from now to the end of time.

Hart's argument and those of several other newly skeptical thinkers put the exobiology and SETI communities into a minor crisis. Then came the most shocking defection: Shklovskii, the Russian radio astronomer with whom Sagan had collaborated on *Intelligent Life in the Universe*. Shklovskii had signaled his impatience with the search for intelligent aliens back at the 1971 Byurakan conference, noting that exobiology lacked the observations and experiments that support other natural sciences. The belief in intelligent extraterrestrial life rested on logic alone. Geometry, he said, was a bit like that, too, but even the most faithful believer in intelligent aliens could never be as convinced of the reality of those creatures as a geometer could be convinced that parallel lines never meet. Then, in 1976, Shklovskii published an article in a Russian journal arguing that "practically, if not absolutely," we are alone in the universe. He wrote that there are no visible "space miracles" erected by intelligent civilizations, and that if there are any aliens out there they're probably too far away to contact. He had a powerful argument that avoided the normal division between "alone" and "not alone." The we-are-alone argument is so hard to reconcile with the size of the universe that it barely survives the modern age of astronomy. But for Shklovskii the distances from Earth to other worlds are of significance. We could be functionally alone. We could be alone *for all practical purposes*. We wouldn't be alone in the technical sense (in fact, there could be millions of civilizations spread across the billions of galaxies), but we would still be condemned to loneliness. Humans, he wrote, are likely to be the "vanguard of matter" in our corner of the universe, with all the weighty implications thereof.

Shklovskii's old friends, such as Sagan and Drake, were taken aback. They figured their Russian friend was disillusioned in part by the foolishness of Soviet politicians. His beef wasn't with the aliens, surely; it was with the idiots who ran his country. Shklovskii, they figured, had lost the dream of contacting aliens because he assumed that the aliens would be like humans—bumblers, incompetents, certain to destroy themselves with the misused fruits of their knowledge.

Sagan and Shklovskii argued over the Russian's conversion every time they met thereafter. Once, Shklovskii made the point that life does not always evolve toward some progressively higher state. There are dead ends. There are creatures who come to the end of their genetic lines. He cited the saber-toothed tiger. The saber-toothed tiger had teeth so large that they were actually an impediment. For Shklovskii, the implication may have been that humans could be a dead end, too. Their gift—intelligence, technology, the manipulation of the world around them—could be an evolutionary adaptation leading to nowhere except, perhaps, to extinction.

The aliens, even during the hard times of the mid-1970s, still retained a grip (three-fingered, to be sure) on the public imagination. The UFO movement had been boosted by a wave of sightings in 1973 and 1974, leading to Steven Spielberg's movie *Close Encounters of the Third Kind* in 1977, with François Truffaut playing a role based on the respected ufologist Jacques Vallée. Erich von Daniken turned his theory of ancient astronauts (*Chariots of the Gods?*) into a publishing phenomenon. *Star Wars* would soon show what kind of mutant outlaws hang out at saloons in deep space. *Star Trek* existed only in reruns, but the trekkies were so numerous in the 1970s that one publisher printed 450,000 copies of a set of Starship *Enterprise* blueprints and "space cadet manual."

Ufology and science fiction, however, were never supposed to be the centerpiece of the space program. We were supposed to go into space for real, with actual astronauts and genuine spaceships. Unfortunately, the Conquest of Space wasn't happening on schedule in the post-Apollo era. The 1970s would go down in history as the decade that saw the rise of the environmental and feminist movements, and space would be reduced to a footnote. The only people who really felt the tug, the gravitational attraction, of outer space were the dreamers, the trekkies, and various Californians.

The military still cared about space, because it was potentially a platform for war. The national-security strategists wanted to keep the aerospace industry humming along in case someday we needed to fight our battles on the high frontier. The race to the moon was famous for its technological offshoots—the classic, apocryphal example being Teflon—but perhaps the greatest creation was the aerospace component of (to use President Eisenhower's phrase) the military-industrial complex. These companies, dependent on government contracts, mastered the art of Space Age rhetoric, often running advertisements in magazines and on television reminding the citizenry that humans are destined to go to the stars (but must first build multibillion-dollar orbiting monstrosities).

The great breakthroughs and technological leaps that had been anticipated by Space Age dreamers were taking a long time to materialize. This made no sense. In the Space Age, things were supposed to happen at accelerating rates. This was a time of bizarre deceleration. It had started to dawn on the average American citizen that we are not going to be dodging Klingon warships or riding giant sand worms on the planet Dune anytime in the near future. The fantastic momentum of Apollo vanished, contrary to all known laws of inertia. Most ordinary people stopped using the phrase "Space Age," except ironically. Those bulky, pressurized moon-walking suits started to look as quaintly out of date as the Space Needle in Seattle or the conveyor belt at the entrance of the home of George and Jane Jetson.

In the 1960s, the human-space relationship had been obvious: Space would become our habitat. We would someday be star creatures, voyaging across the cosmos to discover new worlds and exotic alien races. Every kid coming of age in the Apollo era knew that someday we'd be hopping from Earth to Mars as easily as riding a skateboard to the 7-Eleven. (John Logsdon of George Washington

University's Space Policy Institute remembers paying $5 to Pan Am in 1969 to reserve his seat on a commercial flight to the moon.) Every time you turned around, Walter Cronkite or Jules Bergman was on TV talking about the imminent launch of another Gemini or Apollo spacecraft. TV captured the gantry-eye view of the Saturn V rocket igniting under brave men bound for the stars. This wasn't science fiction, it was real life, actual technological progress taking place in real time and real space. Only a fool would doubt that we were conquering the final frontier. We would not be confined to a single, lonely, seemingly unremarkable planet in one cranny of one galaxy, but instead would cruise the cosmos, traversing the clearest, purest medium anyone could imagine. Space was cold but frictionless. It was clean! You could zoom through the stuff. In *Star Trek* the stars drifted by like little wads of cotton blowing on a lovely breeze. Space was certainly not a hostile environment. As Piers Bizony would later write in *2001: Filming the Future:* "Up to 1970 or so, everybody really *did* think that the future in space would be just like *Star Trek.*"

As with many historical developments the people responsible for putting the Space Age into a tumble probably had no idea they were doing anything of the sort. They made some bureaucratic decisions that seemed reasonable. The NASA budget (in inflation-adjusted dollars) peaked in 1965 in the massive buildup to Apollo. With money pouring into the Vietnam War, there was a sense in Washington— outside NASA, at least—that it would be a mistake to launch another costly Apollo-style project after the moonshot. The obvious mission of choice would be to go to Mars, but that would cost upward of $100 billion. The propulsion requirements for a Mars trip were positively hideous. As soon as Apollo 11 succeeded and the euphoria died down, it became time for practical bureaucrats to seize the moment. The Bureau of the Budget in 1969 celebrated man's landing on the moon by slashing the projected NASA budget for fiscal year 1971. That forced the cancellation of three moon landings, which would have been Apollo 18, 19, and 20. NASA decided that it made more sense to put launch capability into a reusable Space Shuttle, rather than disposable rockets that would wind up at the bottom of the sea. It was perfectly rational and responsible. It was the spirit of recycling! Who could complain about that? NASA spent much of the 1970s preparing to launch the Space Shuttle, but it didn't take a genius to see that it would go only a couple of hundred miles off terra firma—the equivalent of a drive from Washington, D.C., to Roanoke. At that point, mankind's penetration of the cosmos would come to a halt. The new philosophy of NASA seemed to be that the sky's the limit—literally.

The public didn't protest the decline of space travel. The business of hopping around on the moon in space suits simply did not carry much relevance to the passions of the day. The intelligentsia applauded the cuts in NASA's budget. What was the point of spending $25 billion to send a dozen white men to the moon when so many cities were crumbling and rivers full of poison? The astronauts were not remotely representative of the human race. They seemed to be the same person, again and again. The only interesting astronauts were the ones who went crazy or got depressed. To the average professor at America's select universities, the Apollo program had been some kind of military-industrial stunt, a power play, implicitly hegemonic. The mid-1970s saw the rise in academia of postmodernism, which in its worst moments was the antithesis of rigorous scientific thinking. The postmodernists had a serious antipathy to traditional science, with its top-down, authoritarian regulation of truth and reality. To a postmodernist, reality is whatever we decide it is. There's no objective truth to anything. The Enlightenment was over. As the biologist E. O. Wilson would later write, protesting this trend, "Enlightenment thinkers believe we can know everything, and radical postmodernists believe we can know nothing."

The Space Age was written off as a neocolonial historical spasm, a fantasy of nerds with propeller beanies. It represented a psychological flaw—a pathology—of human beings. It was an analogue to war. We should make peace, and stay home, on Spaceship Earth. If we could send a man to the moon (people would say), why couldn't we solve (fill in heartbreaking social problem)?

In this moment of Luddite technophobia, there were still a few subcultures that believed, passionately, in the Conquest of Space. Among them were the advocates of space colonies. A new group called the L5 Society planned to live in massive space stations in a gravitationally stable point near the moon. They had a guru: Gerard O'Neill.

O'Neill had thought up the idea of space colonies while teaching a freshman physics class at Princeton in 1969. Over the next five years, he developed his proposals and grew increasingly enthusiastic, and finally, in 1976, produced a popular book, *The High Frontier.* The

O'Neill space stations would have rivers, clouds, all the comforts of home. *The High Frontier* begins with a description of life aboard a space station, told in the form of an imaginary letter written by a married couple to another couple preparing to join them. O'Neill suggested that such a letter could be written on a space station as soon as the early 1990s:

> You'll notice immediately the small scale of things, but for a town of 10,000 people we're in rather good shape for entertainment: four small cinemas, quite a few good small restaurants, and many amateur theatrical and musical groups. It takes only a few minutes to travel over to the neighboring communities, so we visit them often for movies, concerts, or just a change in climate. There are ballet productions on the big stage out in the low-gravity recreational complex that serves all the residents of our region of space. Ballet in 1/10 gravity is beautiful to watch: dreamlike, and very graceful. You've seen it on TV, but the reality is even better. Of course, right here in Alpha we have our own low-gravity swimming pools, and our clubrooms for human-powered flight. Quite often Jenny and I climb the path to the "North Pole" and pedal out along the zero-gravity axis of the sphere for half an hour or so, especially after sunset, when we can see the soft lights from the pathways below.

So it's quite groovy. Space is sort of like Sedona or Laguna Beach or Tiburon—an artists' colony literally tens of thousands of miles from street crime, bad schools, pollution, corruption, and the New Jersey Turnpike.

For O'Neill there were obvious advantages to life in space. There's no night, for one thing. Sunlight all the time means a constant source of energy. There's no worry about lack of frontier, because there's obviously space in space. A space colony would also be free of the "gravitational disadvantage" of human civilization on the surface of the Earth. The lack of gravity on a space colony (except where artificially created through rotating habitats) would make manufacturing easier. As O'Neill put it in July 1975, when he testified before Congress, "Industrial operations on Earth are shackled by a strong gravity which can never be 'turned off.'" Using materials from the moon and the asteroid belt, the colonies could proliferate and provide an opportunity for eco-

The space colony dream: Just like living on Earth, but cleaner, with more gravitational options.   NASA

nomic growth that would be impossible on the surface of an overly crowded planet with finite resources. O'Neill didn't sell his colonies as a romantic adventure so much as an economic necessity.

NASA decided that it didn't like the word "colonies," because so many of the world's indigenous peoples had had a disastrous experience with colonialism. So NASA adopted the term "space settlements." In 1975, a summer study program at NASA's Ames Research Center—with O'Neill on board as the technical adviser—came up with one of the space agency's most impractical schemes, a proposal to build a colony in space several kilometers in diameter. It would require two thousand construction workers. Ideally, it would be built near the asteroid belt, where the raw materials are abundant, and then propelled by rocket power back toward Earth, to the orbital point known as L5. But the study also sees other possibilities: "They may, however, prefer to go the other way, to strike out on their own for some distant part of the solar system. At all distances out to the orbit of Pluto and beyond, it is possible to obtain Earth-normal solar intensity with a concentrating mirror whose mass is small compared to that of the habitat."

At the core of the space-colony dream was a deep pessimism about the ability of human civilization to survive and prosper on Earth. "In the future, the Earth might be looked upon as an uncomfortable and inconvenient place to live as compared to the extraterrestrial communities."

But the Left never went along with O'Neill's space-colony vision, just as it would initially be reluctant to accept the coming of the personal computer. A typical remark was that of Lewis Mumford in a letter to Stewart Brand, a colony supporter, that was reprinted in Brand's 1975 book *Space Colonies:* "I regard Space Colonies as another pathological manifestation of the culture that has spent all its resources on expanding the nuclear means for exterminating the human race. Such proposals are only technological disguises for infantile fantasies." Wendell Berry wrote that O'Neill's colonies were strikingly conventional: "If it should be implemented, it will be the rebirth of the idea of Progress with all its old lust for unrestrained expansion, its totalitarian concentrations of energy and wealth, its obliviousness to the concerns of character and community, its exclusive reliance on technical and economic criteria, its disinterest in consequence, its contempt for human value, its compulsive salesmanship."

This overt contempt for space travel had even filtered into science fiction. The genre is too broad and inventive to suffer anything as simple as a trend, but by the 1970s it was common to find SF narratives built around dystopian nightmares, around technology gone awry and science gone cruel. Doomsday wasn't much of a plot device because it was already a given. The question was, *then* what? The world as we know it will be destroyed. That's just the starting point of the story. The SF writers could see more clearly than anyone the dark side of the Space Age, the primitive urges, the latent savagery, of our ventures beyond Earth. No longer could anyone talk about space travel with an entirely straight face; this was now the stuff of irony and farce.

In 1969, the year of the moon landing, Kurt Vonnegut produced *Slaughterhouse-Five*, another of his grim, comic explorations of the lethal consequences of technology. The hero, Billy Pilgrim, is kidnapped by a flying saucer in 1967 and is taken to the planet Tralfamadore and exhibited naked in a zoo (where he mates with the former Hollywood film star Montana Wildhack). Someone in the crowd asks what he has learned in his time on their planet, and Billy Pilgrim launches into a breathless speech about how marvelous it is to see the inhabitants of an entire planet live in peace. Humans, he says, have been engaged in "senseless slaughter since the beginning of time." He warns his audience that humans may pose a danger to the entire galaxy. The Tralfamadorians close their little hands over their eyes, their way of telling him that what he is saying is stupid. Billy Pilgrim asks a guide what he said that was so dumb. The guide explains that the Tralfamadorians already know how the universe ends.

"We blow it up, experimenting with new fuels for our flying saucers. A Tralfamadorian test pilot presses a starter button, and the whole Universe disappears."

What few people know is that in 1976, with the Space Age desperately needing something to cheer about, a man named Gilbert Levin found life—or so he believes—on Mars.

I drove out to Beltsville, Maryland, just beyond the Beltway, to hear Levin tell the story. He works in a low-slung modern building with a sign outside saying Biospherics, Incorporated. Levin is the chief executive officer. I found a man who is storkish, a ceiling-scraper, wearing a pink shirt and an expression of amused contentment. I noticed on his shelves a number of books about Mars, which is to be expected, but, strangely enough, there were also books about sugar. He explained that his company has patented the use of a rare, natural sugar that is only partially metabolized by human beings.

The sugar has a Mars connection. In the early 1960s, when Levin was designing experiments for detecting life on Mars—he had a NASA contract—he realized that all amino acids used by living things on Earth have a certain structure that scientists call "left-handed." (Why they are left-handed and not right-handed is an unsolved mystery of life on Earth.) Levin's proposal for detecting life on Mars was to splash amino acids and other nutrients on Martian soil samples and see if any Martian microbes gobbled the stuff up. In thinking up this plan, he real-

7

The Mars Effect

ized that Martian organisms might metabolize right-handed amino acids rather than left-handed amino acids. Eventually he got the idea to make a type of sugar, using "backward" carbohydrates, that humans wouldn't metabolize. It could be sold as the perfect nonfattening sweetener. His product is called tagatose. It might conceivably make a Martian fat but would be almost unrecognizable to the body of a human. Better yet, it would function just like ordinary sugar, unlike products such as NutraSweet (aspartame), which is worthless for baking. (Note that you've never eaten a NutraSweetened cookie.)

The problem, at least when I met Levin, was that his new, radical, back-handed sugar retailed for about $3,000 a pound. Still, his company kept pressing ahead, financed by a lucrative, unrelated business handling calls to 800 numbers. In the meantime, he was spending his life going to conferences and telling people about how he found life on Mars.

"Maybe I'm not completely nuts," he said. He picked up a book that had a photograph of the surface of Mars taken by the Viking lander. "I dunno. Look at the corner of that thing." He indicated a rock with a slightly greenish stain on it. "That has some color in it."

Color, which suggests . . . life.

It's important to understand what people were expecting when NASA sent two spacecraft to Mars in 1976—the famous mission named Viking. There was abundant reason to suspect that Mars did not have a scrap of life, not a microbe. It was Sagan's ex-wife, the biologist Lynn Margulis, who came up with the most astute reason for why Mars would show no sign of life. She said a planet couldn't be just slightly alive. The Gaia Hypothesis, developed by Margulis and James Lovelock, declared that the Earth functions in some respects like an organism, that the living things interact with the nonbiological systems to form an essentially stable, functioning biosphere. Earth is wildly, preposterously alive, overrun with creatures large and small, with elephants and paramecia. The various microbes of Earth are not adjunct phenomena to the physical structure of the atmosphere and oceans, but rather active partners in the creation of the habitable environment. Perhaps that is the secret to life on any planet. Life has to seize the entire joint, and shape it, and regulate it, and dance with it. Mars is a planet where life, if it ever existed, didn't have control of the air and the water. It looked dead from a distance; it would turn out to be dead upon inspection. So predicted the rules of Gaia.

To the extent that Mars held much promise for exobiology, it was largely by default. There weren't many other places in the solar system where life could plausibly exist. In the early part of the twentieth century, there had been some hopes for Venus. Venus is the sister planet to Earth, being nearly as large and composed of the same elements. In the 1920s, astronomers could tell that it has an extensive atmosphere, and many suspected that it is wet and steamy. Spectroscopic analysis found no signature of oxygen or water, however. In 1932 came evidence that Venus has an atmosphere heavy in carbon dioxide. Edwin Martz, a colleague of the Mars-crazy Harvard astronomer William Pickering, suggested in 1934 that the CO_2 is given off by animals and plants. By 1940, astronomers were discussing the possibility that Venus suffers an extensive greenhouse effect and that its surface temperature might be above the boiling point of water. A few diehards, however, came up with explanations that allowed Venusian life. No one had any conclusive answers. Finally, in 1956, came radar evidence that Venus had a surface temperature of something like eight hundred degrees Fahrenheit—broiling! Sagan made a name for himself with his doctoral dissertation showing how the greenhouse effect had driven temperatures to a life-excluding extreme.

And yet even Sagan continued to lobby for possible life on Venus. In a letter to the journal *Nature* in August 1967, Sagan and Harold Morowitz, at that time a molecular biophysicist at Yale, proposed that the clouds of Venus might have liquid water and the organic material necessary for life. They noted that an organism in the clouds would have to live at a fixed altitude, because if it were carried by downdrafts to the lower atmosphere it would be cooked in the high heat of the planet, and if it were carried upward to the high atmosphere it would run out of moisture and get too cold. "We therefore imagine an isopycnic organism constructed as a float bladder," they wrote. ". . . The organism is essentially a spherical hydrogen gasbag. . . ."

Few other scientists were captivated by the possibility of floating gasbag creatures on Venus.

That pretty much left Mars, with its possible signs of seasonal "vegetation," as a candidate for life. But on June 14, 1965, the Mariner 4 spacecraft flew by Mars and took twenty-two pictures. They were transmitted back to the Jet Propulsion Laboratory. The shocking news: Mars looked like the moon. Mars was a cratered wasteland. No canals, no Martians, no John Carter and Deja Thoris. There was neither love nor hate on Mars, barely any air, nothing moving—a planet so static and lifeless it didn't even have erosion.

This was a scientific triumph and a cultural disaster. The nightmare of science is that sometimes you learn things you don't want to know.

Sagan had no choice but to go where the data led him. He theorized that the seasonal markings are caused by giant dust storms. When the dust lifts into the atmosphere, it exposes darker soil underneath. The man who wanted so badly to find life on Mars had produced what would be one of his greatest academic successes, matched only, perhaps, by his work showing that Venus suffered an extreme greenhouse effect. Sagan was killing off these planets one by one.

Still, Mars made a comeback. Six years after Mariner 4's fly-by, the Mariner 9 spacecraft achieved orbit around Mars and began sending back images. The spacecraft found more than just craters this time. It discerned the largest canyon in the solar system, Valles Marineris—as long as the United States is wide. It also found the largest mountain, Olympus Mons, a shield volcano taller than Everest. Most important, Mariner 9 found numerous serpentine channels that appeared to be dried riverbeds. Shades of Percival Lowell! (But of course the channels weren't straight, like canals, and couldn't be seen from Earth.) The channels renewed interest in the possible biology of Mars. They appeared to show that Mars had once been warmer and wetter, and that it had enough atmospheric pressure for water to remain liquid at the surface. Sagan and the other optimists suggested that there might still be oases, hot spots where frozen water could thaw and organisms survive. There'd be no way to know for sure until a spacecraft touched down and surveyed the place.

That was the job of Viking. The Viking project scientist was Gerald (Jerry) Soffen, a skinny, slightly devilish-looking man with a high forehead, black hair, and a goatee. Soffen, I discovered, has a different version of the story of how Gilbert Levin found life on Mars, one in which Levin doesn't actually find it. I found Soffen in room 5-E27 at NASA headquarters, where he was nominally a senior scientist in the new astrobiology venture but was more precisely an *éminence grise*. He remembered with a smile the time of Viking, when the Jet Propulsion Laboratory in Pasadena became the center of the

NASA universe, and when Soffen himself was something of a celebrity. "We were the fair-haired boys," he said.

Viking had one major problem: Everyone assumed it was all about life. The biology investigations got all the publicity. There were geology and atmospheric experiments, but no one cares about rocks and air. People care about life, and only life. Everything else is material for some dull article in *Scientific American.*

With so much talk about possible life on Mars, the mission had a high probability of creating a major disappointment. The planet had no sign of liquid water and hardly any atmosphere. Scientists tried to have an open mind about life, imagining exotic routes for something nonliving to become alive, but they remained chauvinistic about liquid water, carbon molecules, and heat. The place was usually colder than the coldest ice cave in Antarctica. Liquid water on Mars wouldn't even freeze, were some to burble miraculously to the surface; it would instantly turn to vapor because of the paltriness of atmospheric pressure. Only a raging optimist would think that the Red Planet might still be an abode of living things. Such a scientist was, in fact, on the Viking team: Sagan. He, his wife, and their young son had moved to California.

For several years, Sagan had been an official member of the Viking imaging team, but with his multiple careers (professor, researcher, author, twenty-four-hour quote machine) he didn't have time to help as much as team leader Thomas Mutch would have liked. At one point in 1973, Mutch wrote Sagan a terse letter asking if perhaps the funding given to Sagan for work on Viking would be better used by some other scientist. "A great deal has happened over the past year, but your contribution has been limited," Mutch wrote. Sagan defended himself, but it would not be the last time that people noticed how thin he had spread himself. Even so, he was passionate about Viking, in theory if not always in deed. He privately told colleagues that he believed the probe might well find more than just simple life— it might find animated creatures, Martian animals. Sagan complained, in fact, that the cameras were not ideally designed to capture a moving object: They might not notice a lumbering beast.

In his book *The Cosmic Connection,* published three years before the Viking mission, he acknowledged that on Mars, which has no protective ozone layer, life would have to cope with the harsh, germicidal ultraviolet radiation blasting down upon the surface. But Sagan said Martian life might simply adapt to that. "We can easily imagine organisms walking around with small ultraviolet-opaque shields on their backs: Martian turtles."

In the spring before the Viking landing, Sagan authored half a dozen articles that were distributed by the NASA public-affairs office, and again he talked about the possibility of finding Martian creatures. "For some reason it is thought to be a sign of caution to admit the possibility of microbes on Mars but to exclude peremptorily the possibility of macrobes—organisms large enough for us to see unaided were we on Mars. But there seems to be no evidence for or against Martian macrobes, and, for all we know, there is a thriving population of large organisms on the planet. Nothing in our present understanding of Mars excludes this possibility."

Perchance to dream! This was all Sagan wanted from the world, an easing up of negativity, an opening of the mind to amazing possibilities not yet excluded from rigorous thought by contravening evidence. Keep your mind open, he implored (but not so open that your brains fall out, he whispered). As for the other scientists, this sort of talk about Martian animals—turtles?—struck them as bizarre, ludicrous. Sagan's argument seemed to be based on the notion that Mariner 9 couldn't have detected large animals on Earth if it had gone into orbit a hundred thousand years ago, before humans started reworking the surface of the planet. Therefore, the Mariner 9 images of Mars might similarly have overlooked wondrous preintelligent fauna. The problem was, had Mariner 9 orbited Earth a thousand centuries ago, it would have noticed giant oceans of liquid water at the surface and a thick atmosphere rich in nitrogen and oxygen and all sorts of other things that make a planet look habitable.

At one point Sagan went to Soffen with a dramatic suggestion. Let's put a flashlight on board, he said. There could be something like moths that would be attracted to the light. Soffen thought that unlikely, but out of respect for Sagan he relayed the idea to the project manager, Jim Martin. The idea died right there.

The first Viking lander began its descent to the surface on July 20, 1976. As it dropped, Viking found significant amounts of nitrogen. The pulse rate at JPL jumped. Nitrogen had been thought absent from the puny Martian atmosphere. Nitrogen was good. Nitrogen was *excellent.* The element is critical for biology as it exists on Earth. The odds of finding life on the Red Planet seemed to improve.

The spacecraft touched down shortly before noon, Pacific Daylight Time. Soffen ran to his wife and hugged her. The next thing to do was wait for the first picture. It would take several hours to arrive from the lander.

At JPL, everyone crowded around TV monitors. The picture filled the screens one thin line at a time. There was some initial confusion about straight, dark lines seen in the first picture—almost like tire tracks. The lines were just a momentary obscuration of the landscape, caused by dust kicked up by the lander. The rest of the image appeared: rocks and sand. The camera was looking straight down at the surface, as it was programmed to do. The footpad of the lander could be seen, resting comfortably on Martian soil. The next picture was a panorama, three hundred degrees of horizon. Mars looked like the volcanic deserts of eastern Washington State. The rocks were of all sizes, including one that looked like a Midas muffler and another that looked like a Dutch shoe. As Eric Burgess recounted later in his book *To the Red Planet*, the assembled scientists and journalists at the Jet Propulsion Laboratory immediately noted things that looked, at first glance, either artificial or biological: "The rolling nature of the countryside held everyone spellbound. Were those dark markings vegetation? There's a rock that looks like a cylinder! People imagined all sorts of things among the jumbled rocks. But when copies of the pictures became available soon afterward, it was quite apparent that no vegetation existed on the surface of the Chryse basin."

It was a moment of awesome revelation, even though the lack of vegetation had been discussed for years. Before there had been chatter, a lot of plausibility arguments and extrapolations. People had talked about Mars for thousands of years with different degrees of sophistication. Now, suddenly, they were *there*. No matter what happened with the biology experiments, the pictures from the Martian surface would ensure that Viking would go down as a fantastic achievement. Jerry Soffen and his crew had peered into a piece of the Creation as no one had before—a rare, wonderful Galileo moment.

One day, there was a ruckus in the press room about a new image from the lander showing something unusual on a Martian rock. It looked like the letter "B"—like graffiti. Or maybe it was the number 8. Either way, it was tremendously intriguing. In the heat of the moment, no one had time to ask why anyone would use the same letters and numerals on Mars that the Europeans and Arabs began using on

Earth in the Middle Ages. (Though Sagan raised this point later on *The Tonight Show* with Johnny Carson.)

Aboard each of the two Viking landers was a life-detection "package" about a cubic foot in size and loaded with contraptions. These would perform three different experiments, including Gilbert Levin's, which was known as the Labeled Release experiment. The principle was simple: The lander would extend its robotic arm, scoop up some dirt, put it in a container, and squirt a bunch of liquefied nutrients on it. The broth was made of amino acids and carbohydrates, simple molecules that are metabolized by microbes on Earth. The Levin experiment would detect whether the nutrients were metabolized. The nutrients were "labeled" with radioactive carbon. If Martian bugs metabolized the nutrients, they would presumably emit radioactive carbon gas. Another experiment, designed by Norman Horowitz, would test for photosynthetic activity. The third, by Vance Oyama, was similar to Levin's in that it would test for gases emitted or absorbed by the "soil" after being squirted with a complex liquid substance (this was called the "chicken soup" experiment).

The main problem with all three experiments was that they were

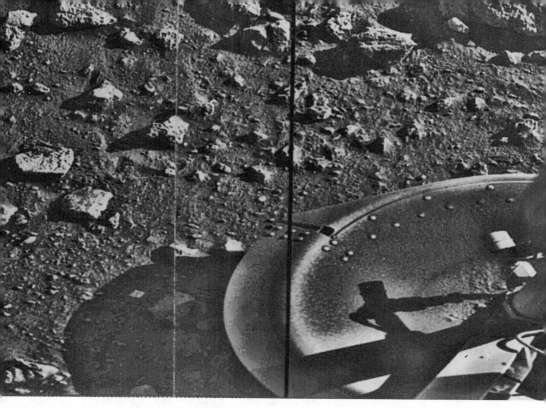

The first image from the surface of Mars: excellent rocks, but no animals, vegetation, or Martians. NASA

designed in the 1960s, when people had a slightly more benign view of the surface of Mars. No one knew in the early 1960s, for example, that Mars had so little atmosphere. Also, the experiments would be conducted inside the Viking craft, which would be operating at temperatures far above the ambient temperature of Mars. The scientists were the first to admit that Viking had gone all the way to Mars to test for the equivalent of Earthlife living in Earth conditions. The tests didn't have much that was *Martian* about them.

But then something amazing happened. Two of the three tests showed signs of life.

Or at least that was what everyone thought initially. Norm Horowitz's experiment, which at least didn't use any liquid water, showed that the soil seemed to absorb carbon atoms in a manner suggestive of something alive. And Levin's worked precisely as he'd drawn it up on the chalkboard. The nutrients were squirted on the soil, and the soil emitted a measurable amount of radioactive gas, just as Levin had hoped. Some of the same soil sample was then baked at sterilizing temperatures. More nutrients were squirted, and this time there was no radioactive emission—again, fitting with Levin's theory that the heating of the sample would kill whatever was alive.

For Levin, the conclusion was obvious: They'd found, and killed, the first known Martian life.

Levin knew that it was an unconfirmed test, and in fact it would be another ten years before he would publish a paper declaring that it was "more probable than not" that his experiment detected life. But at first glance the results certainly appeared to be consistent with biology. Levin remembers being applauded—"Gil, congratulations, you've really done it." People signed a commemorative photograph to mark the incredible achievement. Before the mission he had been told, by NASA officials, that detection of life on Mars would be the greatest scientific discovery in history. And now he was the discoverer. History would remember the name Gilbert Levin!

The news rocketed around the world: Viking had found something that looked a lot like life, and though the results were tentative—ambiguous—unverified and inconclusive—the obvious next move would be to ponder the cosmic implications. On NBC, reporter Jim Hartz asked Sagan what, "in cosmic terms," the significance of Viking was.

"Well, I think its historical significance is immense, because we are, or at least our machines are, breaking the shackles that tie us to Earth," Sagan began. "We're putting our toes in the cosmic ocean and seeing that it's kind of warm, benign, and interesting. And I think the long historical perspective this time will be considered as significant and important as the time between, say, the voyages of Columbus and the voyages of Magellan. . . ."

All this flowed from Sagan effortlessly, the kind of quotes that a reporter simply never got from a scientist—a regular scientist, that is, non-Saganesque. He didn't even use jargon.

"If there are two planets and they both had life independently originating, that's a key clue that life is easy to come by. And then I think it would be a short step to the grand conclusion that we live in a populated galaxy, that there is life throughout the galaxy. And I think the step to intelligent life would then only be a small one," Sagan said.

Some gas has burst from Martian soil, and Sagan had his cue to talk about aliens. But just when he seemed to be on the verge of letting his brain go winging off into deep space, he'd yank it back—and say that there might not be life on Mars after all. Even this, though, would be positive in a way, Sagan told NBC, because it would give us "a sense of the preciousness of our little planet. Either way, whether we find life or not, I think the point of the uniqueness of human beings, the uniqueness of all the organisms on the Earth, will be driven home powerfully, because the clear lesson of evolution is that nowhere else can there be life like here. There may be intelligences elsewhere. They may be smarter and have better music and all the rest of it, but there cannot be human beings."

At JPL, however, the euphoria began to die. Doubts emerged. How did they know they hadn't merely witnessed some kind of interesting chemistry?

One of the Viking experiments, designed by Vance Oyama, had shown oxygen coming from the soil after exposure to water. No one could figure out why that would happen—it wasn't consistent with living organisms. But even more striking were the findings of another instrument that wasn't officially part of the biology package. Viking had brought to Mars a device called a gas chromatograph mass spectrometer. The GCMS could determine, with a fair degree of precision, what kind of molecules were in the Martian soil. The scientists were certain that they'd find organic molecules, those carbon-linked molecules (such as amino acids and hydrocarbons) that had already been found in meteorites, and which appeared to be common in the universe. Organic molecules are necessary for the existence of life as we know it. So the GCMS went to work, fiddling with the Martian soil, warming it up, scrutinizing the emanations. The results: Zilch. No organics. Not a single lonely amino acid.

Now, the doubters came to the fore. With Norman Horowitz in the lead, they produced a new argument: The "positive" results from two biology experiments were the result of chemistry, not biology. The soil, they said, was full of peroxides. (Over time, the doubters developed a scenario in which iron peroxide in the soil, when exposed to water, produced free oxygen, as seen in the Oyama experiment.) Because Mars has no protective ozone layer, ultraviolet sunlight had broken down water molecules (H_2O) into hydrogen peroxide (H_2O_2) and hydrogen. These peroxides destroy organic matter, literally sterilizing the surface in the same way that hydrogen peroxide purchased at a drugstore will disinfect a wound.

No one could know for sure that this was what was happening

on the Martian surface. Levin continued to claim a positive result. Soffen (decades later) told me, "Gil could be right." But he added, "It's so much more plausible that the peroxide in the soil chewed this up, rather than a Martian bug."

In August 1976 the scientists didn't know what to tell the public. Science is complicated. There are moments when you can't get a perfect yes answer or a perfect no answer (in part because you didn't ask the right question).

The press howled: Make up your mind. Weeks passed. The second Viking lander touched down and did more experiments. Same ambiguous results.

Meanwhile, Sagan looked at pictures. He and his second wife, Linda Salzman Sagan, pored over the images every night with their close friends Tim Ferris and Ann Druyan, who were staying with the Sagans while visiting Pasadena. Ferris wrote for *Rolling Stone* and would become, over the years, the best science journalist of his generation. Druyan, his girlfriend, had just graduated from college and wanted to be a novelist. They all knew precisely what they most wanted to find. The first glimpse of the place could tell you that nothing could possibly live there (except those nifty microbes—maybe). The lander rested on terrain that looked like the driest regions of the American West. What they wanted to see in this jumbled landscape was something that looked . . . well, artificial. Some markings, ruins, remnants of previous visitation. Perhaps there would be some parallel lines. Though not the same thing as proof of aliens, parallel lines would be highly suggestive. Mariner 9 had already taken images of "pyramids" (Sagan's word) in the region called Elysium. Sagan had argued ever since that rovers should be dispatched to check out such provocative sites. He felt that Mars was an ideal location for the preservation of evidence of past visits by intelligent creatures. The surface has virtually no tectonic activity and is therefore not recycled. Without water and with little atmosphere, there is little erosion. Alien artifacts from a billion years ago might still be lying around.

They did not see what they wanted to see. The pictures showed rocks, and more rocks. One night Sagan and Ferris jointly examined the panorama picture. They sat on the couch, heads together, and bent the picture above them so that it created the full 360-degree effect. Ferris said he saw something on the horizon: Palm trees. An oasis. Sagan said he saw the same thing. They knew it was an illusion, a projection of human yearning. The human eye is adept at filling in the missing details, and what was missing from Mars, spectacularly, was the slightest visible trace of life.

To Sagan's eyes, Mars still had possible—possible!—signs of ancient civilizations. "Some of the patterns are so odd, intricate, that we cannot be sure they are caused by windblown sand," he said in his TV series *Cosmos* a few years later. The screen showed interesting patterns akin to plowed fields.

Months passed, and still the Viking team could not say one way or the other if Mars had life on it. At a press conference in Washington in November, more than three months after Viking 1 landed, Jerry Soffen, Chuck Klein, and several other scientists ran through the results, step by step, but declined to give the press what it wanted, a conclusion. The scientists were virtually incomprehensible. They talked as though they were giving a graduate-school chemistry seminar. They detailed the various chemical reactions, the outgassing, the absorptions, the theories of molecular interactions, but none of it seemed to cohere into a central point.

Then it was Sagan's turn. He saved the day. He spoke in simple terms that English majors could understand. He implored the reporters to have greater "tolerance for ambiguity." As for his own scrutiny of the lander images, he knew precisely how to frame the results.

"There are no bushes, trees, cacti, giraffes, antelopes, rabbits," Sagan announced. The man knew his audience. These were reporters—in some cases, their knowledge of biology came from what they had seen at the National Zoo. "There are no burrows, tracks, footprints, spoor. There are no patches of color which are uniquely attributable to photosynthetic pigments. In short, there is not the faintest hint on a scale large enough to see at these two places, at these two times, of anything alive."

And then—magically—he offered hope.

"Let me make a few caveats. One is that we have examined one-ten-millionth of the surface of Mars"—he was a master of the impressive number—"with the two landings, and it might well be that these are not characteristic of the other places on the planet."

The unknown, as always, might be far more interesting than the known.

"For example, if I were to drop you into two random places on the planet Earth, those would very likely be oceans, and you would not see any large forms of life in short periods of time, and you would simply disappear from view. In addition, there are many places on the land area of Earth, like the great Peruvian desert, where as far as the eye can see there is nothing alive whatever; and yet, in the oceans and the Peruvian desert, those places are loaded with micro-organisms."

When others lost hope, Sagan was always there with encouragement, inspiration, and helpful ideas about what might well be. His friend and colleague Bruce Murray told me years later that Sagan gradually "assimilated" the fact that Mars had no life—but occasionally had relapses. He would start talking again about how there might be life on Mars. "If anything, that was his UFOs," Murray said.

But not even a voice as persuasive as Sagan's could alter the fact that Viking had failed to find what people wanted. Viking ended a dream that had lasted for ninety-nine years, beginning with Schiaparelli's discovery of the *canali* in 1877. The Mars of Lowell, H. G. Wells, Edgar Rice Burroughs, and the 1938 Orson Welles broadcast of *The War of the Worlds* had vanished. There was nothing that could replace Mars in the imagination. As Norman Horowitz put it, "There is not another object in the solar system with the irresistible allure of the old Mars."

And the old Mars was dead.

Horowitz's 1986 book *To Utopia and Back* concluded with sober words:

> The failure to find life on Mars was a disappointment, but it was also a revelation. Since Mars offered by far the most promising habitat for extraterrestrial life in the solar system, it is now virtually certain that the earth is the only life-bearing planet in our region of the galaxy. We have awakened from a dream. We are alone, we and the other species, actually our relatives, with whom we share the earth.

Gilbert Levin never gave up. To this day he argues that his experiment showed evidence of life on Mars.

"We detected life," he said. At the very least, he said, the results are "consistent with biological hypotheses."

At one point Levin went back to Sagan and tried to persuade him to share authorship on a paper arguing that there was visible lichen or moss or some kind of living substance on the rocks of Mars.

Levin argued that the spots on the rocks, monitored by Viking over the course of several years, had moved. Or at least they might have moved. This required close analysis of a few pixels of data. Levin recalls that Sagan, upon analyzing Levin's images, estimated that there was a 67-percent chance that the spots had moved. But the gatekeeper couldn't bring himself to sign on to the paper, and once again Levin was on his own.

On the tenth anniversary of the Viking mission, the participants gathered for a symposium at the National Academy of Sciences. Norm Horowitz got up and declared emphatically that there was no life on Mars. Then Gilbert Levin had his turn, for what Soffen, who was running the show, assumed would be fairly brief remarks. Levin, however, had a lot to say. He declared, at length, that no one had ever duplicated his results in a laboratory with nonliving materials. After twenty minutes, Soffen got up and indicated that Levin needed to stop talking. Levin didn't stop. Soffen approached Levin every few minutes and said, "Gil, I think it's time for you to stop." Levin wouldn't shut up. After fifty-five minutes, Soffen recalls, he finally could take no more. He was furious. Nearly an hour into the marathon briefing by the man who discovered life on Mars, Soffen managed to dislodge Levin from the microphone and break for coffee. As Soffen recalls, they began yelling at each other outside the hall.

"You bastard! You'll never do that to me again!" Soffen roared.

Levin countered with a string of similar invectives.

Levin has a somewhat different account of what went down—he says he was presenting a research paper and didn't run long in his talk—but he remembers that at one of their post-Viking encounters he and Soffen had an eruption and Soffen called him a "used-car salesman." In any case, the two men didn't speak again to each other for about five years.

Levin remains on the periphery of the Mars establishment, a maverick. But even if his view is in the minority, he still has his sugar. He is optimistic about tagatose. He said he thinks he has found a way to make tagatose for only $2 a pound. If he won't be remembered as the man who found life on Mars, he may yet make the perfect cookie from right-handed sugar.

Do you realize, he said, that the baked-goods industry is four times the size of the soft-drink industry?

8

Course
Correction

A government agency like NASA prefers to schedule its scientific discoveries. The perfect discovery is one that not only changes the world but also can be budgeted. Viking was almost the perfect mission in that regard, but Mars did not cooperate: It insisted on resembling a dead planet. As a technological and scientific endeavor, Viking could still claim to be a stunning achievement. But the fact that would resonate over time, the moral of the story, in a sense, was that Mars was dead. And that was an inglorious result. That wasn't supposed to be. Thousands of science-fiction Martians suddenly died screaming.

So perhaps the breakthrough in the field of extraterrestrial life would not emerge from careful planning by NASA administrators. Another mode of discovery might be necessary: serendipity. To find life beyond Earth might require great feats of patience and calm, a serene open-mindedness. The people with the fiercest passion for the subject of alien life might have to learn to discipline themselves, to erase from their minds any and all presumptions about extraterrestrial organisms. The aliens would have to appear on their own schedule, in their own peculiar habitat.

If life's out there, it'll probably turn up eventually. The discovery may require a billion-dollar investment in a new tool, or it may simply require that some clever student re-examine some existing data. The aliens could be sitting in a filing cabinet, awaiting a new narrative, a new interpretation. They may be discovered through a mistake. When Columbus set out across the Atlantic he felt certain that the Earth was much smaller than the mainstream scientists believed it to be. He knew that China was just a few thousand miles west of Europe. Sometimes you find things that are better than what you were looking for; this may be the model for discovering alien life. Human beings, in one way or another, have managed to venture into the unknown, to see what's out there, to map nature, and there's no reason to think that extraterrestrial life will forever stay so elusive and mysterious. The only way we won't find it is if it isn't there.

Viking, in any case, sent exobiology into a state of shock. It became not so much a dead science as a pale, trembling one. Students knew that they could not pursue extraterrestrial life as a serious subject. (Bruce Jakosky, who would later become one of the country's leading exobiologists, told me that studying extraterrestrial life after Viking "would have meant the death knell of your career.")

Sagan, however, barely broke stride. He banged out a book on human intelligence, *The Dragons of Eden*, that won the Pulitzer Prize. That an astronomer would tackle the brain was a sign of considerable confidence. It established what everyone had long suspected, that Sagan could talk or write about any element in the Drake Equation, from the origin of planets to the rise of big brains to the fate of civilizations. He was the definition of protean. In retrospect, it's amazing he never wrote a cookbook.

NASA, meanwhile, had another planned triumph in the pipeline: Voyager. Two robotic spacecraft, Voyager 1 and 2, would fly to Jupiter and Saturn. Then Voyager 2 would venture on to Uranus and Neptune. Again it was JPL's mission, and again Sagan played a role. In December 1976, the project manager asked him to find a way to include some kind of message on the two spacecraft, to be launched in 1977. Sagan turned to the people around him. He, his wife, Linda, and his friends Tim Ferris, Ann Druyan, Frank Drake, and the artist Jon Lomberg worked together to produce an indestructible metal record, a compendium of recordings of Earth sounds, messages, and music. A copy of the record would go on each spacecraft.

The Voyager Record had a less ambitious antecedent. Back in 1971, soon after the Byurakan conference, Sagan got involved in what he called "Mankind's first serious attempt to communicate with extraterrestrial civilizations." Sagan, knowing that NASA was about to launch a spacecraft, Pioneer 10, that would eventually leave the solar system, asked for permission to put a plaque aboard with a message from Earth. (The idea of including a message, as Sagan acknowledged in a note to Frank Drake, came from Richard Hoagland, who years later would explicate the Face on Mars.) Sagan had only three weeks to devise the message. It might be a timeless and historic missive from Earthlings to the Lords of the Galactic Empire, but it would also have to be off-the-cuff. Basically he'd have to blurt something out. Sagan turned to Drake for help. Working in overdrive, they came up with a message in the presumed intergalactic language of science. It included what Sagan, in a rare lapse into scientific abstruseness, called a "schematic representation of the hyperfine transition between parallel and antiparallel proton and electron spins of the neutral hydrogen atom." Sagan and Drake included a drawing of a cute little spaceship flying from the third planet of a nine-planet solar system. A series of binary numbers and dashes, radiating from a central point, located the solar system in time and space relative to fourteen pulsars. (Pulsars are extremely dense stars that can be detected with radio telescopes and have their own unique signatures, so they serve as landmarks in the otherwise fairly homogeneous expanse of space.) The most striking and controversial part of the plaque was a drawing of two human figures, quite naked. They were drawn by Linda Sagan. Her drawing was simple, and years later hardly looks offensive. But feminists were upset. The woman has her arms by her side, whereas the man raises one hand in greeting. He, therefore, is the figure of action. He's doing something, while she's just a lump. Perhaps worst of all, he has genitals and she doesn't. She's just sort of blank down there in the personal zone. This presumably would confuse the aliens. Sagan himself felt worst about the failure of the drawings to depict the figures as panracial. That had been the intention, but they came out looking white. Indeed, they looked white and suburban. "Hi, we're from Orange County" were the words attributed to the man in one newspaper editorial cartoon.

In any case, Pioneer 10 and a second craft, Pioneer 11, took the

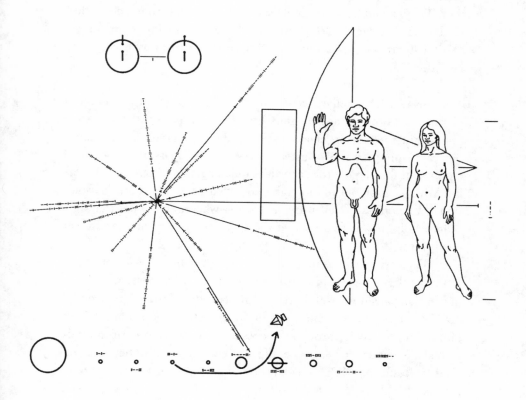

The first message from the human race to civilizations in deep space.
No snide captions please.   NASA

message into deep space, though neither would ever approach another inhabited world. It would take Pioneer 10 fully eighty thousand years just to travel the equivalent of the distance from Earth to Alpha Centauri. Neither spacecraft was likely ever to intercept another solar system. They would both be lonely sentinels passing through a limitless void, their messages undeciphered, the little man waving in greeting at random hydrogen atoms, dust particles, and the microwave background radiation.

The Voyager Interstellar Record would be far more complex and well planned than the Pioneer plaque. Everyone on the team agreed that, once again, the real target audience was the human race, not the aliens out there in space, but—just in case—the Voyagers also contained equipment for playing the records. Linda Sagan later wrote, "Watching Voyager flash out of our sight, and eventually out of our jurisdiction, on its one-way ticket to who-knows-where, one hopes that, like Marco Polo, it will find itself at the gates of some ancient and great civilization."

The basic goal was to capture some representative slice of life on Earth, without succumbing to provincialism. It was a heroic effort, but provincialism is a subtle thing, and two decades later the record may be, more than anything else, a documentary about the 1970s, a glimpse of the world through the eyes of a few 1970s American intellectuals. Religion, for example, did not make the cut. There was no God, no Jesus, no Muhammad, no Buddha. The messengers later explained that there were just too many religions out there—"at least a dozen and probably hundreds of major religions on the Earth." But there was Beethoven, along with Chuck Berry, greetings from the secretary general of the United Nations, greetings in fifty-four languages, and some "whale greetings." Druyan wrote, "As a long overdue gesture of respect for these intelligent co-residents of Earth, we wished to include their salutations among those of the statesmen and diplomats." The record included the sounds of volcanoes, mud pots, wind, rain, surf, crickets, frogs, a wild dog and a domesticated dog, and a kiss. At one point Druyan obtained an EEG of her brain waves as she meditated on important people and ideas through history and sought to avoid thinking of anything inappropriate. Perhaps the aliens would know how to unscramble the static and read her mind.

The messengers worked overtime, frantically. They were gamers, taking seriously their responsibility to represent Earth correctly. Sensitive issues required delicate negotiation. Should they include the Miles Davis recording of George Gershwin's "Summertime"? A black man's version of a Jew's composition seemed a nice multicultural touch, but such matters are never simple, and there were fears that they would be tacitly diminishing the importance of the black musical tradition. They decided to keep Gershwin on Earth. They did find some Navajo night chants. They took a pass on Bob Dylan (why confuse the aliens?). They got permission from the Beatles to include "Here Comes the Sun" but couldn't use it because the Beatles didn't own the copyright to their songs.

Sagan and his compatriots managed to secure a greeting from President Jimmy Carter: "This is a present from a small distant world, a token of our sounds, our science, our images, our music, our thoughts, and our feelings. We are attempting to survive our time so we may live into yours. We hope someday, having solved the problems we face, to join a community of galactic civilizations. This record represents our hope and our determination, and our goodwill in a vast and awesome universe."

Druyan had the job of finding the musical selections and passing her recommendations on to Sagan. She became inspired. She found a song by some monks that, she said in a letter to the Sagans, sounded to her like a joyful transcription of the periodic table of elements. "Imagine finding this in the universe. It would be so encouraging," she wrote. She suggested various Chinese, Indonesian, Japanese, Peruvian, Korean, Irish, and Bulgarian songs. Music, she wrote, might be the most informative thing they could send with the Voyagers. "Thinking about the world's music out in the universe makes me feel motherly, tender towards the earth. What happens if the universe doesn't like our music? What if there's a long silence afterwards? What if the universe says, 'What else can you show me?' "

Ann Druyan first met Sagan at a dinner party at Nora Ephron's apartment in New York City. Her interest in science came primarily from her interest in the philosophy of Karl Marx. Marx had written of history as following certain inexorable scientific principles, and had wanted to dedicate *Das Kapital* to Darwin. Druyan herself had, at the time, rather vaporous standards of evidence for her many and sundry beliefs (as she later acknowledged). She believed, like many intellectuals, that Immanuel Velikovsky in the 1950s had correctly deduced the truth about the solar system, that it was so highly dynamic as to have been rearranged within human history, that Venus was a fragment of Jupiter and had buzzed by the Earth only a few thousand years ago, and so on. She believed in the ancient astronauts of Erich von Daniken. Exotic science had a political appeal, it was defiant, it challenged the orthodoxy of those severe gray men who had found nothing better to do with scientific knowledge than build the bomb. Yet here was Sagan telling her she was wrong. Sagan was clearly not a member of the Establishment. He was a hip professor, an antiwar activist, well credentialed as a liberal if obviously not as far to the left as she was. But he didn't believe any of the cool Velikovskian/von Daniken–esque theories floating through the cafés and dorm rooms. Sagan told Druyan precisely why Venus could not have been formed recently, why the whole *Worlds in Collision* theory of Velikovsky had no factual basis. She didn't like his attitude, his incredible certitude. Yet he was enormously charming at the same time.

She was engaged to Ferris. Something didn't feel right, and she called off the wedding, postponed it. She needed time to figure things out. One day in the late spring of 1977, during a long-distance phone call, she and Carl Sagan figured out that they were in love, and to Druyan it was like the realization of a mathematical truth. They decided, over the phone that day, to get married. They had never even kissed. Three days later, they met in New York City, at the entrance to a tourist boat that circles Manhattan. They got aboard and planned the rest of their lives, deciding to tell their partners of their decision after Voyager launched in August. They would simply try to make the best of a bad situation.

Voyager launched. They broke the news to their partners. It was not a happy time. The legal battle between Carl and Linda Sagan lasted at least eight years—by which time Voyager 1 had already passed Jupiter and Saturn.

The new relationship with Druyan coincided with Sagan's ascendancy to a new level of public acclaim. He made the most important move of his career: He agreed to write and host a TV series about space for

Carl Sagan in 1980 explained the cosmos to the masses. In his universe, there could be untold civilizations more intelligent and advanced than our own.

PBS. *Cosmos* ran for thirteen episodes and became, to that date, the most-watched public-television series in history. Sagan became famous on a global scale. His book *Cosmos* was destined to become an enormous best-seller. He became not merely the spokesman for the community of astronomers and planetary researchers, but also the representative of science itself, the singular promoter of a rational, open-minded, imaginative investigation of (to borrow a phrase from Douglas Adams's *The Hitchhiker's Guide to the Galaxy*) "Life, the Universe, and Everything." No one had ever translated the grandeur of science to a broad audience as well as Sagan. *Cosmos* had everything going for it: a brilliant, imaginative narrator; gobs of money from the corporate sponsor, ARCO; the patience of public television; and, most of all, the hunger among the public for real information about their amazing universe. It was as though they didn't know it was out there! They knew but they didn't really appreciate it. Sagan expanded their horizon by a factor of billions. Years later, people would thank him in the street. A porter once turned down a tip, saying that Sagan had already given him the universe.

In the first episode of *Cosmos*, Sagan manned the bridge of a clean, angular spacecraft with a large viewing screen, not unlike the screen viewed by Kirk and Spock on the *Enterprise*. Sagan led his viewers on a trip across the universe, starting at a point far, far from Earth, billions of light-years distant. The voyage would take the audience homeward. First Sagan showed the realm of the galaxies, billions of them strewn in great superclusters across immense distances of space. Eventually he reached the Milky Way galaxy. He zoomed through the spiral arms of the galaxy, past stars beyond number. "Within this galaxy are stars and worlds and, it may be, an enormous diversity of living things, and intelligent beings and spacefaring civilizations." No scientist was ever more careful to insert the crucial caveat of "may" or "might" or "it is not inconceivable." With proper phrasing, you could countenance almost anything—not just a diversity of living things but an enormous diversity. Finally the starship entered a familiar-looking solar system. The viewers could see a lovely ringed planet. Then they came to a smaller world, blue, with oceans visible, half the planet in darkness, city lights shining. "But this is not the Earth," Sagan announced unexpectedly—it was some other inhabited planet. It turned out we weren't in our familiar solar system after all! We were still far away, looking at a planet reworked by a civilization far more "advanced" than ours. "Are they also a danger to themselves?" Sagan pondered. Only after seeing the world of these superior creatures did Starship Sagan continue onward, toward Earth.

Then the mail started coming. It came from all over the world.

A sixteen-year-old student in Pakistan identifying himself as 1.6 meters in height and 38.6 kilograms in weight wrote: "With due respect, I want the permission to say something. Because I am very afraid of your great personality. . . . Sir! I want your permission to become your student. You will teach me astronomy and its problems in letters, etc. I think it will be very useful for me. . . . O.K. I will wait for your answer in positive."

Rare was the letter that did not ask something of Sagan. A man in Evanston, Illinois, told his entire life story as part of a pitch to Sagan to co-author a book with him—the topic of which was never quite stated. The man just wanted to collaborate, to feed off the Sagan brain and write something.

One young woman wrote to ask, in the sincerest terms, if Sagan would be her date to the prom.

Most of the mail was about science, and it offered stunning evidence that, although people aspire to be scientific, it is not always their natural calling. One day Sagan received a letter from a woman with a graduate degree in anthropology. She lived in St. Petersburg, Florida. She wrote: "A few of my friends and I have longed for the day that we might come into friendly contact with some unknown being, and your scientific notions combined with your wonderful style convey the feeling that anything is possible. To continue along the 'anything is possible' motif, I have a question to ask you. Are you an extraterrestrial? I know this is quite personal, and you have every right to be shocked or even angered at my asking, but it's something I feel could actually be true. After all, it makes perfect sense. For what better reason might you write the things that you do than to prepare us for Contact with another race? I don't know how old you are, but it seems that you have gathered an amount of data and insight on the cosmos incomprehensible to a mere human. . . ."

It was Sagan's own logic thrown back at him, in a sense. Could a mere human—one of the dummies!—have written so many books and expounded on so many mind-boggling theories?

Sagan wrote back: "I like your hypothesis, but I have to confess I'm just another of those humans you can find almost anywhere on this planet."

Everyone wrote to Sagan. A woman in Sebring, Florida, demanded an audience with Sagan so that she could share her revelation that anachronistic instincts of fear and paranoia, operating in "the R-complex and the limbic system," will lead to human destruction.

A man in Richmond, Virginia, wrote: "Your physical body—you?—was once a single cell of protoplasm. How did you become what you are now? What guided, controlled the development? What is life? Can you really believe that 'mind' can be explained as only material structures and electrochemical processes?"

A man in New Delhi wrote: "I have developed a simple theory by which we can move any object at any speed, even more than the speed of light. This theory is flawless but has not yet been practically tested. . . ."

A woman in St. Louis wrote a very short letter: "I am 78 years old. Something has been bothering me for many years. Where does all this noise from our world go? What happens to it?"

A man in Kuwait wrote to suggest that long ago a civilization on Venus came to Earth, and provided the "missing link" between apes and humans. A man in Berkeley wrote to say that the Earth is analogous to a light switch and is about to "snap over" to a condition of lifelessness.

The gatekeeper usually wrote back. He did his part to educate people. But he was bailing out a sinking ship with a pitchfork.

As the Voyagers toured the solar system, they sent back images of exotic worlds and discovered new moons of the distant gas giants. The probes found that Jupiter's moon Io is studded with active volcanoes. They gave scientists their first look at the cracked, icy surface of Europa—and incited the first tentative speculation that it could have a liquid ocean beneath its surface. For the optimists these were thrilling discoveries. One could not rule out life elsewhere in our solar system.

But the pessimists didn't see it. They saw a solar system that is remarkably, strikingly lifeless. They saw dust and gas and rocks. A more or less complete survey of the solar system indicated that only the Earth had life, and indeed only the Earth could possibly have life. A person had only to go down the list of the nine planets to see how forbidding was our little neighborhood. Mercury: A joke, a small planet not much bigger than our moon, cratered, without atmosphere, fried on the side that permanently faces the Sun and frozen on the other. A horrible little world. Venus: About the same size as Earth but hellishly hot, broiled by a massive runaway greenhouse effect. Thick clouds choke the atmosphere, and all liquid water has long ago boiled away. When spacecraft have landed on Venus, they've sent back a few pic-

9

The
Naysayers

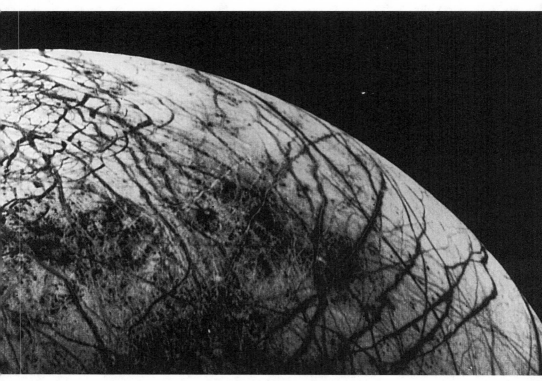

Cracks in the ice of Jupiter's moon Europa.
Is there an ocean down there? And life? NASA

tures in the scant moments before the spaceship has melted. The moon: a cratered wasteland, no atmosphere, no liquid water (only years later, with the Clementine mission, would we find evidence of water ice), essentially an agglomeration of debris from a catastrophic impact billions of years ago. Mars: runtish, frozen, sterile, with an atmosphere too thin to protect anything from the harsh light of the Sun. Jupiter: a giant world of gas with no discrete solid surface. Jupiter's moons: small, scarred, icy. Saturn: more gas. Saturn's moon Titan: covered with thick orange smog and too cold for liquid water. And so on down the list: more gas, more ice, until finally the solar system peters out at the eccentric, pitifully small and slightly ridiculous ice-world called Pluto.

The geologists, atmospheric chemists, and physicists all loved this solar system, because it had an abundance of the qualities of the things they study. Biologists could not love it. Nor, really, could the American people, or the people of other countries with a space program. Where were the Venusians, the Martians, the Jovians, the Saturnians? The aliens who lived on the other worlds of our solar system had gone the way of twice-a-day home mail delivery. That shocking development could be absorbed by the science-fiction writers, because they needed only to shift the extraterrestrials farther away, to distant stars and galaxies. But the failure to find life beyond Earth reduced the incentive to send human beings to other planets. People do not much care for rocks, gas, dust, liquid oceans of hydrocarbons, clouds of methane.

Even as Voyager went into high gear, it was clear that NASA had turned against planetary exploration. A planetary probe was like an oil tanker, a behemoth with tremendous inertia. You could tell years in advance what would be out there beeping its way through the solar system. In the late 1970s, the future looked dull. Most of the energy at NASA was focused on the Space Shuttle, which still hadn't flown.

After Viking, a self-respecting scientist was supposed to study things like cosmic structure, the evolution of galaxies, the abundance of elements in the interstellar medium, atmospheric chemistry, the geomorphology of dead worlds—the boring stuff, the *hard science*. Saganism might be popular, but it wasn't quite mainstream. There were still some people interested in sending humans to Mars, but they were so few they called themselves the Mars Underground. The pessimists ruled the world: They said there might not even be many planets out there upon which life could originate. There had been much publicity in the 1960s about a possible planet orbiting Barnard's Star, one of the Sun's nearest neighbors, but a 1973 study had shattered that analysis and cast a pall over the search for planetary systems. The Barnard's planet had been a big hit in the press and the crowning achievement of a major astronomer who had worked on it for twenty-five years—and now it turned out that it didn't even exist.

Some astronomers were also beginning to think that Earth was in an extremely lucky position in space, a narrow habitable zone in which water could remain liquid. In 1977, the astronomer Michael Hart, who had written the famous reconsideration of the Fermi Paradox, published calculations showing that Earth could not be more than 5 percent closer to the Sun or 1 percent farther away and still maintain a habitable biosphere. In other words, one wrong move and you're dead. Look at Venus—only twenty-six million miles closer to the Sun, and totally scorching. Mars, on the other side, is upwards of thirty-four million miles farther out (it has a highly elliptical orbit), and it is frozen. Earth is in the Goldilocks position—not too hot, not too cold.

A planet, Hart wrote, had to be just the right size, too. A planet whose radius was merely 10 percent bigger than Earth's would spew too much carbon dioxide from its interior and have a runaway greenhouse effect, like Venus, Hart calculated. And yet, if the radius were less than 94 percent of the Earth's, the planet wouldn't hold on to its atmosphere—the fate of Mars. Reinhard Breuer listed many of the obstacles in his 1978 book *Contact with the Stars:*

> It is not enough for the planets to be born within the narrow ecosphere of a star. They must also have the right mass—neither too large nor too small—otherwise either the surface gravity is too high, or the planet cannot retain a suitable atmosphere. The atmosphere must develop a sufficiently thick ozone layer to ward off the life-destroying ultraviolet radiation of the star; and it requires a high carbon dioxide content, to develop a moderate greenhouse effect and so prevent the seas from freezing. Further, the planet must turn on its axis, at the slowest, in 4 terrestrial days, or it will be too hot by day and too cold by night. It must

also travel around its sun in an almost circular orbit, otherwise the seasonal changes will be too great. And even this special orbit must undergo only minimal variations over the millennia.

Constraints everywhere! In 1980, Robert Rood and James Trefil wrote a book, *Are We Alone?*, that argued that the popular, Saganesque notion of millions of civilizations out there was simply nonsense and that the new orthodoxy among the serious scientific community was that the constraints on habitability were far greater than the optimists realized. Rood and Trefil said there might be intelligent creatures out there, but when they crunched the numbers in their own versions of the Drake Equation they got either one civilization in the galaxy (us) or possibly two or three. A small number. The galaxy just didn't have much action, as they saw it.

In 1981, Frank Tipler, a physicist, published another attack on SETI in the *Quarterly Journal of the Royal Astronomical Society*, where Hart's Fermi Paradox article had run six years earlier. Tipler said that if there were civilizations out there we should see signs of von Neumann machines. These theoretical robotic spacecraft, named after the mathematician and computer pioneer John von Neumann, would manufacture copies of themselves and spread across the universe. Tipler pointed out that we were not far from developing such devices and there was no reason some advanced civilization couldn't do it.

All these arguments had flaws, internal loops and triple-twists. On the one hand, the SETI people were making the grand presumption that aliens would try to contact us with radio signals. But the critics made their own presumptions. The Fermi Paradox and the Hart/Tipler arguments are based on the presumption that aliens will build starships or robotic spacecraft and come to our solar system, and therefore that if we don't see such alien starships there must not be any aliens. These skeptics and naysayers wanted badly to be seen as the antidote to mushy thinking, and then they themselves produced a bunch of ideas that, when handled, sort of went . . . squish.

In 1982, Ben Zuckerman and Michael Hart produced another skeptical book, *Extraterrestrials: Where Are They?* They ain't anywhere, the authors said. Zuckerman found himself truly annoyed with the Saganists—they seemed to be bordering on intellectual dishonesty, he thought. Zuckerman had been shocked to hear Sagan give a talk at the

University of Maryland in which he floated once again the notion that Phobos could be an artificial satellite. Zuckerman felt the data on Phobos had long since been revised to preclude the possibility. What was going on? Didn't the visionaries know they had lost? "I don't feel he was intellectually honest," Zuckerman said of Sagan's remarks about Phobos. "I'm sorry to say he was willing to bend the envelope."

Freeman Dyson, a visionary of the highest order, conferred with Sagan during these troubled times. He thought all the various arguments on both sides were fundamentally without scientific weight. It made no sense, he felt, to try to deduce the existence of extraterrestrials, or their lack of existence, from such limited knowledge. He wrote Sagan a letter, chastising him for getting so worked up about Tipler.

> I wonder why you take Tipler so seriously. I think his arguments get more attention than they deserve. I cannot take seriously any of his numbers, nor yours either. Any specific model of the future has to be absurdly narrow and unimaginative. I believe only in limits set by the laws of physics and chemistry.
>
> We all tend to think in terms of metaphors. My metaphor for the future is not homo sapiens but life as a whole. Life has a way of expanding and diversifying to the limits set by physical constraints. So I expect it will be so in the future. Since the laws of physics allow life to spread at speeds of thousands of kilometers per second, it will probably do so.

Sagan wrote back, "I don't myself take Tipler seriously, but I worry when so many others (for example, the editor of Physics Today) seem to do so." He went on: "I think this debate is really about the conflict of two principles, each somewhat plausible: that life (as you say, 'as a whole') is very good at expanding and diversifying within the boundary conditions set by the laws of physics and chemistry; and the Copernican conjecture that there is nothing, in the broad sense, unique about the Earth. Is it more likely that extraterrestrial intelligence, if it exists, has spread to every nook and cranny of the galaxy, or that in a set of natural experiments among [a hundred billion] worlds, we are the first? If we had to bet our lives on it, which of these two contentions would we pick?"

The point being . . . that something didn't quite make sense.

Sagan, apostle of the Assumption of Mediocrity, was not about to start saying that humans were the lone example of intelligence in the galaxy. But on the other hand, the Dyson vision of life spurting through space at the speed of light had no observational confirmation so far. Which was Fermi's point.

Where are they?

And so exobiology remained, through the 1980s, a difficult science. If Sagan wanted to write about contact with aliens, he'd have to make it fiction.

Even as Sagan was finishing up his novel *Contact*, there was a referendum of sorts on Sagan the scientist. He had many supporters, friends, allies, students, people who truly revered him as a thinker and liked him as a person. ("I considered him a god," said planetary scientist John Kerridge.) But there were always some who found him distasteful, a grandstander. Finally, there were a few hardcore specialists who looked over Sagan's work and found it a bit lacking in rigor. Sagan had published millions of words over the years—he was almost in Asimov country when it came to prolificacy—but he hadn't made many significant discoveries.

This all came to a head in 1984. A group of colleagues nominated Sagan for membership in the National Academy of Sciences. Sagan was already middle-aged and had seen many of his friends gain induction over the years. The normal academy procedure is to have specialists in a given field—in Sagan's case, planetary science—make the nomination. But the specialists had declined to do it. Instead, scientists from a variety of other disciplines, among them such luminaries as Freeman Dyson, Lyman Spitzer, Hans Bethe, John Bahcall, Leon Lederman, and Philip Morrison, formed a Voluntary Nominating Group. The effort, led by Ed Salpeter of Cornell, was noble but hopeless. The National Academy of Sciences is as stuffy and rigid an organization as exists in our quadrant of the galaxy. Sagan's greatest achievements—his popularization of science, his landmark *Cosmos* series, his broad engagement of essential questions of life on Earth and in the universe—were almost handicaps, blots on his record, in the context of academy membership. What counted was "hard science." The nominators labored, in a 250-word statement, to argue that Sagan was indeed a hard scientist, as hard as they come. The nomination did not mention that Sagan had arguably done more for the advancement of science in modern American society than anyone else of his generation.

"Sagan is widely known for his effective popularization of science," the statement lamely began, "but this must not be allowed to overshadow his scientific contributions."

It then listed his work on Venus, with Mariner 9, Viking, and Voyager, his study of the dust storms on Mars and climate change on Earth, and his laboratory research on the origin of life.

The nominating letter did not fly. One academy member wrote a blistering response to Salpeter: "In my professional opinion, Carl Sagan the scientist is somewhat above average, but certainly nowhere near the quality that would earn membership in the Academy. I can think of several people in our area (including Carl's student Jim Pollack) who rate much higher. Where Carl is truly outstanding, of course, is as a spokesman for, and popularizer of, science. I have a great liking and respect for Carl, especially because he still works hard at real, day-to-day science. Nevertheless, it is my understanding that election to the Academy is based on scientific excellence and creativity, and not on ability as a popularizer."

The academy member then noted that he had spent time in the library researching the record on scientific inquiry into the atmosphere of Venus and had concluded that Sagan had not originated the idea of a Venusian greenhouse effect. "It is perfectly true that Carl put a lot of effort into trying to resolve these issues; what was lacking was anything that could be called brilliantly original." As for Sagan's work on the origin of life, "it represents a trivial extension of the classic work of Miller and Urey, and is merely a routine tilling of the fields that they opened up."

And so Sagan did not gain membership in the National Academy of Sciences. There was one more try, eight years later, in 1992, but again there were objections, and a debate broke out in the annual meeting, the pro-Sagan crowd arguing again that he was a hard scientist and the skeptics saying his research had not been all that scientifically significant. The other fifty-nine people nominated on the ballot were routinely approved, but because Sagan faced a challenge, under academy rules there had to be a vote on his nomination. He failed to gain the two-thirds majority needed for membership. The most famous scientist in America learned that he'd been officially rejected by the institution of scientific elites. Two years later, the academy, perhaps ashamed of itself, gave Sagan a consolation prize, the Public Welfare Medal, which came dressed up in accolades and made him an "honorary" member with no voting privileges.

This rather specific professional slight, endured over the course of many years, did not diminish Sagan's general optimism and good spirits. He knew he'd had a fortunate life full of incredible rewards. He could even understand why his colleagues would spurn him. He said, "A scientist who devotes his life to studying something arcane like the hyperfine structure of the molybdenum atom, and whose work is ignored by everyone except the world's three other experts on molybdenum, naturally is jealous and outraged to see reporters hanging on me for my latest pronouncement about the possibility of extraterrestrial life."

The academy's actions struck him as fundamentally ungrateful and foolish, because science needed boosters and popularizers, but there were other things in his life that loomed larger. One of them was the specter of his own death. Sagan since adolescence had suffered from a faulty "cardiac sphincter," the valve in his esophagus that keeps food from coming back up. After a meal he had to hop up and down to ensure that he would not subsequently choke. One night in Ithaca in 1983, he went to bed early with a stomach pain, which worsened until finally Druyan drove him to the local hospital. The doctors diagnosed appendicitis and quickly performed surgery. Druyan stayed with her husband in his room. At three o'clock in the morning, Sagan stood up, deathly pale, and fell to the floor. He was bleeding to death from his esophagus. The doctors sent him by ambulance to a bigger hospital in Syracuse, where he underwent ten more hours of surgery, a valve from his intestine replacing the damaged valve in his throat. The operation could easily have killed him. He awoke with an incision from the top of his chest to below his navel.

After that he was a bit different.

Publicly, he became even more political. He dictated a petition against President Reagan's Star Wars plan while still in the hospital. He began his work on "nuclear winter," arguing that a nuclear war would throw so much ash and soot into the atmosphere that the sunlight would be blocked for years, plunging the planet into a period of ice, possibly triggering a mass extinction. This was a political statement encrusted with science, and as such was a dicey business, since in science everything is subject to revision. In the case of nuclear winter, Sagan was using numbers from another researcher, and they proved faulty. The research community determined that the situation wasn't quite so dire; instead of a nuclear winter there might be a nuclear autumn. Politically it might have the same significance, but it didn't help Sagan's reputation as a careful scientist. He forged ahead nonetheless, spending his *Cosmos* capital on politically worthy causes. He began lobbying for a joint U.S.-Soviet manned mission to Mars. Through space we could finally end the Cold War, maybe (and, oh yes, find out if Gilbert Levin had discovered life on Mars or just some odd chemistry).

Privately, he started to experiment with something relatively novel: parenting. He had a baby in his house, his fourth child, but his first and only daughter. For Sasha he would try, finally, to be a fully engaged father. He also paid more attention to his thirteen-year-old son, Nick, who had been living on the other side of the country with his mother. They talked on long drives to the family therapist. They went to movies together, like *Aliens*, after which the elder Sagan complained, "Why do we have to always portray extraterrestrial life as though it's out to destroy us?"

Druyan noticed that her husband learned to do things he would never have done in his earlier marriages—pick up his socks, make the bed, wash the dishes. He would sometimes remember things he had said or done in his first two attempts at married life. "What a jerk I was," he said to his wife.

Sagan had been given another chance, another *life*, and he was determined to do things better this time. Perhaps he realized that we are not, at this point, cosmic creatures, but remain entirely terrestrial. Our responsibilities are close to home, not in the stars. In his novel *Contact*, which he was writing during this period, Ellie Arroway is shadowed by the guilt she feels for ignoring her aging mother. At one point she receives a letter from her stepfather saying that, although she is a big shot who flies around the world and gives speeches, as a human being she hasn't learned anything since high school. At the end of the novel, having made contact with extraterrestrials but having gleaned little information from them,

Arroway finally realizes she had been focused on the wrong things in life:

> She had spent her career attempting to make contact with the most remote and alien of strangers, while in her own life she had made contact with hardly anyone at all. . . . She had studied the universe all her life, but had overlooked its clearest message: For small creatures such as we the vastness is bearable only through love.

10

The Savior

Dan Goldin saved the Space Age, or what was left of it. At the very least, he prolonged the idea that space is relevant to the human future. His achievement was an improbable one, for he joined NASA at a time when the space agency had become a national joke, when it had reached its nadir after a series of failures and blunders, and one spectacular tragedy.

The Challenger explosion on January 28, 1986, had momentarily spurred a verbal rededication to the central mission of spaceflight. The seven astronauts were martyrs to a grander cause. It would have been cruel in that moment of pain and horror to question the underlying goals of the shuttle program, or to note that there was often not a compelling technological or scientific reason to send payloads into orbit with a piloted spacecraft. "It's all part of taking a chance and expanding Man's horizons," President Reagan told the shocked nation. "We'll continue our quest in space."

Soon came the review of the accident, and people were shocked to discover that NASA had failed to pull the plug on a flight even though there were icicles hanging from the shuttle. The contractor for the solid rocket-boosters knew that they were in danger of failing, and the extremely cold temperatures made the launch all the riskier. But government agencies rely heavily on precedent; there had never been a failure of the solid rocket-boosters, and the presumption was that they'd continue to function properly. The accident review made the space agency look like just another gelatinous government bureaucracy. To its critics, NASA had become the equivalent of the Department of Agriculture or the Department of Energy or the Department of Housing and Urban Development.

A brief attempt in 1989 to put more juice into the space program failed miserably. President Bush, newly elected and flush with Kennedyesque frontiersmanship, announced that he wanted the United States to return humans to the moon and then send them on to Mars. Bush's "Space Exploration Initiative" sent NASA into overdrive to come up with a Mars strategy. The space agency sketched out a plan, called the 90-Day Report, that envisioned a bewildering array of aerospace hardware, from giant spaceships to moon bases to space stations and orbiting hangars—what Robert Zubrin, a visionary engineer, derisively called "the parallel universe." Word spread on Capitol Hill that the plan would be a $450-billion boondoggle. After that, there was little utterance of the phrases "Space Exploration Initiative" and "90-Day Report."

Bad things kept happening to NASA. A billion-dollar probe, Galileo, delayed for years already, became delayed further by the Challenger disaster. It was already sitting at the Cape, and the agency decided to transport it back to California pending the resumption of the shuttle program in a couple of years. Technicians loaded Galileo into a truck and drove it at painfully slow speeds on bumpy country roads, and somewhere along the line—in some back-country town like Perry, Williston, Chiefland, or Live Oak—the vibrations may have damaged the probe. No one is sure when or where it happened. The main antenna no longer could be properly deployed—a fact that NASA discovered years later, when the probe was already racing toward Jupiter.

And things got worse still. Nothing could quite compare, for sheer technological sloppiness, with the space telescope that couldn't see straight.

The astronomer Lyman Spitzer had proposed a space telescope in the 1940s. High above the obscuring atmosphere of Earth, the telescope would have the ability to gain sharp images of astronomical structures. This fantastic idea, nurtured over decades by thousands of people, culminated in the launch in 1990 of the Hubble Space Telescope. On May 20, the Hubble saw "first light." Orbiting over New Guinea, it pointed at a star, gathered data on a computer tape, and relayed the information to the ground, where the image was processed and shown on monitors at Goddard Space Flight Center, near Washington. The star was sort of fuzzy, but that didn't bother anyone too much—it would surely be quickly corrected. One astronomer e-mailed his colleagues: "Think what they can do with fine tuning the images."

But a little fine-tuning didn't improve the situation. Within a day, a few astronomers began to mention the disturbing—indeed, rather hideous—possibility that the telescope had a serious flaw. The stars remained out of focus. What did they look like? The phrase that went around was "squashed spiders." There seemed to be a spherical aberration in the images. No one wanted to hear the words "spherical aberration" (much less "squashed spiders"). For the next several weeks everyone scrambled to find a way to make the aberration go away. Maybe, thought some astronomers, the problem was merely in the camera, not the mirror. Or, if there were a spherical aberration, perhaps it could be corrected with tiny "actuators" placed behind the mirror, which could slightly alter the shape of the glass. But gradually the truth emerged: The spiders were there to stay.

NASA announced the mirror flaw at a press conference on June 27, 1990, and the howls of derision could be heard around the world. "NASA's $1.5 Billion Blunder" groaned the cover story in *Newsweek*. Cartoonists and comics adopted the Hubble as the archetype of a screwup, in the same way that a man on a desert island with one palm tree is the archetype of isolation. A massive investigation sought to find out what had gone wrong. The mirror had been tested with a device called a "null corrector." For the null corrector to work properly, a lens had to be in just the right place. To test whether the lens was in the right place, technicians used a measuring rod and a beam of light. But the beam of light hit a tiny cap stuck on the end of the rod. Normally the technicians would have known there was something wrong, because the cap was coated in nonreflective paint and the beam of light wouldn't have bounced off the surface. But there was a scrape in the paint surface. A tiny scrape! And so the light bounced, and the rod was in the wrong spot, and thus the lens was also in the wrong spot, and thus the null corrector was 1.3 millimeters out of place, and thus the primary mirror had a spherical aberration—and thus there would be no pretty pictures.

Into this pit of gloom and doom stepped Goldin. He arrived on April Fools' Day, 1992. The bureaucrats viewed him as little more than a hired assassin. The previous NASA administrator, Richard Truly, had bombed out just six weeks earlier, and Goldin had been yanked in out of the blue to replace him. Hardly anyone wanted the job; even if the agency hadn't been such a mess, few executive-caliber people wanted a position that might last only until the following winter, when a new administration would be inaugurated.

Goldin arrived as a credentialed downsizer. There was no doubt that he would annihilate much of what he encountered. NASA was an agency that desperately needed to be grabbed by the lapels and shaken, a hidebound bureaucracy. There were Code D and Code L and Code M and Code Z and all these other coded departments. There were massive spaceflight and research centers spread around the country. The flow chart of authority looked as if someone had splattered boxes at random on a sheet of paper and connected them with thin gray lines. A given piece of hardware, such as a spaceship, was effectively owned and operated by several separate fiefdoms, each one with a stake in the craft, from the space-science people to the astronautics people to the rocket builders to the legislative liaisons who had to make sure the thing was built in the right congressional districts. The only way an operation like this could survive over time was through the development of routines and patterns and habits. But now came Goldin, who wanted to change everything. He carried with him an abiding interest in the new concept called Total Quality Management, and immediately started something called the Office of Continuous Improvement. He had Red Teams and Blue Teams. He published an analysis of his own "metrics."

Goldin arrived with some scores to settle. For a quarter of a century, he'd been at TRW, an aerospace company, working on highly classified black budget programs. He worked on Ronald Reagan's Star Wars technology, including "brilliant pebbles," a program to develop small projectiles that could be aimed at Soviet nuclear warheads. During the late 1980s, he became a pest to NASA bureaucrats, criticizing their tendency to plan overwrought space projects that couldn't possibly get funding. He specifically criticized a $30-billion plan for the Earth Observing System, a series of huge satellites that would study the planet from space, monitoring, for example, the signs of global warming. (The wags howled: A $30-billion thermometer!) Goldin may have been correct to criticize EOS, but his abrasive style did not help his campaign. A couple of NASA's top officials called Goldin's boss at TRW and demanded that he be muzzled. Otherwise, the NASA executives said, TRW would be cut off from further space-agency grants. This, at least, is how Goldin tells the story.

Thus it was a rather beautiful moment for Goldin when he suddenly became the boss of these people who had disliked him so intensely. He vowed to do nothing to win them over. Goldin couldn't fire anybody—it's the government—but he could shove people into stupid jobs. In the case of Leonard Fisk, the associate administrator for space science, Goldin exiled him to a made-up post called "chief scientist." It meant Fisk went from being in charge of billions of dollars' worth of science to being a man with no authority at all. Fisk soon departed.

Goldin earned the reputation of a bully. His critics didn't mince words:

"He will say or do whatever is expedient at the moment."

"If you're working for him, you don't want to come across a problem that's contrary to his view."

One official summarized the feelings of middle managers: "They all hate him."

Goldin knew he was not, as he put it, "warm and fuzzy." He didn't think he could afford to be. "I'm a driven person. My intensity sometimes scares people," he said when I asked him about the criticisms. "I took this job not to be friends with people. I took this job to be an agent of change at NASA. I recognize that I'm going to upset some rice bowls." When he came, he said, NASA's budget had recently increased dramatically and there wasn't anything to show for it.

"Hubble was blind, Galileo was deaf. . . . We had a few Battlestar Galacticas on the Earth-science platform. . . . And everyone was happy! Everybody was happy! I don't think I'm a bully, but I came with an agenda."

What few people appreciated about Goldin was how seriously

he took the notion of the Space Age, of man's destiny in the stars. "Downsizer" didn't adequately describe him. His mantra—"faster, better, cheaper"—made him sound like an efficiency-demon, but in truth Goldin dreamed of bigger things. He had been a space nut since he was seven years old and living on Boynton Avenue in the South Bronx. His father would take him to the Hayden Planetarium at the American Museum of Natural History. Goldin could barely see anything: He had progressive myopia. An oddball, the one kid who couldn't play sports, he was driven toward books, and by the time he was in his early teens he had homed in on the burgeoning literature on futuristic space travel. Straight out of City College, he went to work for NASA, in 1962, thinking he'd help invent the next fantastic propulsion system. He worked on electric-ion propulsion, and anticipated a Mars voyage soon after Apollo reached the moon. In those days no one was afraid to talk about crazy stuff like antigravity devices and antimatter annihilators and faster-than-light travel. He left the agency after the budget began to fall in the late 1960s. When he had returned to NASA, as the top boss, he insisted that the agency revert to form, that it embrace those radical ideas again, that it start taking chances, inventing new things, dreaming.

And then one day, after all those years of mistakes and heartache, the sun broke through the clouds. In December 1993, the astronauts on the Space Shuttle used a robotic arm to latch onto the Hubble Space Telescope. Donning their spacesuits, they ventured out into the vacuum and installed within the telescope a new, small mirror that had precisely the configuration to counteract the spherical flaw in the primary mirror. The astronauts, just like that, fixed the Hubble. Astronauts! The forgotten people of the space program! They still zoomed into space about eight times a year, but few of them would ever be recognized in a shopping mall. The shuttle had always been the brunt of ridicule from the intelligentsia and here it had galloped to the rescue. People even started to talk about astronauts by name. They knew about this astronaut named Story Musgrave. It was like the glory days.

And then came the pictures. The Hubble not only worked, it worked sensationally. Human beings saw a new rendering of the universe. The Hubble—a joke no more—had made the cosmos come alive.

11

The Pillars of Creation

The essence of science is knowing what we're looking at. We try to figure out this thing in front of us. Early in a baby's life comes the discovery of the hand, the marvelous tool, only slowly revealing itself to be under the baby's mental control. Anthropologists study the emergence of the opposable thumb as a key moment in the rise of a species that manipulated its environment. Biologists and anatomists and chemists all study the nature of living things, how the brain, the most complex system in the known universe, can operate a nervous system in such a way as to command the clenching of a fist. A physicist goes beneath the world of cells and tissues and molecules, and studies the substance of reality itself, the protons and neutrons, the electrons and quarks. We are still, collectively, looking at our hands and wondering what exactly is going on here.

Sagan, for one, knew what he was looking at when he looked at space. He carried with him the knowledge that he had a specific connection to space, that the heavy elements in his body, the carbon and oxygen and phosphorus, were birthed in the cores of exploding stars. Our atoms were once out there. We came from space. We are starstuff, Sagan would tell everyone.

You can't actually see billions of stars in the night sky. On the darkest night, with no moon, in the desert or on a mountain, the human eye can see only about three thousand stars. More to the point, the human eye can't really tell what it's looking at. A star does not look like an extremely distant sun. Even the moon doesn't look like a world. The Sun does not look terribly far away, and reveals little of its true size. (Which is why Icarus was so confused.) Figuring out what these things in the sky are has been an ongoing project for several thousand years. The single most elusive fact of nature may be the motion of the Earth. Try as one might, it is impossible to detect the fact that the planet is moving—spinning on its axis and simultaneously orbiting the Sun. Ignorant of the Earth's motion, a cosmologist of the ancient world would struggle mightily to make sense of what was happening in the heavens.

Aristotle, nicknamed "the brain" by Plato, was a compulsive explainer. In his treatise *De Caelo*, meaning *On the Heavens*, he developed a physical theory for the way the universe is put together. He viewed the universe as a series of spheres, an idea originally proposed by another student of Plato's, Eudoxus. There were spheres of water, air, and fire, each seeking its ideal level and surrounding the central Earth. Fire and air go up, earth and water go down. Beyond these four elements came the spheres, respectively, of the moon, the planets, the Sun, the stars, and then the sphere of the *primum mobile* (prime mover) that caused the motion of the heavenly objects. Last came the sphere of the gods and the heavenly host. Although to our modern ear this sounds one step removed from pure nonsense (you know, did he really believe that stuff or was he just making it up as he went?), Aristotle's cosmos remained more or less intact for two thousand years. It was not a random assemblage of elements but, rather, a carefully ordered structure filled with meaning. Things were put together in a manner that bespoke their underlying value.

For many people the desire to believe in models of the world that are pleasing is more powerful than the desire to verify through experiment. Plato, for one, rejected the argument that reality should be put to scientific tests. His world was literally an ideal one. To this day, the Platonic approach to the world is highly popular, if by other names. Aristotle at least believed that scientific observation should trump a philosophical ideal.

The problem with the Aristotelian universe was that it didn't exactly match what astronomers observed. The planets, particularly Mars, wander in a bizarre fashion. Mars moves eastward among the stars, then slows down, grinds to a complete halt, gets brighter, and takes off westward, dims, stops again, and goes back east. In the second century A.D., the Greek astronomer Ptolemy refined the Aristotelian model and dressed it out with precise observations of the movements of the moon, Sun, and planets. Ptolemy dealt with the retrograde motion of Mars by inventing a model in which the orbit of the Red Planet has an "epicycle," a secondary circular motion imposed upon the original. To fit the theory further into observed Martian motion, he came up with an eccentricity to the epicycle. Indeed, the Ptolemaic system began piling up epicycles and equants and deferents to explain the subtle motion of the heavens.

Ptolemy's work survived over the centuries only because the Arabs kept alive the science of astronomy as Europe descended into its long intellectual decline. In the Renaissance, when the Europeans began rediscovering the works of antiquity, Ptolemy returned to prominence. He was still the authority on the cosmos. And the system worked, for all practical purposes. Your average navigator was able to cross the oceans without needing to know that the Earth went around the Sun. Columbus, Magellan, da Gama, and all the other great explorers labored in a universe that was securely geocentric.

Sometimes a new idea, a powerful vision of nature, comes along that changes the very structure of reality. And sometimes the agent of change is a tool. In 1608, an article in a diplomatic newsletter in The Hague noted that a man named Hans Lipperhey of the province of Zeeland had requested a patent on a spyglass that could detect stars not seen with the naked eye. Naturally the emphasis was on the military uses of such a device. By the end of 1609, Galileo had improved the design. He discovered that the moon had mountains, that it was a world unto itself, a place and not just a thing. Galileo discovered that Jupiter had its own little system of planets. He saw that Venus had phases like the moon. He saw that the Milky Way was made of countless individual stars. The stars, he noticed, were mere points of light, but the planets appeared as disks. This was compelling evidence that the stars and the planets were not the same type of body (the planets until then had been merely wandering stars). In modern civilization we celebrate the achievements of scientists such as Galileo, but we can hardly imagine the thrill, the visceral pleasure, of those early years of telescopic astronomy. If science is the process of knowing what it is that you're looking at, then Galileo experienced the ultimate scientific rush. Every human to walk the Earth had seen those stars and planets, had seen the moon and the Sun, but no one had seen them as Galileo had.

In the 1600s the cosmos increasingly revealed its vastness. One astronomer, Christiaan Huygens, made the assumption that the Sun and the stars were the same type of object, and he tried to figure out how far away the stars would have to be to appear as mere points of light. He calculated that Sirius must be 27,664 times as far from the Earth as is the Sun. A cannonball fired toward Sirius would not arrive for 691,600 years (though perhaps it depends on which cannon and which cannonball). The universe was turning out to be of a scale that dwarfed the human species. Astronomers were still wildly underestimating the true size, but a process had begun. The universe was getting bigger.

Everyone was still perplexed by certain objects, such as the "nebulae." These were wispy structures that could be seen with telescopes throughout the skies and were quite unlike the stars. The philosopher Immanuel Kant argued in the late eighteenth century that the nebulae were separate "island universes," outside the Milky Way, but he had no data, just a hunch. The fact that some of the nebulae were spiral in shape led astronomers to suspect that they were whirlpools of gas of the type that might give rise to a star. In 1864, amateur astronomer William Huggins used a spectroscope (a device for analyzing the chemical nature of light) to discover that a nebula in the constellation Draco was gaseous. A few decades later, the astronomer James Jeans showed that a collapsing cloud of gas in the process of forming a star might obtain the spiral structure seen so commonly among the nebulae. When James Keeler of the Lick Observatory showed, in the 1890s, that there were more than a hundred thousand spiral nebulae, this appeared to be the conclusive proof that these objects are merely clouds internal to our own galaxy, since no one could imagine that there could be that many island universes. It would be absurd, a cosmos so large.

But an absurdly large universe is what the astronomers eventually found. A lowly assistant at Harvard's observatory, Henrietta Swan

Leavitt, decoded the information contained within the fluctuating luminosity of a type of star known as a Cepheid variable. Most stars have a constant level of brightness, but the Cepheids get brighter and dimmer in a regular cycle. Leavitt realized that the period of the Cepheid—its particular, quirky rhythm of brightening and dimming—was directly associated with its absolute magnitude, its true brightness. If an astronomer knows the absolute magnitude of a star, and compares it with the apparent magnitude—how faint it has gotten as the light travels across all that distance—the astronomer can roughly estimate its distance from the Earth. At Mount Wilson, Edwin Hubble started searching for Cepheids in the Andromeda nebula. In 1924, he announced that he had found one. Then he found many more. He did the calculations. The Andromeda nebula was hundreds of thousands of light-years from Earth, a tremendous distance, clearly not in our galaxy. Although some nebulae were, indeed, gaseous structures within the Milky Way (such as the one in Draco observed by William Huggins), most of the hundreds of thousands of spiral nebulae were separate galaxies. The universe had changed scale again. Vaster and vaster it grew. Harlow Shapley, who had argued that the nebulae were internal to the Milky Way, received a letter from Hubble with the groundbreaking news that the nebulae were extragalactic. Shapley turned to an associate, and said, "Here is the letter that has destroyed my universe."

As the tools improved, the universe grew not only larger, but stranger. In the late 1920s, Edwin Hubble and his associate Milt Humason made one of the most stunning discoveries in the history of science, the redshift of the galaxies. The light from the galaxies is shifted to the red end of the spectrum, indicating that they are all flying away from us. The universe moves. It is expanding at a tremendous (though not easily calculated) rate. This spectacular drama implies that long ago everything we see existed in a single ultradense point in space and time. The discovery of the redshift led, over time, to the development of a powerful new model of the origin of the universe, the Big Bang.

So now, in the 1990s, this fabulous new tool, the Hubble Space Telescope, would inevitably change the universe once again. One of its first discoveries was that many stars are surrounded by disks of dust, apparently the spawning ground of planets. The Hubble even saw something planetlike, a "brown dwarf" star orbiting a larger star.

An astronomer named Robert E. Williams of the Space Telescope Science Institute in Baltimore had an idea for a special way to use the Hubble. He wanted to point it into the darkness of the sky, at a blank spot among the stars, and keep it there for two weeks. He picked a patch of space—no bigger than Roosevelt's eye on a dime held at arm's length—just above the bowl of the Big Dipper. The Hubble took 340 exposures, each of forty minutes' duration. The raw data poured into Baltimore. Williams figured he'd clean up the noise and the static and see something like thirty galaxies in that patch of blackness. But he did not see thirty galaxies. He saw two thousand galaxies. The Hubble had discovered that there are far more galaxies than anyone had ever realized. The universe kept growing. The word "vast" no longer seemed adequate as an adjective for this particular cosmos. (*Very* vast? Humongous? Pretty darn jumbo?)

The "Hubble Deep Field" amazed astronomers, but it was only the second-most-popular Hubble image. The winner, by far, was titled "Gaseous Pillars in M16—Eagle Nebula." The Eagle Nebula is about seven thousand light-years from Earth (nearby in cosmic terms) and is so named because in earlier photographs, with terrestrial telescopes, astronomers could discern something like the silhouette of a bald eagle with wings upraised. (I recently stumbled across one of these old shots of the Eagle Nebula while viewing, on video, an episode of the 1960s TV series *Lost in Space*. The nebula is in the background of a prolonged scene in which John and Maureen Robinson are trapped outside the Jupiter II spaceship while a comet comes closer and closer, baking them into unconsciousness with the intensity of its heat [it was an oddly hot comet]. I noted that the same episode features an iteration of the Assumption of Mediocrity. The Space Family Robinson boards an abandoned alien craft that looks something like a giant doorknob, and realizes the aliens have mapped much of the galaxy. "Compared to them we're still in nursery school," says Major Donald West. The aliens then appear, and are rather inscrutable. They look like random debris slapped onto a chicken-wire frame set on rollers.)

The Hubble did not see an eagle in the Eagle Nebula. The eagle was vaporized by the dramatically increased detail of the new image. The picture showed three pillars of gas and dust, bathed in yellow sunlight, and fully three-dimensional. The image would soon be framed on Goldin's office wall, and on the walls of offices throughout NASA. It shows a structure that seems to go beyond structure, to carry meaning beyond astrophysics. It is certainly the most beautiful interpretation of outer space ever produced by the cold hardware of modern astronomy.

"Gaseous Pillars in M16—Eagle Nebula" has come to be known more popularly as the "Pillars of Creation" picture. The cloud is a star-forming region, and so "Pillars of Creation" is not merely a metaphorical name. The first thing one notices about the image is that it looks like a piece of sky—as though someone has taken a three-headed cumulus cloud from Earth and shipped it to a neon district of outer space. Sprinkled around and even within the cloud are pink stars (pink!) of varying magnitude, some adorned with pretty "diffraction spikes," which are caused by supporting struts on the telescope and make the stars look as though they are twinkling. The pillars are rust-red. The backdrop is blue. Everything is illuminated by an unseen, godlike source of light that is above the frame and is pouring sweet yellow sunshine onto the billowing tips of the pillars. The image makes space look friendly, inviting. It looks like a glorious sunset in deep space.

You want to go there.

"Pillars of Creation" soon became something of an unofficial logo for NASA space science. The pillars popped up on brochures, posters, Web sites, and textbook covers, not to mention glossy magazines, the front pages of newspapers, and even the movies (the radio waves from Earth in *Contact* are depicted blasting through the Eagle Nebula). At one point Goldin took a blowup of the image to a town meeting in Bozeman, Montana. There was no NASA business in Bozeman, to be sure, but one of the state's U.S. senators had been a big supporter of the agency, and Goldin dutifully paid a visit. As a prop, Goldin brought the Eagle Nebula. The Montanans stood up and applauded. They clapped for NASA, they clapped for the Hubble, they clapped for this marvelous image, but most of all they clapped for space.

It turns out there is nothing in space that actually looks like the Pillars of Creation. At a meeting of the American Astronomical Society in

The Hubble Space Telescope converted the flat, faraway Eagle Nebula into the inspiring Pillars of Creation, oriented for drama. NASA

Toronto, I showed an 8½-by-11-inch copy of the pillars to several astronomers to get their reactions. John Graham, of the Carnegie Institution in Washington, examined the gaseous structures, admired them, and then said, "This is oriented to look picturesque." He tilted the picture to make the pillars point to the right, about two o'clock. "True north-south orientation is something like that," Graham said. The pillars had looked much more dramatic in a vertical pose, which more closely resembles the shape of cumulus clouds in the pressure gradient of our atmosphere. If the pillars were hanging down, rather than pointing up, they would not be so pretty at all, and might even be borderline repugnant.

Every astronomer praised the technical virtuosity of the Eagle Nebula image, but also acknowledged its cosmetic nature. Geoff Marcy said of the picture, "It's so artificial-looking you wonder what part of it is right." Charles Steidel said, "I think this looks fake." Bruce Bohannan said, "The stars should be white." (As I suspected!) Owen Gingerich said, "They look to me like giant columns of smoke rising up. But when you look at it, you realize it's something better than that, because of this splendid illumination around the edge. There is an otherworldliness about it. You'd think that Dalí, even with his imagination, couldn't have come up with something like this." But then Gingerich pointed out the craftiness of the shot. He noted that it has been contrast-enhanced. In real life, the nebula would be much fainter than what we see in the famous Hubble image. From a starship hovering nearby there wouldn't be much to see other than a dim blob (and the stars wouldn't have those nice diffraction spikes). Most objects in space are much less brilliant than they appear in textbooks. The universe is a dim place, it turns out, and because it's dim there's not really very much color.

And so the "Pillars of Creation" image is to some extent an artifact of the telescope, not an "objective" look at a structure in space. This is hardly cheating, since the human eye, with its limited capacities and its bias toward certain frequencies of electromagnetic radiation, is an arbitrary standard for declaring the "true" brightness of something. We all use imperfect equipment, whether it's a telescope or an eyeball. If there is a single aspect of the image that is truly illusory, it's the general sense that everything in the frame is close together. The depth dimension is flattened out, and all we get is a happy little bunching of objects. Our minds prefer this: We reflexively shrink space down and make it more manageable. The pillar on the left is about six trillion miles in length. The Sun and Earth and all the planets of our solar system would fit snugly into a tiny nub of that pillar.

I kept a picture of the Eagle Nebula on my bulletin board for well over a year. I loved the picture at first. The Eagle Nebula looked like a terrific place to hang out. You'd enjoy that awesome view (let's ignore the skeptics who say the thing is dim). You'd live on a world with beaches and palm trees, and at night, after the sun went down, you'd stare at the pillars and think about your place in the cosmos, and by golly you'd feel pretty good about it.

Over time, the Eagle Nebula sort of flattened out on me. It began to look like an $8^{1}/_{2}$-by-11-inch print. I kept reminding myself of the scale, that one inch equals roughly one trillion miles. And yet the majesty of the nebula began to fade. I wondered: What was the meaning of this structure? Does the meaning exist only to the extent that we insert it into the frame?

One day Goldin, in his office, pointed to his magnificent blowup of the nebula and said, "It's a question mark."

There is a classic novel by Stanislaw Lem, *Solaris*, about a planet that is covered with an intelligent ocean. What the ocean may be "thinking" is unclear. It is utterly inscrutable to the human scientists who are sent to study it. The ocean merely creates structures—sometimes abstract shapes that emerge from the water and later disappear, sometimes imitations of human architectures, such as spaceships. The ocean is able, somehow, to detect memories in the minds of the scientists, and from those memories to generate perfect simulacra of lovers long gone. The lovers tag along with the scientists. The scientists think they're going mad. At no point does the ocean communicate. For all the efforts of human science, "contact" is never truly made.

The ocean Solaris may serve as a useful analogue to the universe itself. Scientists study the structures of the cosmos endlessly, and try to separate the structures into categories. There are stars, comets, dust clouds, interstellar hydrogen molecules, lonely red giants in the middle of nowhere, galaxies, galactic clusters, galactic superclusters, and even a Great Wall of superclusters, millions of galaxies strung together in what amounts to the largest structure any human has ever witnessed. The scientists measure everything. They stare into the Sun. They collect the data.

And no one knows what it all means.

12

The Worm

The worm changed everything. The field of exobiology, or astrobiology, or bioastronomy, could pretty much be broken down into B.W. and A.W. Before the worm, there was a lot of hand-waving. After the worm, there was *evidence*. There were assertions that could be scrutinized and possibly even verified. The age of blah-blah was over!

The worm reclined languidly on a tiny flake of a meteorite from Mars. The meteorite is known as ALH84001, and to a remarkable degree scientists have learned to say "Ayell-aitch-eighty-four-double-oh-one" without gasping for air. I wanted to see the worm, but assumed it had long since been locked up somewhere or destroyed or perhaps donated to a museum. It turned out it was on top of a filing cabinet in the corner of David McKay's office at the Johnson Space Center in Houston. McKay is a quiet man with gray hair and spectacles. He doesn't get animated. He moves the way he talks, deliberately. He took a Tupperware-like tub off the top of a filing cabinet, set it on a table, reached inside, and pulled out a smaller, clear plastic snap-top case not much bigger than a postage stamp. The label said "ALH84001, 198."

Inside this case was a thin black strip, known as a carbon boat. Several flakes of rock, each smaller than a grain of Kitty Litter, rested on the strip. McKay pointed to one of the flakes. That's the one, he said. That flake.

So you're the little flake who started this war.

Down on the surface of the flake was the first specimen of extraterrestrial life. Unless, that is, McKay had gotten it all wrong, and the specimen was actually just a blob of an inorganic mineral. Given all that had come before, all those centuries of philosophizing, the endless debate over a plurality of worlds, the Mars craze of Percival Lowell, the disappointments (amid the triumphs) of Mariner and Viking, this question—life or mineral?—was hardly esoteric. On this flake might turn the future of exo/bio/astrobiology, or whatever it was called. This little thing could reveal the elusive truth about life beyond Earth.

McKay desperately wanted to be right. If he was wrong, his "discovery" would go down in history as a Lowellian misperception, more Mars canals on a microscopic scale. People would compare it to the notorious 1989 announcement of "cold fusion," in which two scientists thought they could get more energy out of a test tube than they'd put in. "It's a detective story. It's a tricky business," McKay said. "And it's not real obvious which way it's going to end up."

McKay and his colleague Everett Gibson saw the worm one night in late 1995. They had gone to the engineering lab after hours to take advantage of a new electron microscope designed to inspect cracks and defects in Space Shuttle parts. This was a piece of disaster-driven technology. The Challenger blew up; Reagan vowed we'd go on with our exploring of the heavens; someone built a gadget that could see a tiny crack in an O-ring. The main function was to inspect the tiles on the heat shield of the shuttle. What few people had yet managed to do was use such a microscope for the study of biology. The device, equipped with something called a "field emission gun," can see the world of the very small in a detail far sharper and clearer than anyone had seen before. We know the rule: New tools change our view of reality. But there's also that shake-out period, when people aren't sure what they're looking at, and have to make an assumption.

McKay and Gibson placed the flake in the microscope. They already suspected that the Mars meteorite contained something extraordinary, for they had found suggestive evidence of biological activity.

But they hadn't seen anything that looked like the fossil of a bacterium—yet.

And then suddenly they saw something. It was just draped on the surface of the flake of rock, a tubular structure that seemed to have segments, maybe a dozen of them. McKay and Gibson, taking turns, had the same thought: It looked like a worm. This was not some "marker" of life, this was life itself, an organism, fossilized, preserved for billions of years until by freak chance it wound up in a field-emission-gun scanning electron microscope at the Johnson Space Center. If this was what they thought it was, they had just discovered something no one had ever found before. This wasn't life as we knew it. This thing was Martian.

Soon they found other little creaturelike things. Some of them were bunched in groups, like colonies. Many were S-shaped. Most had the tubular quality—quite frankly they looked a lot like Chee-tos.

They didn't tell their colleagues, even ones in adjacent offices. Wendell Mendell worked directly next to McKay, a friend for several decades, and had only the faintest inkling that something strange was afoot. One day, for example, he noticed an assistant to McKay reading a paper on micro-organisms outside McKay's office. Why would McKay care about micro-organisms? He studied moon rocks! There was no life on the moon. It made no sense. Only much later, just before the sensational public announcement, did another friend tell Mendell, "They've got this picture, and it looks like a worm."

"A worm? What???" Mendell said.

McKay and Gibson worked in Building 31 of the Johnson Space Center, the sprawling NASA facility about twenty-five miles from downtown Houston. Like everything else at JSC, Building 31 is purely functional, a government rectangle thrown up in the mid-1960s, at the frenzied peak of the Space Age. McKay had been here the entire time, a highly credentialed and respected veteran of the scientific Establishment. A geochemist, he became a leading expert on lunar regolith, the "soil" of the moon. In the early 1990s—discouraged that the NASA decision-makers didn't support a particular mission he favored—he turned his attention to meteorites.

Down the hall, Everett Gibson was a different sort of character. Gibson tended to thunder a bit. He was sure of himself. He didn't have quite the reputation of McKay, and perhaps he sensed that. The

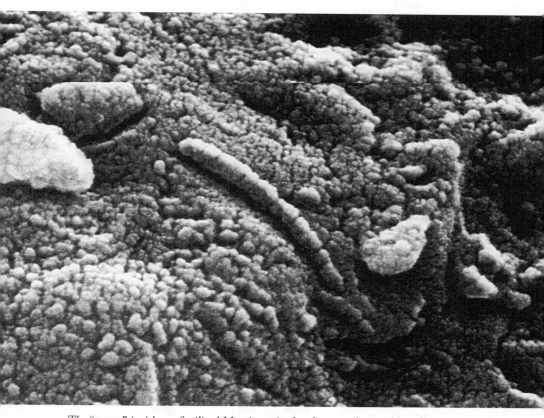

The "worm" is either a fossilized Martian microbe, the severed appendage of same, or just a blob of inorganic material. A possible example of the truism "Believing is seeing." NASA critics of the team would sometimes make a distinction between McKay and Gibson. At least McKay readily acknowledged the possibility that the worm and everything else in ALH84001 were nonbiological. Gibson seemed outraged that anyone would question his conclusions. He spoke in the manner of a machine gun—rat-tat-tatting the facts that bolstered his case.

The idea that meteorites might contain life, or remnant life, was not something that McKay and Gibson dreamed up. There had been a major hoo-hah in 1961 when Bartholomew Nagy and two colleagues announced that a carbonaceous meteorite found in France in 1864 contained molecules characteristic of biological processes. Then Nagy claimed that some of the structures found in the meteorite were "organized" and resembled fossil algae. Some seemed to have cell walls and nuclei. One had something like an appendage. A debate raged over whether the hexagonal shapes found by Nagy and his colleagues were biological in origin, or merely akin to snowflakes, one of nature's lovely but lifeless geometric constructions. Eventually one of the sages of the meteorite community, Ed Anders, figured out what Nagy had done wrong. Nagy had studied the meteorite during the summer, but in winter the same piece of meteorite did not show any of the "organized elements." The reason: Nagy had merely been looking at ragweed pollen. *Oops.*

And so the number of examples of extraterrestrial life, having gone from zero to one, was ratcheted back to zero.

A new category of unusual meteorites, called SNCs (the initials of three cities where these meteorites fell), came under scrutiny in the early 1980s. In 1983, Johnson Space Center scientists Don Bogard and Pratt Johnson found bubbles of trapped gas in one of the SNCs, and the chemical nature of the gas was exactly the same as that of the atmosphere of Mars as recorded by the two Viking landers in 1976. The eventual conclusion: These SNCs weren't ordinary meteorites but were chunks of the planet Mars. Pretty amazing! This was further proof that space, contrary to the Arthur C. Clarke vision of a pristine medium, is a messy arena. Space is full of junk, debris, dust, and even the ejecta of planets. As NASA planetary scientist Chris McKay puts it, for billions of years Mars and Earth have been swapping spit.

In 1984, Roberta Score, a meteorite collector in Antarctica, found the rock that became known as ALH84001. It weighed about four pounds. She found it lying in a field of white snow and ice in the shadow of the Allan Hills. The rock spent the next nine years misunderstood and underappreciated. Everyone assumed it was an ordinary meteorite. Only in 1993 did a closer chemical analysis reveal the signs of a Martian rock. Incredibly, it was 4.5 billion years old, and the material in the cracks in the rock had apparently been deposited about 3.5 billion years ago. This was by far the oldest of the Martian meteorites and a solid sample of a world removed from us not only by space but also by a tremendous stretch of time. It had been ejected from Mars when something slammed into the planet about sixteen million years ago. At least that's what the rock appeared to be telling the researchers, who had developed all sorts of nifty tricks for decoding the marks and spots and blotches on a given piece of matter. When an object drifts in space, it is struck at a fairly constant rate by tiny particles. These particles leave tracks. The Mars meteorite appeared to have about sixteen million years of wear and tear. Through a completely different method, the study of carbon-14 decay, the researchers concluded that the rock had been cooling its heels in Antarctica for thirteen thousand years. Bit by bit, the scientists put together a narrative for ALH84001. This igneous rock had quite a tale.

Inside the rock were some even more intriguing features. There were disk-shaped blobs of carbonate along the cracks—iron, calcium, magnesium, and other stuff that could have percolated through the rock from a hot spring full of living things. These carbonates could have only been formed when water came into contact with the rock. There were also tiny grains of magnetite, which can be a byproduct of bacteria. And there were PAHs, a type of hydrocarbon sometimes produced biologically. They began to see an obvious pattern—a "blanket of biology," as Gibson put it. "David and I went outside the typical thinking box," Gibson said. The typical thinking box would not allow a credentialed scientist to entertain the notion of Martian nanofossils.

"It's bigger than both of us," Gibson told McKay one day as their suspicions grew.

They expanded the team, sought out specialists, and tapped the shoulder of a colleague named Kathie Thomas-Keprta. "We do not want you to tell anybody what you're doing," Gibson told her. "One thing we think may be in this sample is evidence of biogenic processes."

Thomas-Keprta couldn't believe it. She thought to herself:

"These guys are nuts." But she eventually converted. The discovery of the worm gave them the final (and most sensational) link in their chain of evidence. For a year and a half the scientists prepared their case, and then presented their results to the journal *Science*. There was much debate over what to say in conclusion. There were an infinite number of ways of saying maybe—strong maybes and weak maybes, maybes that sounded definitive and maybes that had gaping escape hatches. The final draft said that the lines of evidence, when considered separately, did not prove anything about relic life in the rock. But when they were considered collectively, "we conclude that they are evidence for primitive life on early Mars."

This was a strong, firm, chest-thumping maybe, a scream of discovery. Life on Mars! Run with it.

The Houston team received a summons to Washington. On July 31, McKay and Gibson were ushered into Goldin's office. He grilled them, giving them what he called the "Ph.D.-qualifier" treatment. Gibson had been told it would take thirty minutes, but it took three hours. Goldin didn't know anything about meteorites, carbonates, or oxygen isotopes, but he did understand the basic scientific process. He started with the geology of the rock. How did they know it was from Mars? How did they know how old it was? How did they know when it had been ejected into space? How did they know how long it had drifted in the vacuum before landing in Antarctica? How did it migrate from the place it fell to the meteorite field at the base of the Allan Hills?

After a couple of hours of this, Goldin finally asked the big question: Were they certain they were right?

He remembers that Gibson gave the answer: "We think we have it nailed. We have it four different ways."

After the scientists left, Goldin knew he had to get the White House into the loop. His associate administrator, Wes Huntress, called Leon Panetta, the chief of staff for President Clinton. "You found *what?*" Panetta said. Goldin and Huntress hurried to the White House, handing a photograph of this odd little thing—this worm—to the president's gatekeeper. Panetta was surprised. He had expected something larger, something more obviously Martian, an animal of some kind, not this little squiggly Chee-to. He was thinking it would be more like the prehistoric man who had recently been discovered frozen solid in a glacier in the Alps.

"How big is this?" he asked.

They told him: less than five hundred nanometers. They explained how small that is.

"Whoa," Panetta said. The NASA people impressed upon him the fact that this was the first example of extraterrestrial life (if it was confirmed, they hastened to add). They said this implied the existence of other life forms in space: We weren't as alone as we thought. Panetta told them to wait where they were, and he went down the hall to the Oval Office. A couple of minutes later he was back. The president wanted to see them, he said.

Clinton sat at his desk while the men stood around him and explained the situation. Clinton looked at the worm. He fired questions at them. How did they know this? The conversation quickly shifted to the all-important implications. Clinton wanted to know if the early microbial life on Mars might have evolved later into more complex Martian organisms. Goldin replied that none of the surveying of the planet by robotic spacecraft had turned up any sign of larger creatures. (No turtles.) But he emphasized again that the Martian "nanofossils" could mean that life is commonly distributed in the universe. That had cosmic implications.

This is a day that we'll remember, Clinton said.

The president was very much a Space Age person, someone who had reached maturity in the years of the civil rights movement, Apollo, Vietnam, and Watergate, who had seen America at its best and worst. It would be easy for him to grasp the Assumption of Mediocrity. The baby boomers who had opposed the Vietnam War knew almost by instinct that a society must never presume that it is superior to another. What is not clear is whether Clinton had any particular opinion about, or interest in, exobiology. We recall that he made that specific request of Webb Hubbell to find the answer to the question "Are there UFOs?" (and to the question of who shot JFK). Hubbell claimed in his memoir that Clinton was utterly serious. (Hubbell reported that he tried to get Clinton an answer. He said that, during a trip to NORAD, the military's strategic-air-defense command in Colorado Springs, he asked a colonel whether there were any signs of extraterrestrial visitors. The colonel said there weren't. Hubbell said he wasn't satisfied with the answers he got on either the UFO or the Kennedy matter.)

Unlike Al Gore, Clinton rarely showed much interest in science. His passion was for history, and during the late summer of 1996, as he ran for re-election, he became particularly focused on the Progressive Era, on Teddy Roosevelt and the transition from an agricultural economy to an industrial economy. In his stump speech he would tell audiences that America was going through a similarly profound transition, from an industrial economy to an information-based economy. Clinton was riding high in the polls against a weak opponent, and he could afford to wax philosophical, to massage the big themes, to talk about millennial changes. This was a president who was ready, emotionally, intellectually, to find life beyond Earth.

(Clinton, incidentally, had occasionally tried to follow John F. Kennedy's example and have intellectuals fêted at the White House. On one occasion in 1994, for example, he had hosted a group of scholars for a banquet to commemorate Thomas Jefferson, including Carl Sagan and Ann Druyan. Druyan sat with the president, Sagan with the First Lady. At the end of the evening they compared notes. Sagan told his wife he had spent most of the night talking about the evolution of human beings from earlier primates. Hillary Rodham Clinton, he said, had been enraptured by his discussion of the sexual habits of the alpha male gorilla.)

When Goldin and Huntress finished briefing Clinton, they went to the vice-president, the official administration expert on all things scientific. Soon the word was passed to the president's outside political adviser, Dick Morris, who then leaked the news to Sherry Rowlands, the call girl he'd been hanging out with at the Jefferson Hotel. In her diary entry of August 2, 1996, Rowlands recorded the stunning discovery. Over drinks and dinner, an ebullient Morris had offered a toast. "To us," he said. Then, Rowlands wrote, he said he had something very secret to tell her. Top military stuff. Only seven people in the entire world knew it.

"He said they found proof of life on Pluto!"

Obviously the news would have to be revealed quickly to the outside world. NASA scrambled to put together a press conference. Goldin insisted that an independent voice be included—an outright skeptic, if necessary. He didn't want the life-on-Mars discovery to sound like a done deal, even though he fervently hoped it was true. A naysayer at the press conference would help NASA save face later if the worm and its cousins turned out to be flaky.

There was, in fact, a quintessential gray eminence who would be the perfect outsider at the press conference. His name was J. William Schopf. He was not a geologist or a chemist; he didn't study meteorites or lunar rocks. He was, rather, a biologist who specialized in the remote past—a paleobiologist. One thing had always held true throughout the history of the study of extraterrestrial life: Biologists were the skeptics. Only biologists, it seemed, fully understood and respected the complexity of life. Everyone in the business of extraterrestrial life knew about Schopf. He was an opinionated, brilliant man who wouldn't hesitate to let people know what they ought to think. Time and again he raised his booming voice when idiots and fools and charlatans claimed that they'd found spectacular signs of ancient life. What they'd found, usually, was mud and dirt and clay. Schopf actually believed, fervently, that there is life beyond Earth, probably even in our own solar system. He was a Saganist to the core. But like Sagan he also wanted to keep the field credible, free of bogus or overstated assertions that would stain everyone else when they proved untrue. He was very cautious in his optimism: "I'm certain we will find life elsewhere, probably within our solar system, if you give me 250 years."

Schopf learned about the Mars-rock research months before the McKay-Gibson paper reached publication in the journal *Science*. He'd told the Houston team early on that he saw nothing biological in the rock. The more recent images of worms and Chee-tos didn't change his mind. Some of the purported Martian life forms were only twenty nanometers long, far smaller than any organism ever found on Earth. Size matters when it comes to discerning what's biological and what's not. At some point you literally run out of room—there's not enough space to clamp together a sufficient number of molecules to have an inside and an outside and working internal parts and all that complicated machinery of biology. (The worm measured about 360 nanometers from end to end, which was still about 150 times as short as a human hair is *wide*.) And why weren't any of the nanofossils smushed or flopped all over one another? They looked too orderly. The scientists had overinterpreted the data, he thought.

He later wrote up his reaction:

> I had been brought in to shore up the paleontologic guess of ge-
> ologists schooled in rocks and minerals, but not biology. Their
> guess was wrong. The discs certainly were not remnants of pro-
> tozoans. A number of the objects were simple foramlike discs but
> many of the others merged one into another in a totally nonbio-
> logic way. Their overall size-range also did not fit biology and
> they lacked any of the telltale features—pores, tubules, wall lay-
> ers, spines, chambers, internal structures—that earmark tiny pro-
> tozoan shells. Moreover, the "lifelike" traits they did possess
> (carbonate composition, discoidal shape, ringed rims) could be
> explained by ordinary inorganic processes.

Schopf, invited to speak at the NASA press conference, reluc-
tantly agreed, on the condition that he could say what he honestly felt.
Then everything accelerated. Schopf was vacationing in the mountains when he got a call: The press conference won't be in ten days,
it'll be *tomorrow*. Get on a plane to Dulles immediately!

Schopf flew to Washington, and the next morning, August 7,
1996, went to NASA headquarters, expecting to sit in on a casual conversation around a table with some scientists and reporters. As he arrived, so did eighteen camera crews, a couple of hundred journalists,
and assorted space junkies and gadflies. By midday it would be a mad-
house. Schopf knew exactly who he was: "Daniel in the lions' den."

Before the official press conference there was a practice version.
The Mars-rock scientists ran through their lines. Then Schopf gave
his skeptical account. He felt that, on a scale of 1 to 10, with 1 being
highly doubtful and 10 being a certainty, the Mars microfossils rated a
2. It wasn't completely out of the question that the Houston people
had found life, it was merely highly unlikely. When he finished his
practice presentation, the NASA associate administrator for public af-
fairs, Laurie Boeder, offered some sharp coaching instruction to the
Houston scientists. They had just been routed by this lone ranger
from California. Boeder told the Houston scientists to be more enthusiastic in the real press conference. From Schopf's written account:
"I finished. Utter silence. Then a woman on the Headquarters staff
rose and berated the troops: 'Schopf has just demolished you. Can't

you guys be more positive?!' " (Boeder later said she merely encouraged the scientists to be enthusiastic, not to hype their findings as being conclusive. McKay's version: "The tenor was, You guys are really reticent to say much. Be forceful about it. Project.")

In retrospect, it might not have been a great idea to coach the scientists in the art of public speaking. The natural instinct of most scientists is to admit that there is much they don't know and don't understand, and, indeed, that further review may yield a different and contradictory conclusion. But it's an extremely rare moment when scientists hold a press conference on live international television. This was quite possibly historic. In the bright lights of this great day, it might have seemed permissible to turn up the volume a bit, to stand taller and thrust out the jaw and exude dense vapors of authority. Be commanding. Be excited about the results. *Project.*

And so "maybe" started to sound like "probably."

Kathie Thomas-Keprta, meanwhile, felt ill. She hadn't slept enough. (What she didn't realize was that she was pregnant.) It would be a long day.

As the scientists prepared to go before the cameras, a bigger media star took the stage across town, at the White House. The president went on live television to applaud the scientists. "I am determined that the American space program will put its full intellectual power and technological prowess behind the search for further evidence of life on Mars," he said. If the findings are borne out, he said, "It will surely be one of the most stunning insights into our world that science has ever uncovered. Its implications are as far-reaching and as awe-inspiring as can be imagined." Some of the footage was later lifted by the producers of the movie *Contact.*

Clinton's remarks brought everything to a halt at NASA headquarters. CNN was on Clinton at the White House and wouldn't cut to Goldin until the president had shut up. Goldin retreated from the NASA podium and went behind the stage, pacing in the clutter of a back room. I followed him. He clearly didn't want to be ambushed by a reporter, and an underling chased me away. I managed to ask him one question: Do you know for sure that this is true? Goldin said, "We sort of know."

Finally Clinton stopped talking, and Goldin took the podium. This was surely one of the greatest moments in the history of science—and the sound system failed. A high-pitched whine filled the room. Minutes passed. Goldin had been ready to announce the discovery of life on Mars and instead had to wait while someone fixed the sound system. When he could at last speak, he announced, in a reassuring tone, that whoever was responsible for the sound-system failure would not be fired. He was trying to be sensitive, in his own heavy-handed way. He knew someone out there had to be puckered up pretty bad.

"It's an unbelievable day," he said. He turned to the scientists. "I'm so proud of you, words cannot describe it."

One by one, McKay, Gibson, Thomas-Keprta, Richard Zare of Stanford, and the rest of the team gave their presentations. They had slides, even a video with animation. Their case was slick and compelling. They showed the worm, which looked huge up there on the screen—a spectacular creature, a truly magnificent Martian. The scientists really did appear to have their case nailed four ways.

Then came Schopf. He gave the rock a 2 on the plausibility scale of 1 to 10. He cited Carl Sagan's rule: Extraordinary claims demand extraordinary evidence.

But it didn't matter. The news was so sensational, everyone wanted it to be true. The photograph of the worm hit the wire services and sped around the planet.

The number of examples of extraterrestrial life—verified by *actual NASA scientists*—had finally entered the positive range. One example. And a tiny one. A feeble little smidgen of one. But you needed only one to end our miserable solitude in a cold and uncaring and empty and hostile universe. This was the start of a community. It would mean more money for NASA, a greater sense of purpose, a quickening of the step. The renaissance that had begun with the Hubble repair mission had now reached a peak.

The Space Age was back!

The debate over the worm continued, but one thing was certain: Suddenly there was money for
exobiology, lots of money, maybe as
much as an additional billion dollars
over a five-year span. NASA jacked
up the budget for the new program
called Origins, one of Goldin's pets.
Origins would, among other things,
figure out how to build a Planet
Finder, a telescope that could see
extrasolar planets and remotely detect any spectroscopic signatures of
life. In the months to come, NASA
would start assembling a new program: Astrobiology. NASA officials
met with the top researchers across
the country to figure out where and
how and when to search for life.
Where do we look for life on Mars?
What do we do with Europa? Is
there possibly life or prebiotic
chemistry on Callisto, Ganymede,
Titan, Triton, or any of those other
strange moons out there among the
giant planets? The search for alien
life hadn't felt this legitimate, this
real, since the Viking mission.

Wes Huntress didn't want
people to think that all the money
pouring into Origins and the Mars
program was simply the result of
the meteorite. The Mars rock, he
argued, was merely a catalyst.
Huntress had a chemist's way of describing the situation: "It's all been
sitting there in colloidal suspension,
and you dropped 84001 in there,

13

Star Trekking

and—boom!—it coalesced and dropped to the bottom of the beaker." When he said this, I vowed to remember the phrase "colloidal suspension." It might have enduring metaphorical usefulness for the alien issue in general.

Ed Weiler, the NASA scientist in charge of the Origins program (before being promoted to the top job in space science), had a fairly typical reaction to the findings of the Houston team: If this rock really had nanofossils, it would only confirm what he already suspected. The universe is full of life, Weiler thinks. That's his paradigm, Mars rock or no Mars rock. "It hasn't changed my mind one iota," he told me one day in a hallway at Headquarters. "I believe it *has* to be out there."

Gilbert Levin, meanwhile, had been conspicuously uninvited to the press conference, but he made sure that everyone knew he was the one who had originally found evidence of Martian life back in 1976. This Mars-microbe business was something he'd known all along, he said. In *Space News*, the industry's leading newspaper, Levin declared that the Hubble Space Telescope should specifically look at Mars for the seasonal waves of darkening on the surface. Levin hadn't given up on vegetation.

In Pasadena, Donna Shirley, head of the Mars-exploration program, received a message from NASA headquarters: Put more of a "life" emphasis on her experiments. "There was a life frenzy," Shirley remembers. Goldin seemed particularly impatient to get direct evidence of life. He pushed his deputies to speed up a Mars sample-return program. Just go grab some, Goldin said. The timetable was already ambitious. There would be spacecraft landing on Mars in 2001 and 2003, both equipped with rovers that would pick up samples and stash them. Then, in 2005, another spacecraft would land, and pick up the samples; after two years (when the planets were in the right position again), it would return them to Earth, to parachute into the desert of Nevada or Utah in 2008.

This plan would be revised many times, and eventually a more modest rover would be assigned to the 2001 mission. The rover wouldn't have much range; to find a rock with biogenic features would take a stroke of luck. The whole sample-return program also had a major hitch: It triggered planetary-protection concerns. Sagan in the 1960s had been the most vocal figure in the planetary-protection debate, arguing that astronauts should be quarantined when they returned from the moon. He didn't think the moon had life on it, but he wouldn't rule it out. Sagan was even more worried, in coming years, about the possible contamination of Mars by visiting spaceships from Earth. When life from one planet visits another it breaks the safety seal (albeit an imperfect one) imposed by astronomical distances. Now NASA wanted to bring home a batch of exotic material from a world that, thanks to ALH84001, didn't seem so dead anymore. There was a very slight, bordering on infinitesimal, possibility that a sample of Martian soil might contaminate the Earth. The hand-wringers could imagine dire scenarios.

"Suppose they're pathogenic micro-organisms," Levin said one day. "Suppose it eats up all the wheat in the world. Suppose it goes for people. Suppose it goes for cattle."

He paused.

"It's the Andromeda Strain."

In the movie of that title, based on the novel by Michael Crichton, an extraterrestrial microbe is brought to Earth aboard a crashed satellite, and it quickly wipes out all the humans in a small town. A few intrepid scientists and a lucky random mutation help stave off global catastrophe. In real life, it's wildly improbable that an extraterrestrial microbe would pose a danger to Earth. It probably wouldn't wipe out a single terrestrial germ. The alien life form wouldn't be adapted to our planet, and wouldn't be able to exploit our environment. A human being can't randomly wander around eating things, like grass or tree bark or rocks or dirt. An organism has to practice stuff like that, and evolve, and may eventually become an organism that, for example, finds it appealing to kill a cow and grind up its flesh and fry it in a pan and put it on a bun with ketchup and lettuce and tomato and a pickle.

Most likely a Martian microbe would die instantly when exposed to terrestrial conditions. But the mere thought of global disaster put NASA in a bind. Lawyers began researching all the permits they'd need to bring in a sample of dirt from Mars. Among other agencies, the Department of Agriculture would want to have a sniff at it. The Environmental Protection Agency would throw up some hurdles. It was going to be a legal nightmare, and everyone knew it. Amid the excitement were some who said this was all happening too fast. There were scientists who felt NASA wasn't ready to find life on Mars. No one knew where to look, what tools to use to gather samples, what kind of material would most likely harbor fossils. Should you collect little rocks or big rocks? Should you go to a canyon or a crater? Would it be possible to dig deep into the rock, or would rovers have to pick up whatever was lying around on the surface? "Mars is a dartboard and we're trying to hit the target where life is, and right now the lights aren't even on," Kevin Hutchins, a young geologist at the University of Colorado, complained at a conference in 1997.

The Viking mission had provided a warning of sorts: You don't want to set yourself up for a spectacular disappointment. If everyone had it to do all over again, the Viking mission would have had a different set of biology experiments. No one knew, until Viking, that Mars had an unusual chemistry at the surface. Maybe it was inevitable that Viking would go for broke with its instant just-add-nutrients test for life—it would have required inhuman patience to do otherwise—but the prevailing sentiment in the Mars program, prior to ALH84001, had been to take it slow this time. And slow was less and less an option.

Goldin was bothered by the realization that NASA wasn't really a life-sciences outfit. He was running an agency full of engineers, astronomers, and physicists, but no biologists. He went around asking: Where are the *biologists?* He started grilling people at meetings about whether they had any biology training. At one point he called together all his senior advisers on the Mars-exploration program, and asked everyone with training in life sciences to raise his or her hand. Out of fifty-five people in the room, only one hand went up. Goldin was appalled. This would change. He told me, "Everyone at NASA is going to be trained in biology. We're sending people back to graduate school. We're not just talking the talk, we're walking the walk."

In addition to the specter of the end of life on Earth, there was another nuisance as NASA geared up for the assault on Mars. More and more people were saying that the Mars rock had nothing in it but minerals. McKay and Gibson found themselves the target of severe criticism and challenges from fellow meteoriticists. Ed Anders, the man who had shot down Bartholomew Nagy's hexagonal shapes thirty-five years earlier, blasted McKay in an e-mail, referring to the worm and its cousins as "turd-like shapes." Was this just envy? There weren't many people in the world who had devoted their lives to studying the spew of space, and they had resigned themselves to almost certain anonymity. Suddenly a couple of their peers came along and were the toast of the town. The skeptics were screaming, but no one could disprove the point right away. Hardly anyone could even understand half the data reported in the *Science* article. It was a maze of arguments from numerous distinct scientific disciplines—the *Finnegans Wake* of exobiology.

McKay and Gibson were in a delicate position. Their critics could pick out any weak link in their long chain of evidence. In science, if you don't try hard enough to prove yourself wrong, your friends will gleefully take up your slack. Proving the work of others false is virtually a sacred rite. Some critics said the carbonates in the rock had been formed at temperatures far above the level at which biological processes can occur. Other researchers backed the Houston team on the temperature question. Probably the most damaging claim came from researchers who found that the little squiggly shapes were just flakes of the rock, rendered more biological in appearance by the coating process used in preparing samples. John Kerridge, who had been a prominent figure in earlier NASA plans for the study of Mars, relentlessly criticized the Houston team. Kerridge believed that McKay and Gibson had lost perspective, that they wanted too badly to believe in their conclusion. "There are strong echoes of Lowell," he said gravely.

Gibson seemed to become more emphatic even as the criticism increased. Whenever I saw Gibson, I'd bring up the latest negative finding, and he always had an answer. The critics were looking at the wrong part of the rock. They didn't understand the research. They were refuted by other research. The rock had fossils in it—Gibson remained certain.

McKay was always more cautious.

"I think there's an 80-to-90-percent chance that we're right. I think it's a much more complicated situation than we originally anticipated. There may be contamination that we had not corrected for. There may even be some Antarctic microbes in there."

McKay said he understood what it would mean if they turned out to be wrong. History would not be kind. Even though his team did everything by the book, and tried to be cautious, and tried to check out every other possible explanation, the final verdict would be harsh if the rock turned out to contain only a bunch of minerals.

"We'll be held up as a terrible example of a false conclusion. In some ways, we'll be disgraced."

McKay and Gibson had no choice but to press ahead and try to bolster their evidence. They considered going back to the original flake and searching again for the worm. They had a thought: Maybe it was just a piece of an organism. It was so small, after all. It might be a limb, some kind of appendage, perhaps a tail—tails can get detached. McKay and Gibson figured they could use a needle with some sticky stuff on the end and reach down and try to pluck the worm (or the tail) off the surface of the flake. Then they could cut it open with a tiny knife blade made of diamond. At least that was their idea: They'd never done anything like that before.

Gibson began to worry that he wasn't getting his fair share of the credit. Articles and books were being prepared that cited McKay as the discoverer, but Gibson felt he had been the first to realize the significance of the rock. He wasn't some afterthought!

Criticism, in-fighting, jealousy—the Mars rock was quickly turning into a can of worms, as it were.

But maybe it wouldn't matter, ultimately. McKay and Gibson were rapidly becoming famous for their work. They were getting five or six invitations a day to talk about what they'd done. It was hard to find time to do serious science. Everyone wanted to hear about the great breakthrough, the incredible discovery of life on Mars. The general public didn't know how isolated McKay and his team had become, how unpersuasive the research had been within the scientific community. Scientists usually prefer the simpler of two competing explanations. A mineral is far simpler than a microbe.

One day in the early summer of 1997, Goldin's lieutenants in the Space Science Division were telling him about their long-term strategic plan, all the great missions they'd dreamed up. Obviously, they were going to scour Mars for more of those nanofossils, and even bring back samples. Someday they were going to put a submarine, an "aquabot," into the ocean under the icy crust of Europa. They were going to float balloons through the atmosphere of Titan. They were going to stick probes in every corner of the solar system.

And all the while they would examine the cosmos with a new fleet of telescopes. The Hubble had helped people remember something they had forgotten in the race to send rockets to the moon: Space is full of information that can be obtained remotely. You didn't have to go there after all—you could stay home and observe the damn thing. Although no one was saying it out loud, the Space Age had become just a subset of the Information Age. Space wouldn't have to be a habitat—star trekking and colonization wouldn't necessarily be the way of the future. Instead, space could be the ultimate database. The information is encoded in electromagnetism. There is information in the optical wavelengths, the infrared, the gamma rays, the X-rays, the ultraviolet, the radio. NASA had new instruments that could "see" the universe in all these wavelengths. The space scientists were going to take the measure of the whole universe, top to bottom—exploring the cosmos so deeply they would reach the very beginning of time itself.

In one corner of their list of projects could be seen two words: "Interstellar probe."

"I like that one!" Goldin said.

This was the kind of dreaming that Goldin had been wanting to see at the space agency. Surely it was time to launch a probe that was not just interplanetary but *interstellar.*

Wes Huntress realized Goldin had misunderstood the meaning of "Interstellar probe." Huntress explained that this was not a plan to visit another star, not by a long shot. The idea was to go out to the heliopause, at about a hundred astronomical units (AU), or a hundred times the distance from the Earth to the Sun. At that point the Sun no longer provides a steady breeze of atoms, and is but a bright star in space. This would be very far from Earth, but much farther yet from another solar system.

Goldin got the point, but wanted more. He wanted the star-trek mission. He had the citizens of America to think about, the taxpayers, and he knew they didn't give a hoot about the interstellar medium, about how many hydrogen atoms can be found per cubic centimeter in the voids of space. They would want to go to Alpha Centauri, explore a planet, find some life.

The Goldin plan was logical and incremental and compelling. First you build Planet Finder. Then you build Planet Mapper—an even more wondrous telescope that can see, on an extrasolar planet, oceans and continents and mountain ranges. Then you look at your pictures of these strange and exotic worlds and decide which one you're going to visit. There are a thousand stars within forty light-years of us. One of them, maybe dozens of them, will have a blue marble begging for exploration. You can't phone this one in. You gotta go.

And if nothing else, Goldin would tell people, he wanted to get people thinking, thinking out of the box. This was a chance to recapture some of that old-fashioned gee-whiz antigravity spirit. Goldin said, "I wanted people to start thinking about things other than burning oxygen and hydrogen in a combustion engine."

He instructed Huntress to start cooking up a real, no-nonsense interstellar mission. Huntress, like everyone else, was somewhat pessimistic that a trip to Alpha Centauri could be accomplished. He called Charles Elachi, one of the top people at the Jet Propulsion Laboratory, and asked for help. Elachi turned to one of the most imaginative people he knew: Henry Harris, an obscure scientist working in a windowless office, but a man with incredible ideas. Maybe he would be the one to get us to the stars.

Long after midnight in Pasadena, Henry Harris found himself unable to sleep, his restless mind preoccupied with the difficult voyage to Alpha Centauri. He lived in a simple white house on a quiet street where, at that hour, a few other people may have been awake, but surely none were thinking about sending a spaceship to another solar system. He sat down at his computer and typed some numbers. He stared at the monitor. The numbers were appropriately astronomical: Alpha Centauri is twenty-six trillion miles from Earth, give or take a few hundred billion.

To judge from the floor plan of his house, his study should have been a dining room. It was squarely on the route between the living room and the kitchen. The study was also something of a recording studio. He had two synthesizers, a guitar, piles of sheet music, a piano, an electric organ. Henry Harris had been a lounge act in Las Vegas in the 1970s, and he still recorded love songs, inspired by his wife, Paula. She had met him in a bar in the early 1980s, immediately captivated by, as she put it later, "this gorgeous piece of eye candy." He was now in his mid-fifties, paunchy, and rather bald. His efforts to constrain his baldness with some deft maneuvering of his remaining hair gave the impression of an elaborate engi-

14

To Infinity and Beyond

neering project. No matter, he was still a balladeer, a crooner, and by no means was this irrelevant to the Alpha Centauri project. He felt his singing and songwriting, his days in Vegas and Reno and all those nightclubs across America, gave him the artistic temperament necessary to imagine something as beyond-the-norm as interstellar travel.

Thinking about the future, about far-off technological marvels, about human destiny, was for Harris both a government job and a personality trait. Going to another star—an entirely new solar system—had been a dream of the visionaries since the dawn of the Space Age. Most people in the realm of space science, the hardheads, the number-crunchers, the frowners and groaners and bureaucratic hand-wringers, could not imagine such a voyage and would certainly laugh at Harris if they knew what he was planning. Alpha Centauri? Insane! The presumption had always been that you can't get there from here. Interstellar travel, even with robotic probes—bloodless machines—had too many constraints, too many problems of propulsion, cost, communication over vast distances, all the inescapable rules of energetics. Harris knew these facts as well as anyone, but unlike other people he could not throw up his hands and walk away. He had to engage the issue, come up with some ideas, develop scenarios—for gosh sakes, this wasn't some harebrained inspiration of the Advanced Projects people, this was an assignment from the ultimate boss. It came straight from Goldin.

If he were writing science fiction, he would just make something up, like "warp drive" or "hyperspace." Warp drive was the secret of the Starship *Enterprise*'s ability to explore the universe in a "five-year mission," something that would normally challenge the structure of space-time. Hyperspace was a loose term with various meanings, but, like warp drive, it involved shortcuts through space. If the shortest distance between points on a piece of paper is a straight line, then an even shorter distance can be created by folding the paper and putting one point directly on top of the other. Near the beginning of the *Foundation* trilogy, Isaac Asimov wrote, "Through hyper-space, that unimaginable region that was neither space nor time, matter nor energy, something nor nothing, one could traverse the length of the Galaxy in the interval between two neighboring instants in time." In real life, unfortunately, hyperspace travel was still neither here nor there. There were no magic wands at Harris's disposal, nothing remotely as nifty as the propulsion system used to improbable effect in

The Hitchhiker's Guide to the Galaxy, and called, fittingly, "Improbability Drive."

He knew that on *Star Trek* the *Enterprise* used—according to the clever scriptwriters—canisters of antimatter. A canister of antimatter would not be something you could pick up down at the filling station. Scientists in real life had heroically manufactured a few ephemeral particles of the stuff, the barest trace, submicroscopic, detectable only as a kind of faint echo in the inscrutable interior of massive atom-smashers. It had a way of winking into existence for a fraction of a second and then exploding upon contact with ordinary matter, a mutual annihilation governed by Einstein's famous equation $E = mc^2$. For the moment no one would want to stick that stuff into an engine. It could get nasty fast.

There were other ideas floating around out there. Freeman Dyson had imagined a spacecraft called an astrochicken, part machine, part organism. It made sense in principle, but the science of biotechnology had a long way to go before it could produce a computerized hen that could cluck through deep space. Mark that idea down as ahead of its time. So too was the transportation used in the movie *Contact*. The film had Jodie Foster fall into an artificially created black hole, which somehow fed her into a bunch of wormholes that formed a galactic subway system. That was supposed to be a plausible, scientific, no-magic-wands version of interstellar space travel—Sagan had even cleared the idea with some physicists at Caltech—but Harris knew he couldn't construct an intragalactic wormhole transporter on a government budget.

He had a better idea. He was convinced it might actually work—not immediately, to be sure, but someday. He would argue that NASA should invest its energy in laser-driven light-sails. It was essentially a variation on the *Niña*, the *Pinta*, and the *Santa Maria*. Your spaceship is a sailboat of sorts. The sail is extremely thin and reflective. The "wind" is light itself, fired from a massive laser on Earth. The laser-wind by definition moves at the speed of light. It's a thin, fast breeze. The spaceship would accelerate to tremendous velocities. You could beam that sucker on out there.

Would it work? He crunched the numbers. Twenty-six trillion miles to the star. Four light-years and change. To get there in forty years—a reasonable span—you'd need to go a tenth the speed of light, about eighteen thousand miles per *second*. To propel a thin light-sail to that speed requires a laser producing ten billion watts minimum, and then, out in space, at a distance more or less equivalent to the orbit of the moon, you'd need a giant lens to cohere the laser beam. In fact, the lens would have to be—he roughed out the equations—the size of Texas. The numbers just didn't look good.

He finally went to bed at 5 a.m., knowing he would have to rise again in just a few hours. No one could say he wasn't determined. He felt this was a serious matter, not some scientific lark. If humans didn't find a way to leave the planet, he told himself, their days were numbered.

At JPL, people never said they were "at work" or "at JPL" but rather that they were "on lab." If they were anywhere other than JPL they were simply "off lab." Henry Harris liked to wear a sportcoat and khakis and comfortable shoes, no tie. He looked like a high-school math teacher. Across the world there were people like him, unremarkable in appearance, not particularly animated or flashy, yet bursting with knowledge, minds whirring, ideas rocketing around the brain case at all hours of day and night. They were the ones who wrote letters to the editor catching a sloppy reporter in a minor but annoying misstatement of the arrangement of the periodic table. They did not blink at the sight of scientific notation, exponents, Greek symbols, and many of them knew by heart the latest estimates of the Hubble Constant for the expansion of the universe. They could sketch the DNA molecule on a napkin, knew the principles of microchip technology, and loved to inform children that a spider is not an insect. These were ordinary people of high intelligence and loaded minds, wandering the world in complete anonymity even though they possessed more knowledge than Aristotle.

Henry Harris would check the EurekAlert Web site for the day's breaking science news—one day he discovered that someone had just invented a Tricorder, one of those *Star Trek* gizmos that can detect living organisms at a distance. He followed science fiction and had written a treatment for a TV show about time travel. In one episode, a race of alien reptiles from deep space went back in time sixty-five million years to construct a giant laser to destroy an asteroid heading directly for Earth. They wanted to prevent the catastrophic impact that ended the Cretaceous Era and wiped out the dinosaurs—their reptilian kinfolk.

Harris worked in a bland rectangular structure known simply as Building 135. He had the title of "senior engineer," but was not precisely a scientist. If anything, Harris would actually call himself a physicist. In the early 1970s, he had done all the work to get a Ph.D. in physics at New Mexico State, but never finished his dissertation. He didn't play by the Establishment rules—he was an artist by temperament. For a while he worked in a bunker at the White Sands Missile Range, writing computer programs for the missiles that would carry death and destruction to the enemy. As a civilian, a man of science, and an iconoclast, he didn't really fit in with the military culture. Once, he decided to show his bosses how their supposedly failsafe method of avoiding an accidental launch could be circumvented. A missile would launch only if two people, far apart, simultaneously turned two different keys. Harris brought a pole into the bunker and demonstrated that it could be attached to, and then twist, both keys. His superiors were unamused. Prove out the system, they said, don't criticize it.

In 1973, Harris quit his military job and left academia as well, giving up on his Ph.D. He went west, to Vegas, to take up a new occupation: lounge singer. With his hair grown long, down to the middle of his back, he worked the lounge at Harrah's, took requests, sang the hit songs of the day, "Tie a Yellow Ribbon" and "Mandy" and the best of Neil Diamond and Burt Bacharach. Sometimes he went to Tahoe, sometimes on the road, playing small clubs across the land. He would never be Wayne Newton, much less Elvis.

One night, as he sat at the bar, a gambler struck up a conversation about Nixon. The gambler liked Nixon, thought he was a great man, screwed over by all this Watergate nonsense. Harris elucidated point by point the various obstructions of justice that Nixon had committed. He then went even further, running down thirty years of political history, corruption, and intrigue culminating in the constitutional crisis of Watergate. When he was finished, the gambler huffed:

"What do you know? You're only the piano player."

Eventually the piano player decided to return to the world of science. While visiting a girlfriend in Los Angeles in 1978 he saw an advertisement for a technician's job at the Jet Propulsion Laboratory, and he soon was on board, handling data from the Viking orbiters, still tooling around the Red Planet.

At JPL, Henry could show everyone how smart he was. Strange ideas were welcome, to a point. Before Harris started on the Alpha Centauri project, he worked on a proposal for something called the Ice Clipper, a mission to send a projectile to smash into Europa. Computer models indicated that the moon should have plenty of internal heat generated by the tidal forces of Jupiter. Water and heat—plus complex carbon molecules, which seemed to be everywhere in the universe—were the basic ingredients of life. Europa could have creatures swimming around! Or at least microbes. But it wasn't easy landing a spacecraft on Europa, much less drilling down into the ice, which was certainly many miles thick. That's why Harris wanted to smash the moon with a twenty-kilogram hollowed-out copper projectile. The debris, including subsurface material, would drift into orbit, and it could then be picked up by a passing spacecraft. Though the plan had the attribute of simplicity, it seemed a bit brutish, and NASA took a pass on it.

Now Harris had an assignment to work on interstellar travel. He headed one team to look at a specific type of propulsion known as beamed energy. Harris was also given money to stage a conference on interstellar spaceflight. Another team at JPL would write the broader report for NASA, looking at all types of propulsion options. Not everyone shared Harris's enthusiasm for giant laser beams. Fusion energy still had advocates. Back in the 1960s, nuclear-fusion technology promised to be the energy source of the future. But nuclear fusion, at least in nature, required the unusual conditions found deep inside a star. Duplicating that artificially turned out to be trickier than anticipated. In recent decades, researchers had built these astonishing, frightening, house-sized aggregations of metal with cables spewing in every direction and giant magnets humming and graduate students running around in hardhats. Decades passed. The fusion reactors never quite worked as well as Our Friend the Sun. The closest thing anyone had come to making fusion energy work was annihilating the Bikini Atoll. Fusion seemed applicable only to the limited ambition of conducting a world-shattering thermonuclear war.

A famous study called the Daedalus Report had recommended

that an interstellar mission first stop off at Jupiter and mine helium from the atmosphere. Helium would make a great fuel, but the numbers still didn't look good. Even helium was hard to lug around. Harris felt that no fuel, regardless of its intrinsic energy, would be worth the burden of carrying it. He knew that the ideal interstellar spaceship would be small and nifty, not strapped to giant tanks of fuel. It certainly wouldn't be a heroic battleship, like those seen in Hollywood movies. If you go to a movie like *Starship Troopers*, you immediately notice how heavy, gnarled, and appendaged in metal are the spaceships that travel across the universe. In science fiction, no cargo is too heavy, and no one frets about payload launch costs.

The Harris version of a spaceship would solve the fuel problem by carrying none of it. The fuel would be back on the ground, in the form of the laser, the gadget emitting the powerful beam of light. Harris figured the smart thing would be to invest in the general technology of light-sails and lasers, get the ball rolling, and wait for costs to come down and efficiency to increase. In a couple of decades, maybe the numbers would look better, and a giant lens in space would need to be only the size of Delaware.

Interstellar travel is the toughest nut to crack in the world of spaceflight. Space is not designed—is not structured—in a way that makes human spaceflight or even robotic spaceflight easily managed on the interstellar scale. Science fiction makes space look much cozier than it really is. As the SF writer Kim Stanley Robinson puts it, "The public's sense of the size of the universe has been warped down to something that is small and comfortable."

For all the grand structures we see through telescopes—the majestic pillars of dust, the supernova remnants, the spiral galaxies, and so on—a fundamental feature of space is its emptiness. The universe is so large that it can contain untold trillions of stars and still keep them sufficiently spaced out that from the perspective of one star, any other star is a mere point of light.

That night sky is a grand illusion, because everything looks close together, a manageable terrain. The illusion is fostered by the transparency of space. Light is immortal, and if it doesn't hit a cloud of dust it will travel forever, from one side of the universe to the other. We see not the closest stars but the brightest. The orange star on the shoulder of Orion is the enormous star Betelgeuse, and it is much farther away than most of the other stars we see. An astronomical distance is typically described in light-years, a light-year being the distance light travels in a year. Astronomers would say that Betelgeuse is about five hundred light-years from the Earth, though the margin of error is large in measuring stellar distances. When we see Betelgeuse in the night sky, we know we're seeing light emitted about the time Leonardo painted the *Mona Lisa*. Rigel, below and to the right of Betelgeuse, is even farther away, about eight hundred light-years distant, and we see light emitted several decades before Genghis Khan overran the Asian continent and established the largest empire in the history of our planet (and, by creating a huge trading zone, hastened the dissemination of technology, including the Chinese compass, gunpowder, and paper, all of which in turn led to the Age of Discovery, the Gutenberg Bible, and the Renaissance). The Hubble Space Telescope has captured in its mirror the photons emitted by galaxies so far away their existence is described not as a point in space but as a point in time. These galaxies might not even be there anymore, yet we see them, because space is a time machine.

The unit of "light-years" does not help the average person understand distance. Not only does it sound like a unit of time rather than distance (at least to those who have not made the fateful transition to hardcore astronomy nerd), it also begs for some understanding of the speed of light. Light travels 186,000 miles per second. That's difficult to picture, especially extended over a year. If we accept that a light-year is about six trillion miles, we can calculate that Betelgeuse is three quadrillion miles away—that is, 3,000,000,000,000,000 miles. No one could have the faintest notion what that number means. Nor can we grasp three trillion or three billion. (Truthfully, I can't picture three million, either, but maybe there are some folks with mutant brains who can.) Three quadrillion is three thousand trillion or three million billion. It's gibberish! So we just say the words and hope everyone understands that these are big numbers, and that Betelgeuse is a real hike from here.

The simple approach to interstellar travel is to assume it cannot be done and then return to terrestrial problems, like cleaning up the planet and spreading good cheer. But the visionaries could never tolerate that. Henry Harris, for one, found the prospect of an Earthbound human race not only dull but actively dangerous. The entire race could be wiped out by a single catastrophic impact from a comet or an asteroid.

"Space is dangerous. It's unstable. Given enough time, the universe is unstable," Harris told me the first day I met him. He was tired from staying up late with his Alpha Centauri problem, and he navigated haphazardly as I drove us to lunch. I had been told that Harris would be the man to see about interstellar travel. My book was about aliens, not rocketry, but I felt that somewhere along the line I had to get a better grip on whether an intelligent species would be likely to make trips across the galaxy. I didn't realize that Harris would become a good source on everything—he was tapped into a universe of ideas that most scientists chose to ignore. We quickly went off lab and drove to his favorite restaurant near JPL, the Flintridge Inn. We wound up meeting there numerous times over the course of a year— Henry, I discovered, always had something new to say, novel ideas, unusual factoids, startling leaps of faith. He had a comfort level with the fantastic that I found rare among the scientific brethren. He had a way of slowly unveiling his deepest thoughts, his inner fears and obsessions. One got the feeling that, no matter how deeply you delved into his mind, there would always be something stranger in the layer just below.

On this particular day his tone was grave. There's this idea out there, he said, that if we take care of Mother Earth the universe will take care of us.

"That's bullshit," he said.

Bad stuff happens in this universe. For example, there is the threat of neutron stars. Neutron stars are the astonishingly dense cores of red-giant stars that have exploded. If there are two of them together they will orbit each other at tremendous speeds, but the orbit is not stable, and eventually they will collide, emitting a horrific explosion of gamma rays. Those gamma rays, Henry said, can sterilize a large part of a galaxy.

"There's a neutron pair sixteen hundred light-years away that could destroy all life on Earth."

And there might be others that are closer.

"We've got to go to the stars. That's the point. . . . When you live in a violent, unstable universe, you better be prepared to move." That was the big picture, the ultimate ugly scenario—fragility, vulnerability, the threat of catastrophe and annihilation. Our friend Betelgeuse is expected to go supernova in less than two million years, and scientists cannot say for certain that it's far enough away to keep us out of trouble. There will certainly be some effect from the radiation, they assume. Doom is not probable; total annihilation is not the likeliest scenario. But when Betelgeuse blows it will be as bright as the full moon, and it might make more sense to have some humans tucked away on a distant planet on the other side of the galaxy, to play it safe.

Lots of scientists agreed with Harris that survival of the human species required a long-term colonization program. The dinosaurs didn't master spaceflight, and now look at them—they're nothing but bones. They didn't know about bolides. A bolide is an object from space—an asteroid, a comet, maybe even a chunk of some other planet. About sixty-five million years ago, something the size of a mountain punched through the atmosphere at about eighty miles per second and dug a hole thirty miles deep at the tip of the Yucatán Peninsula. The dust and vapor ejected into the atmosphere plunged the planet into years of darkness and freezing temperatures, and triggered a mass extinction. Seventy percent of the species on the planet disappeared, the lumbering, terrible reptiles among them. So this business of going to Alpha Centauri was quite serious for Harris.

He knew that not everyone shared his sense of urgency. He also knew that the Assumption of Mediocrity implied that we shouldn't be able to master something like interstellar travel. No one else had! Why were we so special? Isn't it far more likely that we will run into the same unavoidable problems of energetic and political will that other (hypothetical) civilizations have faced? If our superadvanced space brothers can't get here, how could we imagine that we'd ever go there?

The Fermi Paradox usually is deployed as part of the argument that there aren't any intelligent beings out there. We have to be the first because we're the only ones that exist. But there's another possible solution to the Fermi Paradox, the one that Henry Harris favors, a solution that simply makes the paradox go away in an instant. The Fermi Paradox is not a paradox at all if its central presumption—that aliens have not visited—is wrong.

Picking through his mango-chicken salad, Harris said the reports of UFOs cannot be viewed as simply a "scientific" question. He emphasized the next sentence: "That's a historical question." He said history is not subject to scientific proof. It is cobbled together from stories, memories, anecdotes. "There's documentation all the way back to the Stone Age," he said.

He began choosing his words carefully, knowing that most JPL scientists would heartily disagree with him. Merely admitting a receptivity to UFO stories could make him a laughingstock on lab. It also put him on the wrong side of the issue from the gatekeeper, Sagan, who'd worked with him on the Ice Clipper team. Sagan, said Harris, didn't grasp the simple fact that history wasn't the same thing as science. Sagan wanted to prove the unprovable. Just because it didn't succumb to empirical verification didn't mean the alien presence was imaginary. "There's thousands of people who say they've seen them. Draw your own conclusion. We've got cases where entire cities have seen them. What about the recent thing in Phoenix?"

I told him I'd read about it in *USA Today*. The newspaper had included an artist's sketch of a giant boomeranglike spaceship. Thousands of people in Phoenix had seen a series of lights over the desert that no one (at first) could explain. Others had seen lights passing over the city the same night. Phoenix had become the UFO hot spot over the past few months.

"No government on this planet is going to tell their people that they don't control their skies," Harris said. That would be too disturbing.

The sixty-four-zillion-dollar question, of course, was what kind of propulsion system the aliens used to get here. Did they use lightsails? Antimatter? How did they solve the safety issue, the hazards of solar radiation and tiny dust particles that can puncture the hull of a starship? How did they communicate with their home base if even at the speed of light it takes years to send a radio message? Were they immortal? What was their purpose? Why did they hide? Harris didn't have the answers. But he had some ideas. He'd heard things about their propulsion system. The key, he'd heard, was Element 115—an incredibly heavy element that hadn't even been invented yet on Earth but in theory might exist. He'd heard it said that such an element would have "strange, relativistic properties." It might have some effect on the very fabric of space-time.

These were exciting thoughts. The technology was out there—way out there.

PART TWO

VISITORS

Henry Harris offered a glimpse of exobiology's parallel universe. The 1990s boom in government-budgeted research on extraterrestrial life—driven not only by the Mars rock and the discovery of extrasolar planets, but also by a more general conviction that the universe riots with life—had its match in the burgeoning mythology of UFOs and alien abductions. The scientific view of extraterrestrial life should have clashed with the supernatural version, but instead it provided ammunition, inspiration, reassurance. The only absolute disagreement between the mainstream and the fringe was over the issue of location. Scientists wanted to find microbial alien organisms on Mars or Europa; ufologists wanted to find bigger aliens much closer to home. There was copious public support for both approaches. This was, clearly, the Alien Decade, a very good time to be in the hunt for life-as-we-don't-know-it.

The parallel universe of ufology is one that does not necessarily feature the same laws of physics that humans have experienced. A popular notion is that the aliens in flying saucers have discovered, and mastered, physical principles that are beyond our imagination. This is one reason they do not necessarily show up on the normal instruments used by mainstream scientists. We may

15

A Different Vibration

lack the diagnostic tools for understanding these creatures. The aliens may possess knowledge that would literally boggle our minds, assuming that a mind is something that can be literally boggled. If an ordinary mortal were to discover the truth about this alternative, alien reality, he or she might begin drooling uncontrollably and collapse into a fetal ball, or begin bleeding from the ears. Paradigm shifting can be a nasty business. In John Mack's book *Abduction,* he quotes a parable about a frog who has lived its entire life in the confines of a well. One day it leaves the well and ventures forth upon the land and suddenly sees, for the first time, the great immensity of the ocean.

The frog's head explodes.

We do not know if our feeble Earthling brain, this blunt tool in our skulls, which may be as primitive in comparison with an alien's brain as an amphibian's is with ours, will ever be capable of understanding aliens. At the moment we're no better able to grasp the truth about aliens than we're able to fillet a salmon with a baseball bat. Our brains are not evolved in such a way as to comprehend the physical or spiritual planes that may exist outside our own perceived reality. (I, for one, can barely comprehend New Yorkers.) It may be that "reason" is an impediment to our investigation of visitors from space. We may have to let go of our traditional, linear, butterfly-net approach to nature (see it . . . catch it . . . examine it). We may need to raise our consciousness. How to raise our consciousness is not entirely clear; it has something to do with vibrating at a different frequency.

These are some of the ideas in circulation. I do not endorse these ideas, I just report them. They move through society with the undiscerning, democratic instinct of a virus.

Henry Harris, for one, just wanted the facts. What made him different from many other scientists at NASA was that he sought out sources of information that others shunned. He drove down the intellectual back roads, the blue highways of investigation. He walked abandoned rail lines. Late at night he would sometimes turn on Art Bell, on whose radio show some of the guests would actually phone in from *the future.* Henry showed that it was possible to be scientifically minded, to be dedicated to the tools and processes of the intellectual Establishment, and yet at the same time to entertain thoughts and theories that on their very face were absurd. This didn't mean that he was wrong, necessarily, merely that he was not lashed down and hogtied by the orthodoxies of his profession; that he was open to absurdities as a potential source of breakthrough science.

"Anomalistics" is the study of things outside the accepted reality of mainstream science, and as an enterprise it's a big gamble. It wagers on dark horses, stuff like telepathy, telekinesis, near-death experiences, reincarnation, past-life recall, astral projection, UFOs, the Bermuda Triangle, Bigfoot, Atlantis, interdimensional travel, the "Mars Effect," and the Loch Ness Monster. The payoff on one of these dark horses, should it turn out to be "true," would be phenomenal. For the anomalistic community, there are numerous historical precedents that provide inspiration. Giant squids were supposedly a figment of the imagination of drunken sailors, and then they turned out to be real. And what about meteorites? The mainstream scientists in the eighteenth century refused to believe that rocks could fall from the sky. Such a thing simply *could not be.* Yet it was so. Andy Knoll, a paleobiologist at Harvard and a leading thinker in exobiology circles, told me, "Ninety-nine times out of a hundred, things outside the canonical envelope are wrong. But the hundredth is Copernicus."

On June 24, 1947, amateur pilot Kenneth Arnold saw nine mysterious "disks" flying at otherworldly speed across the Cascade Mountains near Mount Rainier. He assumed they were secret U.S. military planes. His radio report to ground control incited reporters to go to the airport and question him. Arnold told the reporters the objects moved "like a saucer skipping across the water." He was describing their motion, not their shape. But the headline writers took the story from there and boldly declared that Arnold had seen "flying saucers." Two years later, Major Donald Keyhoe, a retired Marine, published an article in the inaptly named *True* magazine in which he established the standard narrative of alien visitation and government cover-up. He expanded the article into a book, *Flying Saucers Are Real,* and a publishing phenomenon was born.

The aliens in those saucers proved, like their science-fiction counterparts, remarkably adaptable to modern civilization. No amount of skepticism and debunking could erase them as a cultural presence. The UFO narrative managed to grow, mutate, and expand far beyond the initial phenomenon of mysterious disks and flashing lights. The lore began to include elements of direct contact, and then abduction, and then breeding experiments. The aliens who had been invisible inside their flying saucers began, by the 1970s, to take on physical form: Spindly, hairless bodies trucking around a head the size of a watermelon, the biological warehouse of a phenomenal brain. They were not little green men (that's just science fiction!) but rather had gray skin. Their lidless, black, almond-shaped eyes were pools of darkness in which no emotion could be discerned. In the alien bestiary the Grays were joined by Tall Nordics, a humanoid species; by Reptilians; and by a smattering of Praying Mantis creatures.

The emergence of these entities was unconstrained by the rejection of the belief system by the mainstream. It didn't matter that Sagan and his fellow astronomers said the evidence for such aliens was uniformly unimpressive. Nor did it matter that the Air Force, in several investigations code-named Sign, Grudge, and Blue Book, spent twenty-two years reaching the conclusion that the aliens didn't exist. They could not be so easily vanquished. For many ordinary people, the existence of alien visitors seemed by far the most plausible explanation for the strange things happening in the skies. Many of the "official" explanations were hard to swallow. The scientists claimed that people were sometimes seeing nothing more than a weather balloon, or the planet Venus, or a meteor, or "swamp gas." Clearly the scientists had a dim view of the intelligence of the average human being. The scientists were essentially saying that everyone who saw a UFO was some kind of moron. Swamp gas! Now that was a stretcher. People didn't know what swamp gas was, exactly, but they knew for certain that they'd never mistake such a thing for a spaceship full of aliens.

If Henry and his intellectual compatriots were correct, the scientific Establishment in the second half of the twentieth century managed to miss the biggest story of all time. The same people who had discovered the structure of the DNA molecule, found subatomic constituents of matter and energy, and photographed craters on the far side of the moon had somehow—amazingly—failed to detect entire fleets of alien starships buzzing around the American West and even in the airspace above Washington, D.C. Sagan and Drake were listening to Andromeda when, if they'd had any sense, they would have been tuning in military radio traffic around Area 51 in Nevada. The true believers have a sense that this can't and won't go on much longer, this collusion of ignorance and outright deception. There is much talk of the Disclosure. The government has been setting up the big event. The Mars rock? It's all part of the plan, the plan to prepare the public for the shattering news. Many people remember what the Brookings Institution wrote in 1962: that the public might be extremely disturbed by an announcement that intelligent life has been found beyond the Earth, and that primitive civilizations often collapse when they come into contact with technologically more advanced civilizations. It all makes sense in the larger context. Of course the government wouldn't reveal the truth about aliens. People would freak. The Disclosure is a long-term project. People are getting warmed up for it. *Close Encounters* and *E.T.* were just part of the plan.

The UFO enigma can be viewed as an astrosociopolitical issue of great complexity, or, more simply, as a question of human psychology. Why do some people construct their world-views around ideas that other people find ludicrous? Where's the fault line? It's not intelligence or social class. It's not like poor, fat, Velveeta-eating people believe in aliens and rich, thin, Brie-eating people don't. Socioeconomic and educational status don't seem to be factors of great import. Geography is probably one of the few factors that have an influence. Aliens seem to be more prevalent in the West, and in California they're simply taken for granted, more strange guests at the cocktail party. There are also more aliens in America than in most other nations. In a country like India, the aliens never show up on the radar.

If there is one thing that reliably separates believers from nonbelievers, it is probably the attitude toward intellectual authorities— toward the "reality police." To believe in aliens requires a rejection of official wisdom. It requires that we believe that the individuals in power, and particularly the gatekeepers of scientific knowledge— Sagan, Goldin, and the American Astronomical Society—are either wittingly or through ignorance telling us a story that simply isn't true. Old-guard leftists, feminists, right-wing conservatives, and gun-toting militia members can all be found in UFO circles. They all think they're being lied to, and they're probably right in certain cases. Suspicion is a powerful motivator. When we lose trust in authority, we lose trust in every aspect of it, every assertion, every alleged fact. Out for a dime, out for a dollar.

"Trust no one" is a mantra on *The X-Files*. It's a perfect philosophy for the Age of Bad Information. There has never been a time when a lie could be so professionally produced. Even the charlatans have graduate degrees from the Institute of Advanced Charlatanry. Got a dumb idea? Start a Web site! On the Internet the extraterrestrial issue is legendarily the second most popular topic, after sex (though I must quickly admit that this is one of those classic Internet factoids that emerge from the digital universe unsourced, their origin a mystery). The new electronic medium is perfect for the transmission of a diffuse UFO mythology in which there is no central governing narrative but, rather, an abundance of theories and anecdotes and half-baked eyewitness accounts and assorted poppycock (undergirded by balderdash). A person can get online and jump from one paranormal site to another, from the writings of a serious ufologist to those of a pure lunatic, endlessly skipping through an infinite network of intrigue. The anarchy of the alien mythology is one thing that makes it so enchanting.

The writers and producers of *The X-Files* understood this perfectly—for years, millions of people watched the painfully gradual unfolding of a vast government conspiracy to cover up some kind of alien-related agenda. The show began in 1993, just as the alien theme was hitting its cultural stride. Creator Chris Carter used the space-alien theme sparingly, and in fact the actual showing of an alien was initially taboo. One of the chief writers in the early years, Glen Morgan, remembers Carter saying, "The minute you see an alien, this show's over."

Instead there were hints and clues about the alien presence and the government cover-up. The show had a "Mythology," as Carter called it. It was a shaggy-dog story—there were always new twists and turns, and no resolution in sight. Gradually the rule against showing aliens succumbed to creative pressures, and by the time Carter made the *X-Files* feature film there was no more playing peekaboo. The Mythology hardened into a fixed narrative: The aliens, it turned out, had cut a deal with evil white men who ran a secret global government, sort of like the Trilateral Commission. The evil white men were going to infect the entire population of the Earth with a virus that would turn humans into slaves of the aliens, in exchange for the aliens' not wiping us out completely, but in fact the aliens had lied to the evil white men and were using the virus to gestate new aliens in the guts of humans. (At least, that's what I think I was seeing up there on the big screen. Plot clarity did not seem to be the movie's paramount aspiration.) The plot is oddly familiar, because it's essentially just a dramatization of what people have been saying in recent years on the fringe of the UFO movement. An alien presence, a breeding program, a massive cover-up—this is virtually plagiarism from a real-life mythology.

The X-Files was spookier when it didn't tell you what was going on. The movie violated the first rule of anomalies: It's what you *don't* know that's really interesting.

One day in 1997, I visited *The X-Files'* set in Vancouver to interview Gillian Anderson, the actress who plays Scully. She had just won an Emmy for best actress in a TV drama, and amid her congratulatory flowers she was engaged in damage control, having failed to thank any of her colleagues in her acceptance speech. She'd already thanked them, she said, when she'd won a Golden Globe, but that, unfortunately, wasn't as prominent an event. For the Emmys she'd thanked only her family. This made for some tension on the set in the days afterward, and she'd taken the step of putting an ad in the trade papers to make up for her gaffe.

Gradually we got around to talking about aliens. On the show Anderson plays a skeptic, an FBI agent with a medical degree, an entirely rational person who is loath to believe in alien abductions, psychic power, or anything supernatural. Her character follows the rules and teachings of mainstream science. But "Dana Scully" is a purely fictitious entity. The real person, Gillian Anderson, is like millions of other people: She believes that aliens have come to Earth, and that there is a government conspiracy to keep the public ignorant.

"It would shock the hell out of me if the government had never been involved in a UFO cover-up and if there were not life on other planets," she said, conflating two distinct issues (which I interject not to be nasty to Anderson, but because everyone does it, almost reflexively). The government, she said, wants to be in total control of people's lives and does not want them to know that there are beings, creatures, more powerful than humans. The government doesn't want people to be scared. This was the Brookings report, filtered through many channels.

"The concept of other beings being ultimately more powerful than us human beings places the public in a state of fear. And once the public's in a state of fear, the government no longer has the same kind of control."

I asked her why aliens would cross interstellar space to visit the Earth and then spend so much time hiding, skulking around in the desert, whispering in little kids' ears, abducting people, and so on.

"Because that is how we have created them," she said. "I mean, I think that we have subconsciously created the kind of alien beings that we believe there are. And that they operate, vibrate—this is going to make me sound like a complete nut—they vibrate on a different energy level than we do, and they are adaptable to our beliefs."

She was just an actress in a trailer, extemporizing, but she was also speaking for millions of people. She was a medium for powerful cultural beliefs, one of which, an extremely alluring one, is that we don't merely detect the nature of reality, we create it. Our beliefs have direct physical consequences. *We make things happen.* Postmodernism has met the self-help movement. The postmodernist professor at an elite university might say that reality is whatever we decide it is—but even that statement would be hedged by the admission that we don't literally change reality. The postmodernist simply believes that reality isn't a fixed, immutable, objectively real system that is external to our minds. To a New Ager, there truly is something real out there, and we are tapping into it, and altering it, in a continual interactive process. The subjective and the objective are the same thing. The working of our minds, the workings of the universe—no longer are these separate matters.

At least that's how I interpret what's going on—one man's subjective take on an alleged trend in a universe with no absolute truth. Talk about vaporous! Talk about something you can't sink your teeth into. (If you spend enough time thinking about this stuff, you suddenly want to go out and grow some corn or build a barn, to do anything at all that has the superficial appeal of being tangible.)

Gillian Anderson had paused after her attempt to explain the different energy levels.

"I'm speaking as I think," she said.

She meant that she hadn't really thought it all the way through, she was still developing her cosmology.

"Given my belief systems," she said, "this is what follows my thought processes."

Later I spoke with David Duchovny, who plays Agent Fox Mulder. He lived next door to me when we were freshmen in college, a factoid—a conversational nugget—that will serve as hard currency throughout the world for the rest of my life. The young Duchovny was alarmingly brilliant for such a good-looking guy. Shortly after the start of school, he showed me a paper he'd written, in high school, on *The Waste Land*, and it was unlike anything I'd ever seen before, a confident and fresh analysis of a much-analyzed classic, by someone who wasn't even old enough to buy a beer. He went on to Yale graduate school and could easily have become a tenured Ivy League professor in a musty English department instead of settling for being a millionaire superstar with countless adoring fans and a glamorous wife. What a waste.

I gave him a ride home and we talked about the alien idea. On his TV show, "Mulder" is the conspiracy theorist, like the real-life Gillian Anderson. The real-life Duchovny is a skeptic, like "Dana Scully." He gave me his concise thesis about why so many people are attracted to the conspiracy theories of *The X-Files:*

"At the base of it, it gives a very easy answer. Which is that there's bad guys out there, they're all-powerful, and they're making your life miserable. It's the Oliver Stone answer. There's a reason bad things happen to good people. We show why. It's not random. We're more religious than *Touched by an Angel.*"

16

Sagan Is Appalled

Carl Sagan, child of immigrants, got his start in the world of astronomy by going to the New York Public Library and asking for a book about stars. The librarian assumed the boy meant Hollywood celebrities. Eventually he got the book he wanted, and became captivated by space. He went to a public high school in Rahway, New Jersey, and then on to the University of Chicago. He rose in the world through sheer force of intellect and imagination. He carried with him the passion of the Enlightenment—"confidence, optimism, eyes to the horizon," as E. O. Wilson puts it. From a humble beginning a boy could someday know the cosmos.

And then, as he grew old, he watched the world go mad. People continued to believe in things that weren't true; worse, they made up new untruths to believe in. "Pathological science" ran rampant. The bookstores were full of New Age tracts about angels, past-life recall, astrology. Astrology! To Sagan, it was obscene. For his sixtieth birthday, his colleagues and admirers held a bash for him at Cornell, sang his praises, and gave him a victory lap at the podium. During the question-and-answer session, he got prickly and difficult. Someone in the audience asked, "What is your personal religion, or is there any type of God to you; like, is there a purpose, given that we're just sitting on this speck in the middle of this sea of stars?"

Sagan answered, "I don't want to duck any questions, and I'm not going to duck this one. But let me ask you first, what do you mean when you use the word 'God'?"

This may have been the one issue in which Sagan's precise thinking, his parsing of possibilities, seemed like a dodge. And it quickly went downhill. The questioner didn't define God, and Sagan seized the moment to argue that the term "God" is too vague. "It covers over differences; it makes for social lubrication, but it is not an aid to truth." Basically, he didn't believe in God, but he didn't like to say the opposite, either. God remained beyond the evidentiary database. ("An atheist has to know a lot more than I know," Sagan said to me in a 1996 phone interview. "An atheist is someone who knows there is no God." Still, he said there were certain images of God that he found extremely implausible. "If you are talking about an outsized, male, light-skinned figure with a long white beard sitting in a throne in the sky and tallying the fall of every sparrow, then I would argue that there is absolutely no evidence for such a God. But that doesn't mean that such a God couldn't exist—just that we haven't found him.")

Someone asked him about astrology.

"Astrology is a hoax," Sagan said. He wasn't going to dance around this one.

"Excuse me?" said the questioner.

"It's a hoax. H-O-A-X."

At some level he may have felt ever-so-slightly guilty of instigating pseudoscientific notions. There was all that ancient-astronaut stuff back in the early 1960s. Didn't he say Phobos might be artificial? Didn't he talk about alien bases on the far side of the moon? Didn't he write a paper called "Direct Contact Among Galactic Civilizations by Relativistic Interstellar Spacecraft"? Maybe he'd been unclear and imprecise in his presentation. Maybe he hadn't hedged enough at crucial moments, hadn't been a better guide as he demanded that people navigate a delicate terrain between the possible and the probable.

What bothered Sagan was that people refused to be rational in an age of scientific wonders. It wasn't just the alien thing, it was the whole antiscientific movement, from New Age thinking to religious fundamentalism. When he and Ann Druyan protested nuclear testing in Nevada he got a close-up look at the thinking of the antinuclear movement. A lot of people were not merely opposed to misused technology, they were against science in general. They were Luddites. Among the protesters was John Mack, the Harvard professor then developing his ideas about alien abductions. A Pulitzer Prize–winner for his psychobiography of Lawrence of Arabia, Mack eventually stunned and enraged his colleagues by writing *Abduction: Human Encounters with Aliens*, proposing that the alien-abduction reports are based on real encounters with "other intelligences." He believed they may not come from our universe at all. Sagan tried to talk Mack out of his new belief system, but to no avail. The tide had turned. Science was losing the battle for the hearts and minds of tens of millions of people. Many had rejected traditional religious cosmologies only to become adherents of new narratives that were no less supernatural. Sagan seemed to grow increasingly worried as he got older.

As the premier proponent of rational thinking in all matters involving space, he had somehow failed. He had allowed this alien thing to fester, grow, and spread beyond the confines of reason. Squelching stupid ideas and advancing smart ones was his job, his duty. He was responsible because he carried with him the answers, the overarching philosophy, that would satiate the imaginative thirst of the public. If only he could show everyone that the universe was a marvelous place even without alien invaders.

Science, for Sagan, was something of a religion. He was what you might call *scientistic*. In middle age he had increasingly campaigned on behalf of the scientific method. In 1996, he published a collection of essays, *The Demon-Haunted World*. This was not one of the joyous, wonder-filled, cosmos-loving books that had made him famous. Somewhat dark and ominous, it told of the danger of unrestrained beliefs and pathological science. He attacked everything from alien abductions to "recovered" memories of Satanic-ritual child abuse. The reader could sense a note of frustration, as though Sagan was astonished that there remained people who rejected science in favor of myth, superstition, and the paranormal.

He had a theory about why many people don't like science. Science says "no" too often.

"A lot of the most fundamental physics can be written in the terms of prohibitive acts," Sagan told me. "Thou shalt not travel faster than light. Thou shalt not measure the position and momentum of an electron simultaneously to whatever accuracy you want. Thou shalt not build a perpetual-motion machine. . . . A lot of people—New Agers, for example—are annoyed. They think everything can be done."

Sagan had tried to teach people their limits. He had tried to teach them to understand their cosmic mediocrity. They didn't want to hear it. They didn't feel mediocre. They believed in superior beings, and these superior beings interacted with humans. The superior beings even came to the Earth and tampered with the evolutionary process so that apes would evolve into something far more important and intelligent. The alien narrative always doubled back at key moments to the story of human existence, and we were fundamentally the stars of the show. The aliens *empowered* humans. This was an idea against which Sagan could not compete, any more than he could compete against the promises of organized religion.

People wanted "yes."

17

The Anomaly Problem

June 29, 1998, was a very good day for the UFO community. Newspapers and TV networks prominently played a breaking story about the credibility of UFO cases. A press release from Stanford University, quoting a Stanford professor named Peter Sturrock, stated that a panel of scientists had examined the UFO issue and had decided to urge the scientific community to take seriously the tales of flying saucers and other oddities. Sturrock had organized the study and had written, with the approval of the scientists, this new and groundbreaking report. The scientists had determined that there was intriguing physical evidence associated with UFO cases, including radiation burns to witnesses, giant objects tracked by radar, and other inexplicable events. The study, quickly labeled the "Sturrock Report," contradicted the conclusion of the 1968 study of UFOs led by Edward Condon of the University of Colorado. The Condon Report had coldly and cruelly declared that nothing of value had come from more than two decades of investigation into UFOs, and that further research "cannot be justified in the expectation that science will be advanced." The Condon Report exiled ufology from mainstream science.

Sturrock's answer to Condon, offered under the imprimatur of

Stanford and reported straightforwardly in mainstream newspapers, was precisely what the ufologists and anomalists needed. Two problems have plagued the people who study UFOs, crop circles, cattle mutilations, nocturnal abductions, DNA sampling, fetal implants, the Face on Mars, and other mysterious phenomena associated with alleged alien entities in our solar system.

The first is the "giggle factor." Many UFO researchers feel scorned and mocked. At the same time, they aren't blind to the kookiness and charlatanry of some of their fellow travelers. They suffer mightily from their association in the public mind with the tabloid press. They do not think Elvis is on a saucer bound for the Pleiades. The giggle factor goes back to the beginning of the movement, to the heyday of burger-stand operator and alien-abduction pioneer George Adamski. Adamski claimed he'd had an encounter with a Venusian, and had been flown to the far side of the moon, where he'd seen forests and waterfalls. He produced a photograph of an alien vessel in which the spaceship looked strikingly like an electric lamp.

The other nagging problem, which may be even more intractable, is that aliens probably aren't visiting our planet.

As you can imagine, this would make the case for an alien presence much, much harder to prove conclusively and beyond a reasonable doubt. Others may disagree strongly with this assessment. But at least as an intellectual exercise, as a hypothetical scenario, we should think about how truly difficult it will be for the ufologists to prove the existence of a massive extraterrestrial invasion of the Earth if such an invasion is not actually occurring. It would take a heroic act of persuasion to nail down such a case. Only in America is the feat even conceivable.

The Sturrock Report did not really come from Stanford. It came from an obscure organization called the Society for Scientific Exploration, of which Sturrock, an acclaimed physicist, was a founder and the current president. What the national press failed to report is that the SSE is an extremely unusual organization.

A few weeks before the Sturrock Report hit the press, the SSE held its annual meeting, in Charlottesville, Virginia. The moment I entered Clark Hall, the University of Virginia environmental sciences building where the conference was being staged, I noticed a tall, thin, dark-haired man haranguing a stout man who was seated quietly at a table. The shouting person had rolled up his left shirtsleeve and was pressing a microphone against the crook of his arm. A whooshing sound filled the air. The sound came from a speaker on the table— some kind of portable contraption, evidently designed for emergency use when making an extremely argumentative point. The sound was of blood pulsing through the man's veins.

"Do you hear that? Do you hear the spiral?" said the loud man.

The seated man calmly replied he did not hear the spiral.

"That's the sound of the vortex!" the loud man said.

This, it turned out, was Ralph Marinelli, of something called the Rudolf Steiner Research Center in Michigan. He had come to the meeting to give a presentation titled "The Heart Is Not a Pump: A Refutation of the Pressure Propulsion Premise of Heart Function." It is Marinelli's radical assertion that blood is self-propulsive. He believes the primary force that puts the blood in motion is an energy source that he has personally discovered, called "cosmic levity." Our blood is self-levitating. The core of the idea is that physicists have misunderstood the concept of centrifugal force. Marinelli, however, failed to persuade his fellow conferees. He was, one scientist confided, "pre-Newtonian."

But there were more wonders to be heard and seen. In the auditorium, a professor of physics from Texas A&M University, Ronald Bryan, was speaking on the topic of "What Can Elementary Particles Tell Us About the Space in Which We Live?" Bryan noted that there may exist dimensions of space beyond those with which we are familiar. The orthodoxy is that these dimensions are curled up at an unimaginably tiny scale and thus have no direct effect on our lives. Bryan, however, believes that people with near-death experiences (NDEs) may have seen these extra dimensions. The oft-reported "tunnel" that people travel through in a near-death experience may lead to another universe like our own, situated in another pocket in "higher space-time."

After Bryan came an Australian, Charles Berner, who announced that he had developed a Theory of Everything, and wrote an equation on the board that I cannot reproduce using standard fonts.

Then came an earnest young man from Santa Fe, Michael Partridge, who talked about the importance of the human heart in determining what is true. He said the distinction between science and spirituality is dissolving, "and science is at last becoming the servant of the human heart and soul." A member of the audience asked him what he meant by "the human heart." Partridge said, "That is a very good question." He paused. "I would have to say I don't really know. The human heart is such a vast part of our being. . . . I feel it needs further study and is not just a physical organ by any means."

We heard next from a Princeton researcher, York Dobyns, who studies whether human subjects can psychically affect the numbers randomly generated by machines. He presented research that tentatively indicated that the positive or negative mental state of the subject affected his or her ability to influence the machine. It was a carefully stated presentation that, were it not for the paranormal subject matter, would have fit into any science conference.

But then came Eugene Mallove of Cold Fusion Technology Inc. Mallove was clearly an angry man. He had dedicated himself to cold-fusion technology. Cold fusion has been a fringe obsession for years (many brilliant people are captivated by the dream of free energy) but remains outside the mainstream. Cold fusion is the phenomenon, Mallove said, "that gives rise to excess power and nuclear effects in metals in contact with liquid and gas media that contain hydrogen and its isotopes." It is an unexplained process, but "it is real, whatever it is," he said.

"I like to call this electro-alchemy," he said.

He showed slides, including (and this was odd for a scientific presentation) a couple of photographs of his critics. One slide showed the former editor of the esteemed science journal *Nature*, John Maddox. "His magazine routinely lies about everything," Mallove said.

After Mallove came a Frenchwoman who spoke of the Mars Effect, which is the theory that the position of Mars in the sky influences the birth of exceptional human beings, such as great athletes.

A review of the meeting's program indicated that I had already missed many other provocative talks. There had been a presentation about Distant Healing, in which "healing practitioners" scattered around the United States remotely improved the condition of AIDS patients in San Francisco (or so it was reported). Another speaker spoke of "experimental birthmarks," in which the body of a dying or deceased person is marked with soot or paste to help identify the person when he or she is reborn into another body carrying a birthmark in the same location. There was a talk about Nibiru, also known as Planet X, an unrecognized planet that exists in the solar system, as revealed by scrutiny of ancient tablets.

In the spirit of straightforward discourse, I should confess that my impression was that I was at some kind of festival for cranks. What was most perplexing, though, was that many of the participants were reputable scientists. Charles Tolbert and Larry Fredrick, two of the prominent members of the society, and credentialed faculty members at the University of Virginia, made it clear that the organization had changed over the years. "It was almost entirely scientists when it started," Fredrick said. The group was forced by financial considerations to have a more open membership, he said. Hence the arrival of the pre-Newtonians.

There was some sentiment that the group needed to reach out to the mainstream and scrape up more speakers with nonparanormal themes, but it was easy to see how the group was trapped by its own mission. People wanted mainstream credibility—they wanted their ideas treated seriously—but they would never have flown across the country to the annual meeting (or across the Atlantic, as in many cases) if the group did not indulge in extreme strangeness. It was the strange that lured them, that they found so interesting, so compelling. They wouldn't be here if this stuff were accepted by the National Academy of Sciences.

Maybe some of it would turn out to be true after all, and they could say, Hey, we called it early, we braved the criticism, we stuck to our guns. Copernicus never quit. Remember the meteorites! Remember the giant squid!

The conference showed how difficult it is to separate the legitimate fringe researcher from the mere crank. A smarter person than I would have to figure out where the "serious" ufology ends and the stupid, gullible, silly ufology begins. A case can be made that we are all doing our best to understand the world and we shouldn't categorize people in a negative way. If you say the cutoff is a belief in the efficacy of science, you'd set a very low standard, since even people with extremely fringe beliefs will avow an interest in remaining scientific. Distancing oneself from the truly lunatic edge of the fringe is the first order of business for anyone wishing to be taken seriously as a scholar of anomalies. It never occurs to anyone that he or she might actually be one of the wackos everyone is constantly referring to. No one ever says, "Yep, I'm basically out of my mind. Someone must have left the door open at the funny farm!" Instead, what you hear is, "I look at this from the perspective of science." And: "Actually, I consider myself a skeptic."

Strangely enough there appeared, in the midst of this day of bizarre talks, a clear-eyed analysis of the pros and cons of research into anomalies. The speaker was Henry Bauer, a professor of chemistry at Virginia Polytechnic Institute. He had a pointed beard and extravagant eyebrows and a way of sounding infinitely patient about topics of a sensational nature. One could imagine him as a psychoanalyst, helping someone cope with the realization that the Mars Effect has not helped him become a better basketball player. Bauer said that, compared with traditional science and social science, the study of anomalies offers a low probability of useful results. Little is known, and there is not much agreement about what is worth pursuing. In the field of anomalies, hoaxes and fraud are common. There are few true "experts" in anomalies. Those who claim to be experts often are in wild disagreement. Moreover, science addresses the "known unknown," whereas anomalies are part of the "*unknown* unknown."

All this sounded extremely cogent and wise. It is worth noting that Bauer, despite detecting the difficulties with anomalistics, has nonetheless spent decades studying the case of the Loch Ness Monster. He considers it an unsolved mystery. "I can't keep up with all the Loch Ness Monster and sea-serpent groups, because everyone wants to publish their own newsletter and wants me to pay $15 a year to get it," Bauer said.

During a break in the proceedings, I heard about the upcoming UFO report from the society. Larry Fredrick said Laurance Rockefeller, a frequent supporter of paranormal research, had given Sturrock a lump of money to organize a conference to examine the "best evidence" in support of the extraterrestrial hypothesis for UFOs. In the fall of 1997, the ufologists presented their evidence to some scientists who had been gathered up by Sturrock. But the "best evidence," the panel decided, did not include any persuasive evidence of aliens or any events beyond the known laws of physics.

This, then, was the backdrop of the news stories in late June 1998. The stories could easily have been a dash of cold water on the UFO community. Instead, they focused on the one positive element of the report—that the scientists thought there might be something useful to be found in UFO cases. The report was written by Sturrock, with contributions from the scientists on the panel, all of whom put their names on Sturrock's final product. The report is, in fact, quite neutral and factual. Sturrock summarized some of the conclusions at the beginning of the report:

> Concerning the case material presented by the investigators, the panel concluded that a few reported incidents may have involved rare but significant phenomena such as electrical activity, but there was no convincing evidence pointing to unknown physical processes or to the involvement of extraterrestrial intelligence.
>
> The panel nevertheless concluded that it would be valuable to carefully evaluate UFO reports since, whenever there are un-explained observations, there is the possibility that scientists will learn something new by studying these observations.
>
> However, to be credible, such evaluations must take place with a spirit of objectivity and a willingness to evaluate rival hy-potheses.

In short, it was hardly an astounding or groundbreaking report. It did not make a strong push in favor of greater investigation of UFO reports. The tone was more permissive than prescriptive. It allowed that someone studying UFOs might not wind up wasting his or her life on a foolish obsession. It recommended "modest" investigation and some institutional support (though it didn't specify any government funding). Go ahead and study these things, it said, because "there is the possibility that scientists will learn something new."

Mark Rodeghier, scientific director for the Center for UFO Studies, and one of the ufologists who presented evidence to the panel, told me that the ufologists were hoping for something more positive from the scientists. "They really said that all the past research didn't amount to much. And that was disappointing to us." I also spoke to one of the scientists, Jay Melosh of the University of Arizona, who said he was disturbed by the press coverage of the report. The thrust of the report was the absence of evidence of aliens, he said. "We were pretty negative about it. The evidence we saw—and it was promised to be the best evidence of the reality of UFOs as extraterrestrial spacecraft—did not support that conclusion." The conclusion that scientists might learn something from studying UFOs "was not very strong," he said. "The stronger conclusion is that there is absolutely nothing that we saw in the evidence that there is extraterrestrial life on Earth."

He cited the photographic evidence. The ufologist Richard Haines presented a Top Ten of flying-saucer photographs from over the years, and the number-one picture, the piece Haines singled out for its persuasiveness, was an image of a mountain with a small object in one corner of the frame. The object is faint, but has the classic flying-saucer shape. There was, however, a shocking flaw in the evidence.

"The photographer never saw this thing. The lady that took the picture didn't see anything. It was first noticed when the film was developed a year later," Melosh said. "Nobody ever looked at the possibility that it was a film defect or a processing defect."

With the type of camera used, anyone looking through the viewfinder would see whatever ended up on the film. But the photographer had somehow failed to notice the flying saucer cruising by the mountain.

This doesn't mean that there aren't more compelling pieces of evidence out there. In fact, there are hundreds of other UFO researchers who weren't invited to speak to the scientific panel. No one invited Budd Hopkins, David Jacobs, or John Mack, among the pioneers of research in alien abductions. Richard Hoagland, the man who believed that Sagan had whispered encouragement about the Face on Mars, was not invited. No one invited Linda Moulton Howe, a journalist who has extensively studied cow mutilations. No one invited Philip Corso, the aging military man who, before his death, claimed that the government "reverse-engineered" the technology from a flying saucer that crashed in Roswell in 1947. Perhaps these other people could have convinced the scientists of the extraterrestrial hypothesis. And perhaps a different group of scientists would have been more impressed by the evidence presented by the ufologists.

But we can't be sure. It's another one of those dadgum unknowns!

18

The Outer
Limits

Laughlin, Nevada, is a gambling town, not much more than a strip of bright lights along a sleepy stretch of the Colorado River. You get there by driving south from Vegas for about seventy-five miles on a road so straight it might have been laid down along the edge of a giant ruler. Along the way you have plenty of time to survey the landscape for mysterious aircraft and secret government facilities. What you notice, in this part of the West, is the "planetness" of the environment. The expansiveness of the terrain gives a motorist going eighty-five miles per hour a sense of traversing a sizable chunk of the planet Earth. There are distant mountain ranges, ancient seabeds, a palpable sense of tectonic forces throwing up slabs of crust. When the Sun goes down, the stars come out so brilliantly, it's as though there is no intervening layer of atmosphere. You drive down that highway and sense the truth that eludes us in the big cities—that we really are space travelers, all of us. On this rock we zoom through the universe.

I was driving to the Sixth Annual International UFO Congress. If the Society for Scientific Exploration represented a relatively elite, serious, respectable community of anomalists, their less skeptical counterparts could be found in scattered conferences with a more New Age feel, gatherings that make little pretension to academic rigor. The UFO Congress was one of the best such meetings, because it had an extremely democratic, inclusive policy of inviting speakers who were truly incredible in every sense.

The congress gathered at a hotel/casino right on the river. It looked like a good place for a convention of folks who don't have the luxury of traveling on a corporate expense account. Rooms cost $18. A sign advertised a $4.95 all-you-can-eat lunch buffet, and another promised a nightly "Tribute to Reba McEntire" starring a woman who was not technically Reba McEntire. There were tables taking $1 bets at blackjack. This was paradise for a low roller.

First stop: the UFO exhibit hall. Vendors lined the periphery, hawking flying-saucer T-shirts, crystals, plastic aliens suitable for dangling from one's rearview mirror, and cassette tapes describing the secret world government. In such a room you can see the many sides of the alien phenomenon, how it is simultaneously enlightening, terrifying, and cute.

At the entrance to the hall stood a man wearing an eyepatch and chewing on the blackened stub of a cigar. He looked like trouble. He was speaking to a man in a jumpsuit. I noted that Jumpsuit Man had bleached hair. I introduced myself and, just trying to break the ice, asked Jumpsuit Man if he wore a jumpsuit because he sees himself as some kind of astronaut-type person.

"No, he *just wears jumpsuits,*" interjected Eyepatch Man, clearly feeling protective of his friend. He made me for a wiseacre, a snob who thought it was a giggle that someone would wear a jumpsuit.

They began talking about a report the night before on CNN, about a flying-saucer landing in Elmwood, Wisconsin. The report didn't actually air—it was a feed, monitored by someone with a satellite, who then passed on the word to various ufologists. Now they were all trying to track down the story and confirm that CNN had said something about a spaceship. The Elmwood angle raised a red flag. Eyepatch Man told me, "Running through Elmwood is the largest electromagnetic anomaly, a lapline, on the surface of the Earth."

Before I could fully reveal my ignorance of the phenomenon of laplines, he redirected my attention.

"There's probably the greatest stigmatist in the world," he said.

I quickly followed the world's greatest stigmatist, leaving behind (forever, as it turned out) the story of the spaceship landing in Elmwood. The stigmatist was Giorgio Bongiovanni. On his forehead was a hideous, bloody scab in the shape of a cross. I introduced myself, and we "shook hands" with only the tips of our fingers: I had to avoid compressing his bandages. At my request, he pulled aside the gauze and revealed his horrible palms. He had the wounds of Christ.

Through an interpreter, Bongiovanni said he had received the stigmata at Fátima, Portugal, on September 2, 1989. His UFO newsletter, *Nonsiamosoli* (which means "We are not alone"), gives the whole story, saying that the Virgin Mary appeared to Bongiovanni and revealed that Jesus Christ has visited civilizations throughout the cosmos, where He has been accepted—unlike on our planet, where He was crucified. The Virgin asked Bongiovanni if he was willing to carry a part of the suffering of her son, and when he said yes, two beams of light came out of her chest, he said, and struck his palms. The bleeding cross on his forehead, he said, arrived a few years later, late at night in a hotel room in Uruguay.

Aliens, Bongiovanni said, have been coming to the Earth from an advanced civilization for thousands of years. Some of them accompanied Jesus in his travels.

"Our forefathers called them angels," he said.

Bongiovanni had just offered up a central premise of the skeptics, as it happens. For the skeptics, aliens are angels (and angels, needless to say, don't exist in the skeptical universe). In a secular era, many people find it hard to justify a belief in immaterial, wispy entities like angels and spirits and demons, because by definition such creatures are outside the normal laws of physics and carry the odor of pre-Enlightenment thinking. But aliens don't have that problem. They have mastered the laws of physics but are still bound to them. Aliens are angels with a scientific veneer. They don't need wings because they have spacecraft. Some can walk through walls, but only because they can manipulate their molecular structure. It's still science.

The theoretical physicist Paul Davies, in his fine little book *Are We Alone?*, summarized what's going on: "What we see in the UFO culture seems to be an expression in the quasi-technological language appropriate to our space age of ancient supernatural beliefs, many of which are an integral part of the folk memories of all cultures."

The equation of aliens and angels is, unfortunately, a bit too simple. If we are to say that people believe in aliens for the same reason they believe in angels, we leave unresolved the question of why they believe in angels. The temptation, for those of us who have not scared the door of a church in a long time, is to give the topic of religion a wide berth. The truth is that many of us are not capable, intellectually, much less spiritually, of engaging the issue of faith, of knowing how to discuss such things without revealing our ignorance. I had managed over the course of many months to become fairly adept at discussing the structure of the universe, the nature of the interstellar medium, the dimensions of the Eagle Nebula, and so on, but I certainly didn't have a clue about the structure, nature, or dimensions of the human spirit. Yet as I spent time in the UFO world it became apparent that, as a cultural phenomenon, the study of aliens had nothing to do with astronomy, chemistry, biology, or planetary geology. It didn't have anything to do with Henry Harris–style propulsion systems or the potential solutions to the Drake Equation. This involved the even more exotic and incomprehensible territory of the human mind, or, if you prefer, the human spirit.

That guy back at the SSE conference who talked about the importance of the "heart" in understanding the world may have been a bit fuzzy in his presentation, but he was onto something. To be up to speed on the topic of aliens, a person needs to be fluent in science *and* in spirituality. But spirituality isn't something in which you can take a crash course. There are books dealing with every imaginable aspect of spiritual belief, spiritual improvement, talking to God, conversing with your dead relatives, unifying "sense" and "soul," and so on. These books don't fill a shelf, they fill entire bookstores (a fine one being The Bodhi Tree in Los Angeles). One does not merely dip into these writings. You can't "bone up" on religion and spirituality. You can't decide that over the weekend you are going to (in addition to washing the car and mowing the yard and cleaning the gutters) raise your consciousness. To fully appreciate the UFO belief system, I would need to die and be reincarnated as an entirely different person.

Nonetheless, I searched as best I could to understand why these folks in Nevada were believing in things that in many ways seemed medieval. One element of the belief system stood out: The aliens solve problems in a world where problems are abundant. It's a puzzling fact that many people, even in a relatively affluent society, are in psychic agony. They're confused, depressed, anxious. They're overwhelmed. They feel as though they don't belong. They're alienated, in a word. Many are also reluctant to ascribe their problems to internal mental processes, mere neurochemistry (although many others find the chemical explanation perfectly satisfactory). It is somehow more reassuring to view their problems as caused by others (which was David Duchovny's point). Pain and suffering provide the energy for creative cosmology.

One day in Las Vegas (if I may be permitted to engage in hyperspace travel for a moment), I met a former city planner named Mark Williams, who had written a book called *Ascend!* His book was unusual in that it had a blurb from God. ("Mark has been more dedicated to absolute truth and accuracy in what is written in this book than any other modern times author that I and the angels have worked with," God wrote in uncharacteristically stilted syntax.)

Williams had a joyous message. Mankind was becoming a peaceful, loving life form. "The only question," he told me in his trailer in Miracle Mile Park, "is how soon are people going to move out of suffering and into joy." The Earth is the only place in the universe with "full free will," he said. The Creator made Earth as a kind of experiment in moral self-governance. He wanted His divine children to learn to become gods. "We had to learn to choose love just as the Creator did." There was still the possibility of self-destruction. It had happened before, with Atlantis and Lemuria. In the late 1960s, he said, the angels and the ETs—separate groups—concluded that we weren't going to make it, that we weren't going to complete the transition to peace and love, and they gave people the opportunity to leave the planet, which explains the rise of "abductions." Most of the ETs are loving creatures from the Pleiades, and the Grays are in retreat. As Williams explained this elaborate situation, I could not help noticing the context: He was a smart guy living alone in a trailer park. He told me he'd suffered for fifteen years. He'd just gotten divorced for the second time. He'd had massive headaches and severe illnesses. Indeed, he had a tendency to gasp. He couldn't breathe very easily, because of a respiratory problem whose origin I refrained from exploring. His message of joy emerged gasp by gasp. He blamed the ETs for his run of hard luck. The bad aliens had stripped people away from him to make him as weak and isolated as possible, he said.

In his story the part that rang truest was the pain.

After only an hour at the UFO convention, I already needed a break, and went to the coffee shop for some lunch. At the counter sat a man whose badge identified him as Steven. He was a thick fellow, with a big, kind, open face. He said he was unemployed, drawing disability, because of a neck problem. There was an awkward moment as I pondered whether to engage further and he no doubt considered whether to continue telling me about his problems. Finally, he volunteered that he had come to the conference because he was hoping the aliens might be able to help him.

"I need an operation on my neck. I hear they have amazing medicine," he said.

Steven said he was also interested in the propulsion mechanisms used by these creatures. By training, he's an engineering technician. He said the shape of the UFOs is what provides the propulsion. He sketched a craft with a blunt nose and a hollow back, somewhat like a fancy golf putter. When I asked what the energy source was, he said, elliptically, "Frequency, I think."

Or maybe it was related to gravity. The Sun, after all, creates gravity by distorting the fabric of space-time. The alien ships might do something similar. And we must remember the profound nature of the alien mind.

"Their starships are telepathic. They are produced telepathically by their minds," he said. "You ever heard of Ashtar Command? That's another entity that has unique starships. They're supposed to be the good guys. They're supposed to give us the technology within five years."

He went on to talk about a fireball he once saw in the sky, and also about his theory that the aliens are here to improve mankind. I asked why some people believe in aliens and some don't.

"Some people don't have a background in science and engineering," he said. "They don't realize a system like that can come together. It's ignorance, actually."

President Clinton, he said, had been in a starship with some aliens in Arkansas.

"He went from Arkansas to New York in about fifteen minutes and back again," he said, as I asked for the check.

The Sixth Annual International UFO Congress was a fringe event even by the standards of people who have dedicated their lives to the study of flying saucers. The mainstream people at MUFON (the Mutual UFO Network) and at the Center for UFO Studies (founded by J. Allen Hynek, an esteemed scientist who went over to the other side) considered this sort of gathering to be a freak show.

Although there's no official story line in the UFO world, several ideas keep popping up, including the theory that humans did not evolve on Earth through the normal process of Darwinian natural selection. We are, rather, supernaturally created, just as it says in the Bible, only in this case the creators are extraterrestrials who know how to manipulate our genetic material. Where these aliens come from varies, though the Pleiades, also known as the "Seven Sisters," are often invoked, as is Nibiru, a hypothesized undiscovered planet in our solar system. (Nibiru is the subject of many writings by Zecharia Sitchin, who believes he has discerned a series of astonishing planetary phenomena recorded in Sumerian tablets thousands of years ago.)

The supernatural and New Age beliefs have largely superseded the traditional study of mysterious aerial phenomena. Seeing a flying saucer just doesn't get you anywhere anymore. You must have contact. The intrepid Philip Corso, in his book *The Day After Roswell*, didn't merely claim that a spaceship had crashed near Roswell and that bodies had been recovered. No, he wrote that he, personally, had entered a darkened building, snooped around with a flashlight, and pried open a crate. Sure enough, there was an alien inside. "At first I thought it was a dead child they were shipping somewhere. But this was no child. It was a four-foot human-shaped figure with arms, bizarre-looking four-fingered hands—I didn't see a thumb—thin legs and feet, and an oversized incandescent lightbulb–shaped head that looked like it was floating over a balloon gondola for a chin."

That's the kind of story the market demands in the 1990s. The field has become preoccupied with mystical, personal, subjective encounters with aliens, the abductions, the matings, the carrying of alien fetuses, the channeling of glib, pedantic extraterrestrials.

To believers, there is a logical explanation for the shift in focus. Budd Hopkins, the author of *Missing Time* and *Intruders*, says that, for the first several decades of the UFO era, ufologists didn't even think about what might be inside those mysterious flying saucers, or why the aliens were here in the first place. "We were busy trying to get the license-plate number on the getaway car without figuring out what the crime was," Hopkins told me in his New York artist's studio. In the 1980s, a rash of books, such as Whitley Strieber's *Communion*, told the terrifying tales of people abducted by aliens as they slept in their bedrooms. The abductions had a fairly consistent narrative and a common description of the aliens: They were gray creatures, hairless, with skinny limbs, large heads, no chins, and, most memorably, huge almond-shaped black eyes with no lids. Could this common description be anything other than conclusive proof that something real was happening? The only hitch was the timing. The descriptions of the aliens tended to echo the appearance of that little wimpy critter that walked out of the mother ship at the end of *Close Encounters of the Third Kind*. Spielberg's alien was based in part on the tale of Betty and Barney Hill, the pioneers of the abduction field, who were nabbed by a prototype Gray in 1961. A 1975 TV movie about the Hills, *The UFO Incident*, showed millions of people what one of these aliens looked like; this, skeptics believe, is the source of the descriptions of aliens given by alleged abductees. (The description by the Hills was inconsistent, however. Betty said the aliens had huge noses, sort of like Jimmy Durante. Barney said they had no noses at all, just a couple of slits for nostrils. Betty said they had dark hair. Barney said they had no hair at all. It's a typical UFO story, with contradictory data, even as the witnesses are emphatic in their accounts. Both, however, cited the large heads and the gray skin.)

Although the consensus is that aliens are conducting breeding experiments with humans, beyond that no one really knows what's happening. Are the aliens malevolent or benign? Are there "good" aliens in conflict with "bad" aliens? Do the aliens portend our doom or our salvation?

One of the most intriguing speakers at the Laughlin conference was Lee Shargel. No one knew much about him; the convention program described him as a recently retired NASA scientist. Shargel began his presentation slowly, portentously. A slide of the Mars rock. A slide of the planet Mars. A slide of the Mars rover, designed for exploration of the surface. So far, nothing but science. Then he spoke of the Gravity

Wave Highway. A gravity wave, he said, moves as fast as "the speed of thought." Our interstellar brothers are using these gravity waves to traverse the universe.

People began murmuring. This was good stuff. Shargel showed slides of an alien spacecraft secretly photographed by a military jet over Roswell in 1947. It wasn't saucer-shaped but looked more like one of the bat hooks that Batman throws around. (Later, Bob Brown, the organizer of the convention, passed around his own skeptical interpretation, which was that the spacecraft looked like the hood ornament on a Cadillac.) Shargel had lots of slides of this spacecraft: buzzing small towns, parked in a crater on the moon, flying next to Comet Hale-Bopp. The audience was transfixed, oohing and aahing. The problem was, in every picture the spacecraft looked the same—it was seen from the same angle. The viewer in each case was to one side and slightly behind and below the craft. There was not a single shot of it from head-on, or from directly behind, or directly below. Even the shots of it on the moon or in deep space were from the same angle. This was quite obviously a set of fakes, and not even very good fakes. It looked as though the faker had just plopped the "spacecraft" down on a photograph and claimed that the object was flying through the frame.

But that was only the left hemisphere of the brain acting up. The right hemisphere was digging this. Shargel was on a roll. He said there are good aliens and bad aliens. The flying hood ornament belongs to the bad aliens, but there are good aliens trailing in the wake of Comet Hale-Bopp. The good, loving, kind aliens are from the planet Chulos, orbiting the star Vega. Amazingly, he had a slide of Chulos. He had a picture of an extrasolar planet! Goldin would have been thrilled. Why build a multibillion-dollar contraption in deep space called Planet Finder when there are people like Shargel who already have the pictures? (*Down*, left brain.)

Shargel said the good aliens will "evict" the bad aliens who have been lurking around Earth since the Roswell crash in 1947. "You have friends out there. And they love you. Believe that. I do," he said.

People clapped.

Shargel then predicted that early the next morning a gravity wave would pass through the Earth, possibly accompanied by a musical tone. People would feel it, he said. One of the conference organizers turned to me and said, "Tomorrow, Uranus and Neptune and Pluto line up with eight constellations and Hale-Bopp."

A fidgety man next to me said quietly, "Nine-thirty-five a.m. our time." He looked happy. "It's rumored that twelve strands of DNA are activated tomorrow."

Which twelve strands? Unknown. What would this new genetic explosion do to us? Unknown. But it was very exciting.

The speech went on and on and on, and before it was over Shargel revealed how three alien beings who looked exactly like dolphins had appeared in front of him in a bubble like Glinda the Good Witch in *The Wizard of Oz* (his comparison). He said these dolphinoids somehow touched his heart, leaving a heart-shaped mark on his chest. When he was finished, someone in the audience asked him to show the mark. Shargel looked hesitant, then relented. He opened his shirt, like Superman baring his "S." Right there, sure enough, was an extremely faint, pinkish blob.

His story was checking out.

I asked a woman named Pam if she thought the photographs looked fake. They did, she said, but added that she believed Shargel anyway, "because of his credentials." She said, "I think this man's right on. Therefore, I want to believe everything he says is true."

I noticed one woman in the crowd, a tall, lanky blonde, who was particularly enthusiastic about the speech. Her name was Miesha Johnston. She had come to Laughlin from Las Vegas with her roommate, Jan Bingham. Miesha and Jan, I learned, were on an incredible voyage of self-discovery. Miesha ran a support group for people who had had "experiences" with aliens. She was gradually coming to terms with the theory advanced by Jan, which was that they themselves were aliens in human bodies, or "Starseed." Jan knew with absolute conviction that she was from the Pleiades. Miesha seemed to be wavering. If she was from space, she was from somewhere in Orion. But she had doubts. It was a bold belief system even by UFO standards.

Jan told me she occupied a human body that used to go by the name of "Val." She believed that a decade earlier Val was transported to a spaceship during the middle of a nap. Val voluntarily gave up her body, and she, Jan, an indigenous Pleiadian, entered it, and returned to Earth to live among humans and continue to raise Val's children. (You can see how this would cause havoc with pronouns—"Jan" would refer to "Val" as "she" and to her alien self as "I.") Jan told me she loves Val's children as dearly as if they were her own. She said she has retained all of Val's memories and much of her emotional foundation. She also completed postgraduate work initiated by Val. Val, said Jan, has also returned to Earth in another body and is very happy.

This was not an abduction but, rather, a "walk-in" situation, an abduction without the hostility. It's copacetic with all the interested parties. "It's not an easy thing to be a walk-in, initially. You have to learn to walk again. You have to learn to talk again," Jan said. Bingham's twenty-two-year-old daughter, who asked me not to identify her, said she was surprised but not distraught when she learned about her mother's new identity.

"She sat me down and explained to me that she wasn't—I don't know how to say this without its sounding really weird—that she wasn't really my mom, that she was a walk-in." But she didn't reject the idea. "I think that it's quite possible that things like this happen, because for people to assume we're the only race out there is just stupid," she said. The water-cooler paradigm, in this case, turned out to be useful—it helped a young woman cope with her discovery that her mother is an alien.

Miesha, meanwhile, wasn't sure if she was literally a walk-in like Jan, or merely a "wanderer," someone with an alien soul at birth. A wanderer typically endures an unsettled, difficult life, until gradually the awareness of the extraterrestrial origin becomes clear. Although Miesha said she feels "resonance" with Orion, she couldn't pinpoint her home planet. She only knew that she had an "off-planet origin."

She said of the aliens: "They're part of our family. We're like their little brothers and sisters. A lot of species have a lot of love for us."

In the universe of Jan and Miesha, there are no distances too vast to be conquered. There is no daunting chasm of space and time separating the islands of intelligent life. In their universe, loving creatures can come together. Within ten years, said Jan and Miesha, the aliens will be walking among us—human-looking aliens as well as dolphinoids. They promise that we'll all get along beautifully.

Soon I was in Las Vegas furthering my interspecies contact with the Starseed. The serious UFO people would not be pleased with this move, for the Starseed are precisely the kind of New Age figures that the traditional ufologists have complained about for years. Ufologists look outward, toward the universe, for answers to the alien enigma. New Agers look inward. It is not clear which technique yields the best information. Neither the serious ufologists nor the New Agers could actually produce an alien-made wristwatch or an alien-solved mathematical equation. In the news-you-can-use category, both sets of investigators came up short. The New Age people at least offered the hope that all this energy being invested in the contemplation of aliens would actually improve a person's life.

Miesha Johnston invited me to sit in on a meeting of her "Starseed Contactee Group." She convened it weekly in her apartment, which was in one of those just-add-water housing developments found in every suburb in America. There were a series of two-story stucco buildings arranged in modules around a nucleic swimming pool, each apartment blessed with wall-to-wall carpeting, a cramped kitchen, and sliding glass doors leading onto a small balcony.

19

Hypnotized

Miesha's place was pleasant, clean, comfortable, but I also sensed the lack of permanence. This was a residence, not a home. She'd be gone in a matter of months. Nothing was rooted here. There are millions of people who live transient lives, with shifting beliefs. A few months pass and they have a new job, a new address, a new religion. In the fluidity of our lives, the novel notions of the millennium are easily transported.

Miesha didn't publish the phone number of the Starseed Contactee Group (which seemed to have different names at different times). Jan explained that they didn't want any nuts calling up. The group's printed guidelines specified that the meeting would be private, that all comments would be kept confidential, and that everything discussed must relate to UFOs, contacts, experiences, and Starseed. This was not to be a forum for debate or investigation. This was about support. Rule 10 made this clear: "We will not invalidate a member's beliefs, opinions, or experiences. You cannot love or hate something about another person unless it reflects to you something you love or hate about yourself." The second sentence in that rule was conceivably a bit argumentative. The first sentence, meanwhile, was a concise and elegant expression of what distinguishes New Age philosophy from scientific methodology. A scientist considers it a sacred obligation to try to invalidate the spurious claims of others. Scientists are trigger-happy invalidators. Some would be happy to explain to a five-year-old kid that Santa is merely an invention of parents who use the fat elf as a proxy disciplinarian.

Another of Miesha's handouts provided members with a list of physical anomalies that could be associated with alien abduction, contact, or experiences. A sample:

> Having seen what appears to be a star in close proximity.
> Familiarity with a certain star or star system in the sky.
> Dreams of UFOs and alien beings.
> Dreams of being in an elevator or going underground.
> Dreams of being under water, breathing under water, and/or breathing other liquids.
> Dreams of doctors doing medical procedures on you or others.
> Dreams of sexual encounters with people with unknown faces or no faces.

Dreams of flying or floating above your bed.

Having the feeling that you have been watched throughout your lifetime.

Being compelled to take a drive or go out by yourself with no real reason why.

Buzzing, ringing, humming, or pulsating sound in one or both ears.

Consistently waking up at the same time in the middle of the night.

As a child, had nightmares and feared sleeping alone. Often wanted to sleep with parents.

About two dozen people showed up. The meeting began when Deborah, a friend of Jan and Miesha, tinkled a bell. Jan put on a CD with serene synthesized music, dimmed the lights, and turned off the phone. Seated on the floor cross-legged, she led us in meditation. She instructed us to see our breathing: "Every cell in your body is filled with pure white light." She told us to visualize our "crown shocker"— that's how I heard it—which is apparently something on top of the head that controls the energy route into the body. (I learned much later, from my yoga-literate wife, that it's a crown "chakra.")

"The universe and all the loving extraterrestrials are bringing beautiful pure white light through that crown chakra," Jan said.

This went on for a while.

"Now say to yourself, I give myself permission to remember my purpose of being here on Earth."

She talked of going forward, forgiveness, love.

"Now imagine a beautiful being of light. . . . That being is about fifteen feet from you. . . . You know that this being is going to bring you a message—a message of higher consciousness. . . . As you get this information you know intuitively it is appropriate to you, it brings peace of mind to you and brings you closer to your brothers and sisters in the sky."

She told us to open our eyes. Everyone seemed to have entered some improved state of mind, except of course the reporter, who had found it hard to pick up the information from the beautiful being of light while taking notes.

Then came the highlight of the evening. A woman I'll call Maddie (she wanted anonymity) produced a cassette tape for us to listen to. She had taped a man named Darryl Anka channeling an alien

named Bashar. Anka had been doing this for years, mostly in Las Vegas and southern California, and he'd written books filled with
Bashar's insights into our world. This particular channeling had been
conducted recently in Vegas and had required a $20 admission fee. On
the audiotape, Anka began by saying—rather straightforwardly, I
thought—"You'll have to believe he's a separate being from myself to
get some benefit out of this."

For the next forty-five minutes, no one in Miesha's apartment
moved or talked. We just listened. Anka, on tape, took a deep breath.
After a pause he suddenly began speaking in a loud, fast, manic voice,
slightly mechanical, with extremely precise enunciation, as though he
were part human and part robot. "We would like to begin this transmission with the following ideas . . ." Anka/Bashar said. What came
next was an eruption of what can only be called erudite-sounding gibberish. Bashar's genius is that he sounds brilliant, even omniscient, although the words themselves do not really add up to anything
meaningful. "See beyond your patterns! Look at the obstacles and
thus expand your reality!" he said—his accent bringing to mind that
of Rex Harrison. "This is physics. Everything is energy. Everything is
energy and motion."

Someone asked Bashar a question about a problem she had in
her personal life.

"You have to function as a totality! As a holistic system!" Bashar
said.

He advised one questioner to tell his parents that he is an alien
from another world and then cut all ties to them. Otherwise his advice
was nothing but bland self-help nostrums, like "Move toward pleasure
and away from pain."

At one point Bashar discussed his own species. "Genetically
speaking, we are a hybrid species in what you call your future time
frame." One of his habits was to note the shortcomings of the human
view of reality—such as the way we think of something called the "future," which apparently is more readily accessible to an advanced
species.

Next came the strangest part. Maddie (on tape) asked Bashar if
he could channel a Gray. In other words, Anka would be channeling
Bashar, who in turn would be channeling a scary alien. It was like hy-
perlinking on the alien Internet.

"We will see if it is allowed," Bashar said, speaking, as always, in the first person plural.

"One moment. One moment. One moment," Bashar said frantically. There seemed to be some discussion between Bashar and the Gray. Then the Gray took over, his voice colder, reptilian, a controlled hiss. The Gray explained why the abductions are taking place. The Grays are in contact with the leaders of our government. The Grays are after our emotions.

"We now through hybridization are beginning to conceive and understand what you label as love energy," the Gray said.

Maddie told him that the abductions must end. The Gray didn't like this and soon left, leaving just Bashar, who said the Gray belonged to one of the "more primitive factions" of aliens.

We stopped the tape, and Maddie talked of the terrifying event. "I felt tremendous fear when the Gray came through. I do feel they were feeding on me at that time."

We went around the room giving our reactions to the tape. A man named Lee said he thought this might have been a first in the UFO movement—the first time a human had directly challenged a Gray. The Gray, Lee said, sounded like someone in the military—the military, he meant, of the Gray empire. When my turn came, I asked if anyone was willing to entertain the possibility that they hadn't just heard a Gray alien or an alien named Bashar but, rather, had been listening to a guy named Darryl Anka as he pretended to be an alien.

No one thought this possible.

"He'd have to be an incredible actor," said one woman.

"His wisdom is far beyond what a human would have," said Lee.

Many months later, I spoke to Anka on the phone and asked him if he really channeled an alien. He said that he didn't know for certain, suggested that he definitely entered an altered state, but allowed that the alien might be an artifact of his own mind. He also sent me a catalogue of his many channelings. He had just made the important decision to quit channeling Bashar so that he could work on an entirely different project. He was designing a UFO theme park.

They all began talking about their abductions, the shadow government, and fetal implants. Jan said that four years ago the Galactic Command had brought in some high-level Grays and Pleiadians to run the abduction/hybridization program so it wouldn't be as traumatic to the human abductees. A college student named Jen said, "I know I birthed a child for them, but I don't remember it being negative at all." She seemed unsure what to think. "I'm just confused, because we read Karl Marx in school today." An older man then started talking about how he was part of something that spans ten dimensions, and he mentioned a group of unusual beings who will return to Earth in eighteen hundred years in silicon bodies—though at this point I had lost all sense of what anyone was saying. The meeting had become as incomprehensible as a discussion of the carbonate globules in ALH84001.

Finally, one member of the group said she wasn't 100-percent certain that Anka was truly channeling an alien. But she did want to say she was impressed with the way Maddie had handled the scene. She added, "We're so unevolved. I'm almost embarrassed to be human. Actually, I hope I'm not."

It takes one simple presumption to be a Starseed or an alien experiencer: that truth is best accessed through an altered mental state, such as when dreaming, dropping off to sleep, or hypnotized, or when being counseled by someone who is at that moment channeling a being from another dimension.

"What people call dreams I call experiences. To me they're just as real," said Jan's friend, David Easler.

Once, he said, he was in an alpha state, about to fall asleep, when Jan's Pleiadian father, Zachary, popped through a portal into the room. Easler remembers saying, "What's up?" (What *do* you say to an alien who pops into your room?) Zachary then left, and another alien, with a much larger head and huge insectile eyes, came through a different portal, scaring him. This was an alien of the praying-mantis variety. The alien helped Easler recover memories of his childhood experiences, including being taken on a ship and having DNA sampled. He believes he has been genetically altered.

"What I feel that they've done is that there was some latent genetic code that they were reactivating," he said.

The genes, he said, gave him special access to space-time manipulation and remote viewing. He now studies alien technology—techniques for superluminal travel. (Henry Harris, check your e-mail.)

His parents are unhappy. They raised him as a Jehovah's Witness. They completely reject his assertion that he has been visited at night by extraterrestrials.

They think the visitors are devils.

"It's like the Church Lady on *Saturday Night Live*," he said. He affected a screechy, hysterical voice: "Could it be *Satan?*"

The next day I went to a vortex in the mountains west of Las Vegas. The Starseed said the Pleiadians had some kind of underground base nearby for their spaceships. Miesha agreed to show me the place. I drove. Jan couldn't go but her friend Deborah could. Miesha and Deborah had different cosmologies within their mutual belief in an alien presence. Miesha felt the aliens were probably positive, benign creatures; she did not buy into some of the more terrifying elements of the abduction scenario. But Deborah had a darker vision.

I could see her in the rearview mirror. She wore shades. She was by any conventional measure a pretty woman, forty-something, blond, but with a serious, brooding face, as though she were contemplating an unpleasant appointment. She told me, "In the early eighties, I started getting hit with feelings of gloom and doom, like the world was coming to an end." It was, as it happened, her "ordinary" life that was ending. She began having abduction experiences, contact with aliens of a malign intent. They would come every night at 3 a.m. and walk around her bed. She couldn't move. "I've had some really horrifying experiences—being in bed and feeling like I'm being thrown out of bed, feelings of being paralyzed. I've visualized entities."

I asked her what she saw.

"What I saw was the devil. What I know today"—her memory corrected by a realization of the alien presence—"is, what I saw was reptilian energy." She didn't realize what she had experienced until she read Barbara Marciniak's book *Bringers of the Dawn*. After much inner work, she came to realize that she herself was an alien spirit from the Pleiades. The Pleiadians are humanoids; this explains her ongoing problem with the reptilians. Different types of aliens don't get along. There are also dolphinoids out there, and serpent people who go *ch-ch-ch-chhhhh.* . . . Some of them, said Deborah, may come from the center of the Earth.

I asked her why the aliens were coming here.

"I think there are some out there that are seeking power over our planet. I think there are some who can't evolve themselves until *we* remember who we are so *we* can evolve."

These theories were also evolving. Everything was up in the air. There wasn't a single narrative that everyone agreed on. Many aliens, many agendas.

We reached the Red Rocks area, west of the city, and drove down a gravel road into an area being graded and platted for a subdivision. The desert air was so clear it was hard to tell the dimensions of the hills around us. We walked through a picnic area. Deborah lingered in one spot, clearly stirred.

"Do you feel something?" Miesha asked.

"Yeah. Chaotic energy."

"A little nauseous?"

"Yeah. You know what this reminds me of? Reptilian energy."

Miesha repeated the rumor that this was a Pleiadian base.

"Maybe it's the *reptilian* Pleiadian base," said Deborah.

We walked across a flattened spot where someone would soon build a house. Chalk lines prefigured the rooms to come. Las Vegas is racing into these mountains as fast as the developers can pour concrete. The population of the county hit a million in the mid-1990s, and the projection is for two million by 2006. The Pleiadian base will not have the same energy when it is covered with houses and lawns.

The city and its expansion show the best and worst of the human race. Our creative energy and adaptability let us invent an entertainment megalopolis where there had been nothing but a small town in the desert. But it is all powered by greed, by a gangster's vision of free-flowing cash. Now they wonder where they'll get water. Do these things happen on other worlds? Do civilizations last for millions of years, sticking around long enough to be detected by radio telescopes, if they have a Las Vegas somewhere in their psyche?

Deborah and Miesha put their hands on the dry, powdery surface of the rust-red cliff at the edge of the basin. They felt it, the energy. I climbed onto an enormous rock vaguely shaped like a spaceship that looked as if it might take off for the Pleiades any second. When we touch a surface, I announced, we feel not the solid matter of the thing but merely the electrical repulsion, the energy of the electron shell. In other words, I was trying to impress them with knowledge with which I have only the faintest acquaintance.

Einstein, I said, was the person who figured out that matter is the same thing as energy. It's the same stuff in a different form. Matter, the saying goes, is frozen energy.

Miesha said, "Did you hear that the feeling is that Einstein was a Starseed alien?"

Then something wonderful happened. Deborah came upon a plaque buried in the ground. It said, enigmatically, "Lizard Lounge Cabana."

Apparently the local Rotary club was going to build a cabana in that spot, named the Lizard Lounge.

"Reptilian Lounge! No shit!" said Deborah.

She laughed—vindicated.

Deborah suspected that my interest in the vortex might signal something. She felt my book project might not be a random event.

"Are you feeling *pulled* to go to the vortex?" she had asked earlier.

I felt no such pull. If anyone asked, I would say that I felt the topic of extraterrestrial life to be a rich one, with both scientific and cultural elements, and that it would be a good chance to see how people manufacture a plausible and rewarding narrative about reality, nature, life, and our place in the universe. That was my boilerplate.

But how could I be sure it was that simple? There's no way to prove one's impulses are "natural" and not inspired by an entirely hidden, unknown, repressed, or undetected energy source.

And as it happened, I did have a little secret. I shared it with Deborah and the other Starseed. For many years I have had a sleep disorder known as sleep paralysis, which is fairly common, I've been assured by doctors. In my case, it seems to be triggered by excess fatigue or too much chemical stimulation, such as when I drink coffee late in the day. I had the problem most severely when I was in college and drinking coffee at night as a way to stay awake and study. I usually have what is called sleep-onset sleep paralysis. As I am drifting off to sleep, I suddenly lose the ability to move. It literally happens in a snap—sudden, complete atonia of the muscles. (You knew it—the author would turn out to be a weirdo after all! A totally predictable last-minute mutation into craziness!) When paralyzed I can't open my eyes or lift a finger. Alarmingly, I'm still conscious. It's particularly troubling on a plane, a train, or a bus, because I can hear people talking around me yet cannot move, and might as well have been zapped by a paralytic raygun. Sometimes I have trouble breathing. To stop these attacks, I have to shake myself awake by moving my legs as best I can. The whole thing is quite disturbing. Once I went to a sleep clinic and spent the night with electrodes attached to my head. The doctors said my on-off switch for waking and dreaming isn't quite normal. My body thinks it needs to go into the dream state—and the muscles become atonic—before the brain is fully asleep. (One thing I learned is that, when a person dreams, his or her body is paralyzed, an ancient adaptation meant to prevent a person from acting out the particulars of a dream. Sleepwalkers have a defect in that system.)

Making the situation all the odder is that during the paralysis it is common for a person to experience auditory or visual hallucinations. I've never seen anything—no alien sightings—but I've had auditory hallucinations of people entering my home or coming into the room. There is also a sense that the paralysis is caused by these intruders. It is hard during the paralytic experience to keep the neurological cause in mind. Rather it seems as though some Other, some Entity, is doing it. In past centuries these entities were assumed to be spirits, such as succubi or incubi. In our day and age they are assumed to be aliens. Joe Nickell, who writes a column for *The Skeptical Inquirer*, has pointed out that a great many of the supposed abductees who are quoted in Whitley and Ann Strieber's *The Communion Letters* report symptoms that sound like sleep paralysis. The nagging problem with this skeptical view is that people who report being abducted by aliens are usually quite adamant about what happened. They don't blame the Other. They blame real aliens. My experience suggests that they would have to make a fairly big leap from having a sense of the Other to having a full-blown abduction fantasy. I never saw an alien or anything like an alien—so why do all these other folks? Can it be that they are truly that suggestible?

Naturally, the Starseed had a different thought. Maybe *I'm* the one who's deluded. Maybe the reason I was working on this book was that I was on a journey to discover my true nature. Maybe I was one of *them*. It made perfect sense! *I am from space.*

There was no choice but to hypnotize me. Deborah came to my hotel room at the Riviera, an aging casino on the Las Vegas Strip. Deborah told me to lie on the bed, close my eyes, and stay as still as possible. She announced that she would put me in an alpha state, somewhere between consciousness and sleep. I briefly thought of all the science lectures I'd been to, and how they had also brought me to this condition.

Imagine your body being weightless, she said. Relax. See yourself breathing. As she talked to me in hushed, soothing tones, I gradually lost the sensation of contact with the bed. I had a sense that I was hovering in space. I felt calm, and vowed to do my best to follow through with Deborah's experiment. Could it be she was right? No, it couldn't be. I knew I was not an alien.

Though, if I were an alien . . . well, I had a truly terrible feeling that I would turn out to be one of the reptilians.

Let's go back, she said, to the moment when you were first paralyzed. Where are you? What do you see?

I told her: I'm in a car going across West Texas. We're on Highway 40 (the road to Roswell, in fact! Creep me out!). There's a Dylan song on the cassette player. I'm in the passenger seat, my buddy Trey Furlow is driving his Honda Accord, we've been road-tripping for days, have just come from New Orleans, and are heading into New Mexico. I'm reclining in the seat, trying to get a few winks, and suddenly I can't move, I am frozen, paralyzed, yet am still awake, can still hear Dylan, still hear every verse to "Lily, Rosemary, and the Jack of Hearts"—a fun song, but a long song, and a very long song when you are completely paralyzed and can't even move your eyelids and are condemned to spend the rest of your life as a presumed vegetable, communicating if at all only through binary twitches.

"Do you see anything?" she asked.

I strained. I tried to see. I saw nothing.

"What part of your body remembers what happened?" she said. "Feel the part of your body that remembers."

"I . . . I . . . I think it's just my *head* that remembers."

"Your head?"

"I think it's just in my brain."

I was failing miserably! My poor old rational self couldn't adopt the necessary dualism to recover a memory of the alien from the memory storage organ in, say, my left foot. Deborah continued to work with me, but eventually I got discouraged and sat up and the trance was gone and I had found no alien. I told her I thought I merely had a neurological quirk that caused the sleep paralysis. I subscribed to the paradigm of the doctors. She seemed highly skeptical. I knew exactly what she was thinking: This guy is so *gullible*.

The disaster may have been inevitable. Cultural movements have an organic quality: They ripen, they swell, and at some point they burst. The grotesque rupture in the UFO world happened near San Diego in March 1997, when, in a matter of hours, thirty-nine people merrily killed themselves so that they could link up with a spaceship following Comet Hale-Bopp.

Reporters covering the story were shocked to find out that the cult had been advertising its philosophy on the Internet, that it hadn't made any secret of its plans. Moreover, this group was the very same UFO cult that had made national headlines twenty-two years earlier. A survivor named Dick Joslyn told me the story:

In 1975, Joslyn lived in L.A., working on and off as an actor and a model. He was so all-American looking, so clean-cut, that he got a job posing in a jumpsuit for a photograph on the cover of Kellogg's Corn Flakes. When not working, he studied with Clarence Klug, a mystical scholar. Klug taught his pupils that the Book of Revelation could be seen as a guide for building something called a Light Body. The group meditated daily. To Joslyn these were the grooviest of times, a moment of wondrous exploration. One day the meditation group went to the home of a woman named

20

Heaven's Gate

Joan Culpepper, who had invited two special guests—a man and a woman from outer space. That's how the event was advertised, how people referred to it: They were going to meet two people from space, and that was certainly weird, but not really that much weirder than the stuff they thought about normally. (Maybe the people from space were, themselves, Light Bodies.)

The man called himself Do. His partner was Ti. They were from the Next Level, they said, the level beyond human, from which Jesus had also come. Their real names (as the whole world would learn two decades later) were Marshall Herff Applewhite and Bonnie Nettles.

Applewhite had been a music professor, and Nettles a nurse. He was married with two kids, she was married with four. There were vague reports years later that he could not reconcile his traditional domestic life with secret yearnings, including homosexual liaisons with students. He moved out and lived for a while with another man, then had a mental breakdown—a time of "severe upheaval and mental confusion," as he would say later. In the hospital he met Nettles. She has a passion for astrology and a deeply spiritual nature. They decided they had known each other in previous lifetimes.

She abandoned her family, and soon they hit the road. They tried to open a restaurant in Taos, and when that didn't succeed they made crosses in a shop in Las Vegas. Soon they returned to the highway, and one day their sports car broke down in front of a yoga center in Portland, Oregon. They begged for money and food, and for a while dug septic tanks along the Rogue River. One night, in a campground at Gold Beach, Oregon, they realized they were the two witnesses mentioned in the Book of Revelation: "And I will grant my two witnesses authority to prophesy for one thousand two hundred sixty days, wearing sackcloth." They realized they were not humans at all but were creatures from the Next Level, in outer space, merely occupying human bodies until the mother ship took them and their "students" off the Earth.

At Joan Culpepper's house this did not go over big. "Do" and "Ti" were obviously bonkers. "If they're from space," Joan Culpepper said out of earshot of the aliens, "why do they talk with a Texas accent?" But Dick Joslyn wasn't so sure. After the session he approached Do and Ti and said, "I think you're either who you say you are, or you're nuts." Ti said, "Which do you think we are?" Joslyn said, "I don't know. I know you're not con artists." He sensed that they were at the very least sincere about their alien status.

He followed. So did dozens of others. Do and Ti, for all their strange talk, inspired people to drop everything and tag along. They migrated to Oregon, Wyoming, Kansas, Texas, New Mexico, camping out and begging for money on college campuses and recruiting more followers. The flock grew to more than one hundred, mostly young people, hippies and runaways.

There was grumbling about the rules of participation, specifically the prohibition of sex. The idea was to overcome all human impulses. Certainly there should be no falling in love or making phone calls to parents or siblings or anything like that. Transcending human weakness had its drawbacks, and soon there were defections. The press caught the scent, and for a few weeks national stories appeared about the "UFO cult." Do and Ti promised that soon a spaceship would arrive and take them all away. They scanned the skies. Everything they saw was an alien vessel. They saw meteors—"meteor craft," they called them. They saw satellites—"satellite craft." They believed that science-fiction shows like *Star Trek* were actually documentaries, that the members of the Next Level could put ideas into the human atmosphere for visionaries to absorb. (This was a more anarchic version of ufology's later theory about the Disclosure.)

The spaceship never came. Some members lost faith in Do and Ti. Dick Joslyn did not. He believed that it was just a matter of time, a matter of looking in the right place, picking up the right vibration. Such a life might look insane to other people, but to him, as he wandered the country, staring at the sky, waiting, it made as much sense as anything else.

At one point the cult watched the movie *Cocoon*, in which the aliens come to Earth not to contact humans but, rather, to bond with the most advanced and cuddly species on the planet, the dolphins. The cult became convinced that contact would require a marine adventure, and they got the notion that if they bought a houseboat they'd be picked up by the spaceship somewhere off the coast of Texas. They bought a used houseboat, fixed it up, and waited. No aliens.

Bonnie Nettles died of cancer in 1985—a premature voyage to

the Next Level. The cult kept moving, and at one point bought expensive electronic equipment to produce a TV show that it planned to call *UFO Update*. Joslyn would have been the anchorman. Everyone went to a UFO convention in Eureka Springs, Arkansas, to bone up on the latest theories. The ufologists didn't know quite what to make of the flying-saucer cult. The TV show, like most of the cult's endeavors, never got off the ground.

Joslyn packed up and quit the cult several times, but always came back. This was his family. The problem was, the cult didn't allow sexual contact. Everyone had a partner, but the relationship had to be platonic, and every six weeks the partnerships were changed by order of Applewhite. Eventually Applewhite and six other male members of the cult elected to be surgically castrated. Joslyn, however, elected in 1990 to return to the real world.

The group finally emerged from its long isolation in 1993, when it began advertising its beliefs on the Internet. Several people heard the promise of Heaven's Gate and dropped everything—marriages, children, jobs—to join the cult out west. The attraction of the cult was that it did not claim that its attempts at consciousness-raising were metaphorical. The ascendancy to the Next Level would be dramatic and irreversible—everyone would literally leave the Earth. Several members of Heaven's Gate had gone to the conference in Laughlin that I'd attended. They heard Lee Shargel give his talk, and bought a cassette tape and a book from him. Shargel signed the book: "To my brothers and sister. We stand together in the circle of light forever. We have been called to the same light."

One thing long noticed about cultists—documented in a classic work of sociology called *When Prophecy Fails*, a book written in the 1950s about a UFO cult—is that they find excuses to keep believing in their leader even when there's abundant reason to be skeptical. Any scrap of validation is cherished, and the Heaven's Gate cultists could find validation everywhere. The energized UFO marketplace gave the group a vast supply of supporting information (however bogus it may have been). In a sense, they believed what they were told. The cult's library, according to the San Diego Public Administrator's Office, included Mack's *Abduction*, and *Secret Life* by Temple University professor David Jacobs, who believes the Earth has been invaded by hostile aliens. They got their information delivered wholesale, by TV, by movies, by books and magazines. They read best-sellers like *The Celestine Prophecy* and *Communion*. They read the works of Zecharia Sitchin. They soaked up revisionist histories of Christianity, like *The Lost Gospel* and *The Dead Sea Scrolls Deception*. They subscribed to *Wired*, *PC* magazine, the *San Diego Union-Tribune*, and Shargel's UFO newsletter, *The Galactic Observer*. On Art Bell's show, they could hear an Emory University professor, Courtney Brown, say that through the technique of "remote viewing" (a form of telepathy that had also been briefly employed by the CIA to "spy" on Soviet military installations) he had determined that there was indeed a massive companion object to Comet Hale-Bopp.

You don't have to be insane to adhere to an apocalyptic ideology centered on extraterrestrial visitation. You just have to be wrong. And it is easier and easier to believe something utterly and tragically wrong. If you are in the market for a wrong idea, you can find a seller.

Heaven's Gate got a $1-million insurance policy to cover the cult members in case of alien abduction. The cult obtained the policy by paying $1,000 to Goodfellow Rebecca Ingrams Pearson of London, a company that also insures clients against conversion to werewolves or vampires. It is quite possible that the cultists feared that, before they could leave with the benign aliens aboard the spaceship, they would be kidnapped by Grays.

They went to their deaths in an orderly fashion. A letter from the cult to a former member explained, "By the time this is read, we will have shed our containers." The cultists made a videotape where they calmly and happily discussed their decision to kill themselves. One member, a woman, said, "Maybe they're crazy for all I know. But I don't have any choice but to go for it, because I've been on this planet for thirty-one years and there's nothing here for me." They literally packed their bags for the trip. Each person pocketed a five-dollar bill and some quarters, as though he might come across a tollbooth. They ate a lethal mixture of phenobarbital and applesauce, went to their rooms, reclined on their beds, and waited for the wonderful moment when they would join the spaceship.

After the suicides I spent a few days making the rounds of UFO cults to see what they had to say for themselves. The Starseed, though obviously unusual in their beliefs, did not have any truly cultlike qualities, because people could drop in once a week and share their stories without having to give up any other part of their lives. The Starseed had a couple of facilitators but nothing like a cult leader. The Heaven's Gate group was truly a mind-control cult. Other UFO "groups" in California were something less than cults but something more than just support groups.

I started with the Aetherius Society, a spiritual organization in Los Angeles, not far from the heart of old Hollywood. A small, tidy man named Alan Moseley seemed to be firmly in charge. He clearly was proud to be a lucid and logical member of society, and he said he has no patience with people who claim they have been abducted by aliens from outer space, or any such nonsense. "I consider myself a rational person," he said.

The Aetherius Society, he went on, teaches that there are intelligent beings living on Venus, Mars, Jupiter, and Saturn. This information was revealed in 1954 to a man named George King, the "chosen Primary Terrestrial Mental Channel" of the extraterrestrials. "George King communicates with intelligences from *this solar system*," Moseley said, putting emphasis on the last three words to highlight how dramatic the information is. This is not like picking up a signal from the Andromeda galaxy—this is direct communication with creatures on planets plainly visible in the night sky. The Aetherius aliens are our neighbors. "I'm sure it would be easier for George to say he was communicating with intelligences from the other side of the Milky Way galaxy. But you have to stick to the truth."

The members of the Heaven's Gate cult had it all wrong, he said.

"They're out to lunch."

The West Coast is not alone in being alien-obsessed, but what makes California special is the density of UFO groups. They tend to disagree on the specifics, but they all point to the more or less imminent arrival of beings from other worlds. The Aetherius Society believes the aliens will come sometime before the year 2035. Down near San Diego, the Unarius Academy of Science teaches that the Space Brothers will land in 2001 (amending, by necessity, an earlier and embarrassing prediction that they would come in 1976). The Raelians, another group with a Los Angeles chapter, put the arrival at some point before 2015. The Raelians believe the aliens will come if someone builds them a special embassy in Jerusalem.

Many UFO groups do not appear to be cults per se but, rather, harmless, almost quaint 1950s-style flying-saucer clubs. The goofiest may be Unarius, in the town of El Cajon, about twenty miles from the Heaven's Gate house. The official vehicle of the Unarius Academy is a Cadillac with a flying saucer mounted on the roof. Painted on the sides are the words "Welcome Your Space Brothers!" The academy itself is a storefront operation, dominated by a single large room with artificial flowers hanging from the ceiling and various dioramas showing futuristic cities. A painting on one wall shows the lost city of Atlantis. Another wall has pictures of Ruth Norman, the cofounder, wearing a crown and carrying a scepter. Strikingly, there are reproductions of the *Mona Lisa* with Ruth Norman's face superimposed. The school teaches that Ruth Norman is the reincarnation of the woman who posed for Leonardo's famous portrait.

Or was the reincarnation. "She has transitioned," says Carol Robinson, an administrator of the school. Norman, as some would put it, is dead.

Longer transitioned yet is Ernest Norman, her husband and the other founder of Unarius, which stands for "Universal Articulate Interdimensional Understanding of Science." The leader of the school is their former student Charles Spiegel, himself an old man. "We're not a cult," he told me. "We're a highly respected school. We teach a very sound curriculum."

He then launched into an explanation of how human beings are the components of an infinite creative intelligence, the reflection of a spiritual brotherhood known as Shamballa, possessed of minds that are not three-dimensional but, rather, are part of a "fourth-dimensional energy system." When humans come to the end of their lives they simply "change frequencies." Spiegel said he was a self-taught physicist and had many degrees. He read *The Voice of Venus*, Ernest Norman's 1960 account of his trip to that planet, and didn't believe it at first. After reading it for the tenth time—at the fringe, people have a tendency to be admirably thorough in their research—he finally said to himself, "My God, this is not science fiction."

The Spiegel spiel has many dimensions (he is the reincarnation of Pontius Pilate, for one thing), but what is most revealing is the room he speaks in. It's painted on three walls to give the impression that one is in a tower with many windows on an island amid a beautiful blue sea. This is the view from the Star Center, an "astral planet."

"This is what it looks like. Beautiful. Peaceful . . . You're looking at a world that does not have smog," he said. "This is what people would call Heaven."

But Heaven is just something people believe in. This is different. This isn't a matter of faith, it's a matter of *knowledge*.

"It's real!" he said.

I kept moving, looking for patterns, for some emergent phenomenon that would link these diverse dogmas and possibly redeem the small but precious chunk of my life spent in LaLa Land.

Up a dingy, stained stairwell in a two-story building on Victory Boulevard in Van Nuys, I found a door with the name Dr. Frank Stranges. Inside was a man of sixty-nine, bearded, with a round face, sitting at a desk with one of the world's most impressive nameplates. The nameplate was nearly the size of an automobile dashboard. His name was in Gothic script, followed by "Ph.D., Th.D." He had two business cards, one for his church, International Evangelism Crusades, and one for NICUFO, the National Investigations Committee on Unidentified Flying Objects. It was hard to tell which hat he preferred to wear, but he claimed to have more followers as an evangelistic crusader—ministers in fifty-five countries, tens of thousands of members, he said—than as a UFO investigator.

"I'm rationally minded," he said (in what was becoming as automatic a comment as "Good morning"). "I'm also spiritually minded. The most difficult task is to separate fact from fiction. But once you get the fact, you hang on to it like there's no tomorrow."

He believes the aliens are mostly humanoid, "except for the ones with the giant heads and the slits for mouths and the huge eyes." Some ETs are "created beings," just like humans. They evolve on planets and then cross the interstellar void in spaceships. But there are also angels and demons that come from the "heavenlies," the literal Heaven, which he believes is a real place on "the rim of the universe." In the Frank Stranges cosmology, aliens aren't angels—these are distinct entities that work side by side. The Grays, meanwhile, are fallen angels who live inside the Earth and emerge from caves near the poles. "These are evil-demon powers who are throwing a monkey wrench into the whole UFO picture."

Stranges is a throwback to the 1950s, the heyday of George Adamski. Stranges has an Adamski-like tale of his own. He says he met a Venusian named Valiant Thor at the Pentagon in 1959—Venusians were hot in the 1950s (literally so, given the research by Sagan and others on the surface temperature of Venus—though the scientific conclusion apparently took a while to reach the general public). As he spoke, a colleague named Jamie, who wouldn't give her full name, listened intently. She said she'd first heard Stranges give a talk twenty years ago and did not believe his Venusian tale. But then a man in a military uniform—it was covered with medals, she recalled—loudly declared that Stranges was telling the truth. This endorsement erased her natural skepticism. Since then, she said, Stranges has offered prophecies every year, and they always come true. I asked for an example. She drew a blank.

"About the stock market," Stranges suggested helpfully.

"Stock market going up and down, that's an example," she said.

We want to divide the world between the rational and the crazy, just as we divide information between the true and the false. Needless to say, each of us is perfectly rational; other people are the crazy ones. What the visitor to the alien territories notices is that even the "kooks" want to get their facts right. Everyone is groping for truth in an age of confusion, information overload, and false prophets. We all know that extraterrestrial life is a huge, hairy unknown, that it's one of the greatest mysteries of our time, but what many people forget is how easily a data-poor topic can be infiltrated by preconceptions and biases and wishful thinking.

It's a dilemma for everyone, the hardcore scientists as well as the people who meet Venusians at the Pentagon. The measure of your scientific impulse is not how hard you labor to prove a point; it's how thoroughly you search for hidden biases and unsupported assumptions that may have skewed your conclusion. To be truly scientific you have to try (as the Houston scientists claim they did) to prove yourself incorrect.

The downfall of the true believer is the intensity of the desire for wonderful truths. We all want the same thing: A shot at life after death. A special role in the universe. Someone to believe in. We want the universe to be orderly and not random. We want the fabric of the Creation to be pulsing with meaning and purpose. No one wants the cosmos to be a giant, pointless, randomly assembled machine.

Human beings are believing creatures. The members of the Heaven's Gate cult went to their deaths with smiles on their faces, giddy, blissed out. They believed, and that was all that mattered. But even believers seek verification. Even Heaven's Gate had an impulse to double-check. Two months before the suicides, several cultists, including Applewhite, paid $3,600 for a computerized telescope with a ten-inch mirror. They used it to look at Comet Hale-Bopp and search for the "companion object." They were following a scientific impulse—seeking direct observation of the vehicle that would rescue them from our doomed planet. They didn't want to rely on rank hearsay.

They saw the comet perfectly. But they saw no spaceship. The aliens were being their usual elusive selves.

And so they returned the telescope to the store, and asked for their money back.

The Heaven's Gate tragedy was a buzzkill for Roswell. The timing couldn't have been worse, coming just four months before the fiftieth anniversary of the most famous saucer crash of them all. The town had planned a truly intergalactic celebration. Organizers had hoped for a hundred thousand people, a Super Bowl of ufology. There was talk of arranging a concert featuring Sheryl Crow, who had used a true-believer UFO song as the first track on her latest CD. But corporate sponsors got nervous after Heaven's Gate, and pulled out. The celebration would still draw a respectable throng, maybe thirty thousand people, but it wouldn't be Woodstock with an alien theme.

The anniversary was planned for the July Fourth weekend, which was more or less fifty years after the incident in question. The reason it's hard to say more precisely is that there actually was no singular "incident." The closer you look at Roswell the more complicated and contentious it becomes, and the basic questions of what, when, and where remain in dispute.

A few facts are generally agreed upon. One is that in early July 1947, the nation was in a flying-saucer frenzy. Kenneth Arnold had seen the nine mysterious disks zooming through the Cascades on June 24. Hundreds of saucer sight-

21

Roswell Crashes

ings had followed. One person, however, didn't know anything about these invaders from space: William "Mac" Brazel, a rancher north of Roswell whose house lacked electricity, radio, or a telephone. He heard about the phenomenon when he went to town on July 5 and stopped at a bar. He told his buddies: He, too, had seen something strange. He'd found some unusual debris. It had crashed on his ranch a couple of weeks earlier. When he found it he didn't think much of it—it was just a bunch of foil, metal, and sticks. He mentioned it to some neighbors, but that was all. The possibility that the material was somehow connected to a flying saucer from outer space did not occur to him.

Brazel talked to the local sheriff, who in turn contacted the Army. Enter Major Jesse Marcel of the 509th Bomb Group intelligence office at Roswell Army Air Field. Major Marcel went to the ranch, retrieved some of the debris, and took it back to headquarters. And that's when things started jumping. Marcel, for whatever reason, felt that this material looked so unusual that it might be from another planet. A certain Lieutenant Walter Haut, who hadn't seen any of the debris, heard about the case and promptly put out a press release announcing that Roswell Army Air Field was in possession of a crashed disk. He drove the press release to the offices of Roswell's newspapers and radio stations.

"The many rumors regarding the flying discs became a reality yesterday when the intelligence office of the 509th Bomb Group of the Eighth Air Force, Roswell Army Air Field, was fortunate enough to gain possession of a disc through the cooperation of one of the local ranchers and the Sheriff's office of Caves County," Haut's press release said. The story ran in newspapers around the world on July 8, including the evening edition of the *Roswell Daily Record*, which featured a now-famous headline, "RAAF Captures Flying Saucer on Ranch in Roswell Region." Considering that the case is emblematic of government conspiracies, it's worth reiterating the obvious: The ordinary citizen (Brazel) did not initiate the story of a flying saucer. Rather, it was a military man (Marcel) who thought he had something extraterrestrial, and then an Army lieutenant promptly announced the great discovery to the whole world. So, as cover-ups go, it was off to a terrible start. And there was no mention of any bodies.

That same day, the Army retracted Haut's press release and declared that the material was debris from an innocuous weather balloon. That was a lie. The debris (the Air Force now says) came from a balloon-borne sensor used in a classified program called Project Mogul. Project Mogul used "balloon trains" to carry aloft sensors that could potentially detect atomic explosions. The initial story worked well enough, though, and Roswell pretty much disappeared from the UFO lore. What didn't disappear was the idea that a saucer had crashed somewhere in the Southwest and that the military had recovered bodies. As Philip J. Klass points out in *The Real Roswell Crashed-Saucer Coverup*, the first crashed-saucer story appeared in 1950 from the pen of Frank Scully, a columnist for the Hollywood trade paper *Variety*. His book, *Behind the Flying Saucers*, became a best-seller. Scully claimed that the military recovered a crashed saucer and four bodies in New Mexico in 1948. His sources were someone he named only as "Dr. Gee" and a businessman named Silas Newton. It didn't take long for Scully to be debunked. Dr. Gee turned out to be Leo GeBauer, who was no doctor at all but, rather, the owner of a radio-parts store in Phoenix. Newton and GeBauer were later arrested and convicted of running a "confidence game" for selling a bogus oil-deposit detector called a Doodlebug.

Only in the late 1970s did Roswell reappear as an important case. Leonard Stringfield, a UFO researcher, began re-examining reports of crashed saucers, and dug up some eyewitnesses (anonymous, unfortunately). Jesse Marcel, the major, was still around and played a key role in the renaissance. In 1978, the UFO researcher Stanton Friedman was in a Baton Rouge, Louisiana, TV station, waiting for a late reporter, when the station manager mentioned that he was buddies with Marcel. Friedman tracked down Marcel and started gathering more information. Marcel told everyone that the material couldn't have been man-made. "The metal was as thin as newsprint and as light as a feather. It was slightly flexible but very strong. He tried to dent it with a sledgehammer but couldn't. Marcel and the CIC agent tried to burn it but it would not burn," report Kevin Randle and Donald Schmitt in their book *UFO Crash at Roswell*.

Using much of Friedman's research, Charles Berlitz *(The Bermuda Triangle)* and William L. Moore wrote the first Roswell book, *The Roswell Incident*, published in 1980. Sagan was on TV on *Cosmos*, and this small band of researchers was telling an alternate narrative of potentially cosmic significance. It is television that turns niche obsessions into cultural landmarks, and the big break in the case came in 1989, when the TV show *Unsolved Mysteries* devoted an episode to Roswell. Suddenly there were eyewitnesses everywhere. People popped up left and right with sensational information about a case that was forty-two years old. There was, for example, Gerald Anderson, who became a key source for Friedman and Don Berliner's 1992 book, *Crash at Corona.* Anderson saw the *Unsolved Mysteries* program and came forward with the story that, as a boy of five, he had seen a crash site with four aliens, one of whom was still alive. Klass, the tireless debunker, later spoke to Anderson's first wife, and she said her husband had never once mentioned to her this amazing tale of a flying saucer and four aliens. She told Klass that Anderson "likes to tell tall tales and he can actually make people believe them."

But there are other witnesses. The case never pivots on any single, ridiculous individual. There were also the MJ-12 documents, which surfaced in 1984 (more specifically, they supposedly were dropped through someone's mail slot), purporting to show that Presidents Truman and Eisenhower knew of "extraterrestrial biological entities" (EBEs). The MJ-12 documents have certain odoriferous qualities that have made them suspect even within UFO circles. It has been noted, for example, that Truman's signature appears to have been lifted from a different letter, a cut-and-paste job. One idea in circulation is that if the documents are fakes, they still might be the work of the government—an intentional hoax as part of a disinformation campaign of deeply enigmatic intent.

Suffice it to say the Roswell case is not airtight. It is hard to avoid looking at the much-reproduced photograph of Major Marcel holding the crash debris. It looks like a downed kite. Somehow this thing crossed hundreds of trillions of miles of interstellar space—surviving asteroids, solar radiation, bulletlike particles traveling near the speed of light—before crashing in the clear desert skies of New Mexico, no doubt downed by an unfortunate puff of wind.

The biggest problem in nailing down what happened is that everyone has a different story. Were there four bodies or seven? Where exactly did the crash happen? Is there anything that we can say definitely did happen? The UFO Enigma Museum in Roswell puts the crash site at a ranch thirty-five miles north of town (with debris spreading to the Brazel ranch), whereas the International UFO Museum said it happened out near the mountains about fifty miles west of Roswell. Two museums, two competing narratives. Even the debunkers disagree. *Popular Mechanics*, disputing the notion of a crashed saucer, declared in a 1997 article that what crashed at Roswell may have been an experimental aircraft containing Japanese pilots: "*PM* suspects the craft that crashed at Roswell will eventually be identified as either a U.S. attempt to re-engineer a second-generation Fugo, or a hybrid craft which uses both Fugo lifting technology and a Horten-inspired lifting body."

After much prodding, including from New Mexico Representative Steven Schiff (who later died—a fact that my sources find suspicious), the Air Force commissioned its own study of the Roswell case. *Report of Air Force Research Regarding the "Roswell Incident,"* released in 1995, declared that there was no hint of an alien spaceship, and that the Project Mogul device was almost certainly the object that crashed at the Brazel ranch. A second, longer, and more argumentative Air Force report, *The Roswell Report: Case Closed*, authored by Capt. James McAndrew, and released just prior to the fiftieth anniversary celebration, floated the novel notion that any "bodies" found at a crash site were anthropomorphic dummies being used in aeromedical tests. These dummies were dropped from weather balloons, and their odd features may have fooled the desert folk into thinking that aliens had crashed.

Unfortunately, the "dummy" explanation proved a real howler in UFO circles. Everyone pointed out that the tests were conducted many years after the Roswell incident. The Air Force theory—that these things get conflated as memories dim—didn't fly. It's certainly an idea worth pondering, but in a matter as wildly and hysterically contentious as UFOs, McAndrew may have been guilty of overexplaining.

It's two hundred miles from Albuquerque to Roswell across land that is dry, wide open, simple—the kind of terrain where anything slightly strange, artificial, or geometric leaps out from the surrounding landscape. At the junction of I-40 and U.S. 285, I saw a man sitting motionless on the tailgate of a white truck next to a contraption on a tripod. Not far away was a tall antenna with peculiar cup-shaped formations near the top. I thought: It doesn't look right.

The road went through Encino and then Vaughn. There were as

many gas stations abandoned as still in business. A sign said "61 miles to Billy the Kid's grave." They promised to be sixty-one lonely miles. On either side of the road were ranches that stretched to the horizon. Snakes crossed the highway; indeed, one crossed directly under my rental car. I did not slow down to assess casualties. It is hard for primates to feel compassion for reptiles.

Near Roswell, a strange structure appeared just above the horizon—globular, white, motionless. It was hovering. As a trained journalist, I was able to determine quite quickly that the structure was a municipal water tower. *Everyone stay calm.*

Then came the signs. The Roswell Mall had a sign offering a million "big ones" to the first "certifiable alien" to walk through the doors. At the truck stop there were UFO anniversary pins, and mini-cookies that were "Out of This World." A seafood restaurant had a sign saying merely "They Came for Fish." Every storefront on Main Street had aliens in the window. At the "trade show" in the Civic Center you could buy UFO T-shirts, UFO lollipops, UFO wind chimes, UFO snow domes, UFO soup ladles. The celebration felt like a county fair, only with a UFO theme, right down to the $39.95 alien embryos floating in a jar. One man promoted antigravity technology (where was NASA?); another hyped *The Urantia Book*, "written by 21 space aliens."

In the coming days, many people voiced their concerns about the commercialization of the anniversary. This seemed to ignore the serious substance of the UFO situation—the *issues*. The gimmicky consumer items would only play into the hands of those who would treat this thing as a giggle. I ran into a man named Jim who had driven more than a thousand miles from Minnesota to commemorate the event. Jim, forty-one, a building official, took note of the carnival atmosphere and said, "I think the serious side is being overlooked a little bit."

He said this even as he wore, on his head, a three-foot triangular aluminum-foil spaceship.

The first night of the festival, there was a debate at a local hotel. It was an odd event, in that it broached the possibility that Roswell UFO Encounter '97 was based on a fictitious event. The "skeptic" was the ufologist Karl Pflock, and he squared off against one of the Roswell case's biggest promoters, Kevin Randle. Pflock's calm and reasonable presentation had one flaw: He had to admit that he was once a civilian data collector for the Central Intelligence Agency—a red flag that provoked one audience member to question Pflock's credibility. Randle was loud, indignant, even huffy at times, and bedeviled by a fly that had gotten into the room. A couple of times he appeared to pound the lectern not from conviction but simply in an attempt to kill the fly.

Pflock said the military is telling the truth when it says that the only thing that crashed near Roswell was a test balloon from Project Mogul. Randle rejoined, "To me there is no doubt that it was a craft constructed on another planet and piloted by a crew not born on this planet." Randle admitted that the crash debris bore a "gross resemblance" to the materials used in Project Mogul, but said it wasn't an "exact match," noting that the Mogul material was made of ordinary tape and sticks and plastic, whereas the stuff found on the ranch, according to Jesse Marcel, could not be burned or destroyed or crumpled, and had other odd properties. Shouting, Randle pulled Scotch tape, string, and aluminum foil from his pockets and began manhandling the material. He cut the string, he crumpled the foil, he showed a stick that was obviously burned.

"That's Project Mogul!" he bellowed.

The audience clapped. They understood his tortured logic. This tape, this foil, this stick, were all so terrestrial. The eyewitness had described the debris as made of bizarre materials. How could Pflock say that the bizarre materials were such ordinary items as the things he'd pulled out of his pockets?

"It does not make sense!" Randle roared.

The stars of the UFO world came pouring into town. John Mack, Whitley Strieber, Budd Hopkins, Stanton Friedman—they all came to Roswell. But the biggest star was someone who hadn't been seen much in recent years: the paleo-ufologist Erich von Daniken.

Von Daniken, short, stout, tanned, undoubtedly rich, now sixty-two, claiming to have sold fifty-six million copies of his books in twenty-eight languages, spoke for ninety minutes before a packed auditorium at the New Mexico Military Institute. He delivered the same pitch that he did two decades ago, in his prime: The poor, half-naked, primitive peoples in places like South and Central America couldn't possibly have built those impressive temples and pyramids without the help of space aliens. "There is absolutely no way for a Stone Age society to cut this stone in such a way," he said after showing a slide of some delicately cut granite slabs. "This is not technology of primitive Indians four thousand meters above sea level in Bolivia."

Von Daniken discussed his latest idea, which is that in the Great Pyramid of Cheops there's a secret room holding secret information about ancient visitors. This hasn't been confirmed, of course—the room is sealed. On more familiar ground, he showed cave paintings of figures that looked a bit like spacemen, though the paintings are somewhat rubbed-out and unfocused, an ancient version of the infamously blurry UFO photographs. He also showed more of his famous aerial shots of the Plain of Nazca in southern Peru, which is crisscrossed with lines that vaguely resemble landing strips and taxiways. He believes he's found an ancient airport. (Sagan, among others, derided Von Daniken for alleging that alien spacecraft would need a landing strip like conventional human planes.) Von Daniken couldn't leave well enough alone, and showed runwaylike lines that left the flatlands and traveled right up into the mountains and across jagged peaks.

He addressed the nagging question of whether his philosophy implies that primitive people are too backward to build their own temples and monuments. "I do not think our ancestors were so stupid and so primitive [that they decided] to construct temples for their entire lifetimes for nonexisting gods," he said. In other words, they wouldn't have built these things if the gods were imaginary. The reason they called the alien spacemen "gods" and failed to describe any actual alien spaceships is, he said, that they "misunderstood technology."

Von Daniken may once upon a time have been the most influential UFO writer in the world. He was not a ufologist per se, because he focused not on alien artifacts but on human ones, like the big stone faces on Easter Island. He lost luster as it became clear not only that he was shy of any scientific credentials—"charlatan" is a label that he endured—but that his philosophy was implicitly elitist, built on the refusal to believe that ancient peoples could stack stones so neatly. His first book, *Chariots of the Gods?*, which he published in 1968, when he was still a hotel manager, sold millions of copies and set the standard for an entire industry of books whose titles jumped off the cover and screamed of paranormal doings. In one book, *Gold of the Gods,* he claimed to have discovered a vast underground cathedral of gold in Ecuador. After his talk in Roswell, he held a brief "press availability," and I was thrilled to have a chance to ask him about that cathedral of gold. I'd been such a fan in the old days, and had been a bit unnerved by the realization that the ancient astronauts were a modern-day projection on the past. I asked him: Did he ever find his way back to the underground cathedral of gold? He said no, he'd never known how to get there, but had been led to the spot by a Hungarian.

And the Hungarian, he said, has died.

That's the problem with anomalies. The Hungarian is always the one with the proof, and the Hungarian is always dead.

Von Daniken told the reporters that he believes the ETs are observing us again. Someone asked where they are, exactly. *(Where are they?)* The great man had heard silly questions like this for decades, and he mustered a tolerant smile. He said, "We don't make diplomatic contact with chickens."

The remark dangled in the air, an invitation to think outside the box. *We don't make diplomatic contact with chickens.* The chicken remark was another variation on the Assumption of Mediocrity. We are the dummies of the galaxy. Compared with the aliens, we're just *poultry.*

There were those in Roswell for whom this UFO subject was a deadly serious business. They seemed to be worried that they knew things that could get them killed, their bones bleaching in the desert sand. The hot story one day was the announcement that UFO researchers had come into possession of a walnut-sized piece of metal, purportedly from the Roswell crash, that in laboratory tests was proved to be of extraterrestrial origin. It remained unclear, however, who had dug up the wonderful alien nugget, and where it had been for half a century. When reporters demanded to see the fragment, they were told it was not on the premises. When they attempted to ask questions of an obscure California scientist who backed the story, the scientist fled.

Otherwise there were no breakthroughs, merely some interesting assertions.

"It doesn't behave like a dream. It behaves like real experiences," John Mack told an audience at the New Mexico Military Institute. He was talking about people who get abducted in their sleep. "The only problem is that, according to our world-view, this is not possible. But this is not my problem."

Not his problem! The problem belongs to the mainstream scientists who say that abductions can't possibly happen. Mack said we need a fundamental change in how we look at the world. We need to change our view of what is real, and break down the barrier between the physical realm and the nonmaterial realm, the realm from which the aliens come. "Frankly, I think the cosmos is a lot more interesting if there is this traffic between that realm and this realm," he said.

A simple form of bias is the desire for a more *interesting* cosmos. But of course it's debatable what, precisely, is an interesting story. There are those of us who would say that the known universe, the official universe, the universe of Edwin Hubble (huge! expanding! once smaller than a pinhead!) is pretty darn interesting. And perhaps I'm grading generously but I'd say the alien narrative, particularly the abduction element, is still rather astonishing if it turns out that there are no abductors from another dimension.

After one of the lectures a woman in the auditorium asked me to feel the implant the aliens left behind in her leg. There was a little red bump. If it was, indeed, an alien implant, that was surely a startling situation. But it was also interesting to think that a crucial piece of supporting evidence for her unorthodox world-view was something that looked to me precisely like an ordinary blemish. (Extraterrestrial implant . . . or zit?)

Everyone made the pilgrimage to the crash site. This was a complicated matter, since there was no agreement on where the crash site actually was. The Brazel "debris field" is a good two-hour drive from Roswell and is tarnished by the fact that no one ever saw alien bodies there. People want to know where the aliens were, and that's where the rival narratives start bumping into each other. The site favored by the International UFO Museum is in a national forest, pretty much in the middle of nowhere. The only crash site within reasonable shuttle-bus distance of Roswell is the Corn Ranch crash site, favored by Randle and Schmitt, which is just half an hour's drive away, at the end of a gravel road. Visitors paid $15 to a woman under a canopy. They were greeted by the high-school football coach, Mike Whalen, a friend of the ranch's owner, Hub Corn. Whalen told me he was amazed at how this flying-saucer thing had taken off. Before UFOs, he said, Roswell hadn't been too exciting.

"The biggest thing I can remember," the coach said, "is a feed company burning down, and watching screaming chickens and roosters fly out the back, on fire."

The tourists were pouring in on yellow school buses. The ranch has twenty-four sections—square miles—and the crash site is a bluff, an arroyo, that curves to form a bowl, sort of like an amphitheater. There's a ceremonial rock beneath the bluff. On the rock is inscribed:

> We don't know who they are
> We don't know why they came
> We only know they changed our view of the universe

Hub Corn was there, talking to the tourists. He was thirty-six years old, with red, leathery skin, a mustache, a purple fishing cap. He talked slowly, unexcitedly. "I believe something did happen," he said. That's about as far as he wanted to go.

The actual spot where the saucer hit is in the most striking portion of the bluff, where the rock has a geological deformation, a V-shaped indentation. A dyspeptic cynic from some big East Coast city might marvel at the coincidence that a spacecraft from Zeta Reticuli (or wherever) crashed in such a geologically interesting spot, amid so flat and unremarkable a landscape. Corn's family has been in these parts for 150 years, and bought this parcel in 1976. Not until March 1994 did he find out that the saucer had crashed here. He was given the news by Randle and Schmitt, who heard it from alleged eyewitnesses in the early 1990s. (The International UFO Museum calls the Corn Ranch the "Alleged Crash Site" and posts an affidavit from the son of the former owner of the property, saying that no one around there ever saw a flying saucer or any military types taking away bodies.)

Sheila Corn, Hub's wife, took over the tour-guide duties as another dozen people showed up from the latest school-bus delivery. She said the spacecraft was black and triangular, and contained five occupants. Depending on which version of the story you follow, she said, all of the alien astronauts were dead, or most were dead but one was alive and sitting next to a rock holding a black box. Someone hit this alien in the head with a rifle butt to get the black box away from him, she said. (Our big moment of contact with an alien and what do we do? Whomp it upside the head with a rifle butt. Humans can be such jerks.)

On the bluff, in a notch, were some silk flowers—fuchsia and gold and lavender. Someone had donated them as a memorial. An attached note said: "In Memory of the Brave Explorers who found peace on Earth far, far from home. From the People of Earth."

The royalty of Roswell gathered one night out at the golf course, in a pavilion that looked like a carport, a mariachi band playing as the sun went down and illuminated the billowing tops of thunderheads roving across the high desert.

A few decades ago, Roswell went through a rough time. The air base closed in 1965—rumor had it that LBJ retaliated for not carrying the congressional district—and the closure left Roswell with four thousand empty houses. The population in 1997 was forty-five thousand, and the locals wanted to see that grow. The 1947 incident has been great for the town, above and beyond whatever it might mean about contact between terrestrial and extraterrestrial civilizations. "We're into the groove. It's great for the economy," said caterer Millie Whitaker.

One man drew a small crowd. It was Jesse Marcel, Jr.—one of the few living people who actually saw the wreckage. Now an ear-nose-and-throat doctor in Montana, he was eleven years old in 1947 when his father showed him the sticks and foil and string of the mysterious debris recovered on the Brazel ranch. He described the debris in a foreword to one of the many books about Roswell:

> What I saw was a bewildering collection of metal-like foil, bakelite or plastic shards and beams. To complete the picture of weirdness, there were even symbols of strange geometric forms and designs of a metallic violaceous hue printed along the length of the I-shaped beams.
>
> What this meant for me is the realization that we are not alone in our galaxy and that there are advanced civilizations using principles of which we have no concept.

The words were a fine example of the logical pivot found in most UFO stories: When seeing something you don't understand, assume it is from another part of the Milky Way galaxy. Now I had the writer in front of me, and I asked directly what made him think the materials were of ET origin.

"It's hard to say. It's just a feeling, a gut feeling. No scientific basis, which is a mistake," he said, seeming a tad embarrassed that this was his answer. He said he's a member of the National Space Society and closely follows developments in space science. He spoke of the Hubble Space Telescope's location of planetary disks around other stars. All "class G" stars like the Sun probably have Earth-like planets, he said. The Mars rock, with its possible ancient microfossils, indicates that life may arise quickly wherever there's liquid water, he said. So he believes the universe is full of life. Seeing the debris gave him the proof he needed: "I know we're no longer the only ones."

I asked how a spaceship could manage to cross such unbelievable distances from whatever world the aliens hail from. Marcel answered by citing something he'd read about NASA. The space agency, he said, can draw energy from the "false vacuum" in space. The otolaryngologist had gotten ahead of the facts a bit, but there was, indeed, a fantastic theory hovering out there, a notion that it might be possible to manipulate the vacuum of space and make interstellar flight easy and cheap. Marcel explained: "It cancels out inertia and mass." He indicated that this explained a central paradox of the Roswell debris. It didn't look as if it could survive a trip of trillions of miles, but if it had no inertia or mass it wouldn't matter. "It can be a very flimsy structure because it has no mass," he said.

Even something flimsy can go somewhere. It's the promise of the future, and the history of the Roswell case. The American spirit runs strong through this part of the world. With grit and imagination and a little commercial enterprise, it is possible to alter one's place in the world, to create a new reality—and to bring life to the desert.

22

The Hole in the Story

After Roswell, I found that I was rapidly losing interest in UFOs. In fact, I had obviously lost interest even at the height of the Roswell celebration. I had been a terrible guest. I think I was struggling to cope with the realization that human beings are fundamentally illogical. In the case of the UFO narrative, they make the same mistake again and again, which is that, given a choice, they will always choose the more implausible (but more colorful) explanation for a given mystery. In Roswell I had to come to terms with the fact that I was a freak, an anomaly, someone who would never accept the reality of the alien presence even when shown the actual spot where researchers believe a flying saucer crashed. My position had hardened: I would not believe in aliens if there were several of them gnawing on my leg.

If I learned nothing else in the alien world, I learned that for the sake of social harmony it is advisable to avoid trying to debunk anyone else's belief. It's not a winnable debate. The issue won't be resolved in a lab. Aliens are a *personal* matter. As such, to argue with a true believer is to risk being rude and invasive, like challenging someone for believing in God. Many skeptics seem truly angry and outraged that religion has survived the modern age. One gets the sense that these skeptics lose sleep at night knowing that someone, somewhere, is believing in something that is unsupported by an evidentiary database. (At Roswell, I felt this same kind of face-empurpling outrage coming on. My gorge was rising, and I didn't even know I had a gorge. This was the signal to move on.)

The Roswell case also demonstrated a recurring problem with UFO stories: The aliens are AWOL. The aliens fail to show for their own party. Even in many "abductions" the aliens aren't seen directly. Instead there are suggestions, remnants, suspicions, "missing time." There are tiny pits in your skin that may be gouge marks from the alien scientists. There are those implant-pimples. In so many cases involving aliens, all we have are the residues, the memories, the tremendous stories. But we don't have actual aliens.

Where are they?

The lore is constructed around the various gaps in the Official Story, gaps that a hypothetical alien could neatly fill. The skeptic Michael Shermer has noted that the human brain is designed to fill in gaps; we can, for example, read words even when some of the letters are missing. The desire to round out reality, to fill in the blanks, is extremely human, even noble. But it requires discipline, lest we insert into the gaps the artifacts of fantasy. Every good conspiracy theory is built around documents that are missing from the files of the Official Story. There are situations that don't quite make sense, and by gum, we expect the world to make sense even if we have to invent a myth to nudge it to the condition of sensibility.

There are responsible, intelligent people in the ufology community who understand the difference between a goofy case and a more plausible one, but the sanity and seriousness of the researcher do not by themselves validate any results. The biggest mistake people make is accepting the "argument from authority." Journalists in particular make this mistake, relying excessively on quotes from "experts," as though this is the same thing as printing established facts. I am as guilty of this as anyone else. (When a story starts to turn sour, when the facts begin to disintegrate, we say: But my expert source told me so. I have it on the record.) Professional persuaders have stunk up the Information Age. A rule of thumb: Any book that puts the author's credentials on the cover—for example, "John Mack, M.D.," or "Dr. Jacques Vallée"—should be viewed with suspicion. Einstein didn't have to hype himself. For that matter, Sagan didn't add "Ph.D." to his byline.

We are all searchers, and the distinction between the "serious" and "silly" searchers is a false one. Only in a few isolated cases did I ever feel that in talking to a UFO believer I was in the presence of a crazy person. Most of them are merely wrong. In fact, the assumption that there could be aliens visiting the Earth is an entirely reasonable one. Except for a relatively small fraction of cases that are outright hoaxes, people really do see and experience unusual things that they can't explain. But the Sturrock Report made the correct call on the critical question of an alien presence: There's no good evidence for it.

Aliens therefore lack scientific stature. They are not as "real" as, for example, dinosaurs. Dinosaurs are a fairly new and highly exotic concept, dating only to about 1840, but their reality is supported by abundant fossil evidence. For aliens to gain scientific stature, we'd need their bones, or their cartilage, or whatever goo holds them together. It would take more than an eyewitness account. It would require some mechanism by which the hypothesis potentially could be falsified (if it's not true, we need a way to prove that it's not true). If a man tells me that an alien popped through a portal into his bedroom and had a giant praying-mantis head, I can't disprove that. So I nod politely, jot down a few notes, and move on. It is impossible to disprove the existence of a covert phenomenon. One of the central premises of aliens is that they are able, through unknown means, to elude detection. They vanish into thin air. Of course they're unprovable—that's one of the great charms of an extremely advanced entity. Unprovability is a sign of superiority. When you reach a certain level of the game you don't have to obey those silly protocols espoused by fretting scientists. John Mack says the aliens aren't even of the "material world." An immaterial physiognomy makes for one slippery critter.

The compelling piece of evidence could turn up at any moment—a deep-sea submersible could happen upon an extraterrestrial spaceship mired like the *Titanic* at the bottom of the ocean. The aliens are ripe for what is known among scientists as an "existence proof." An existence proof involves something that is considered impossible, that violates the known rules of physics or is in direct conflict with a paradigm, but which suddenly manifests itself in irrefutable form. Aliens could land on the White House lawn, in the courtyard at the heart of the Pentagon, or next to the hot dog vendor outside the Department of Agriculture. The existence proof of intelligent extraterrestrial life would put to an end the thousands of years of lame, inconclusive, argumentative philosophizing about possible alien creatures.

The most serious failure of UFO stories is that they aren't really that interesting. The aliens rarely have anything novel to say. They're a bunch of motivational speakers ("You have to function as a totality! As a holistic system!"), urging humans to make something better of themselves. They advise us to overcome our warlike tendencies and polluting habits and rise to a higher level of consciousness. You want to say to these aliens: You crossed half the galaxy to tell us that? You can go through the tales of contact and find almost nothing that could be called "information." Aliens impart helpful nostrums, but has any alien ever provided the answer to an unsolved mathematics problem? When has any alien ever given us a good photograph or videotape, or even a census record that documents the life on another world? I'd like to read an in-depth history of the past thousand years of civilization on any other world. Do aliens also go through periods of feudalism, revolution, enlightenment? Were the aliens ever a bunch of superstitious dumb-asses, like us? We never find out, because the aliens are too busy giving us advice or conducting experiments on us. Aliens have no *history*.

This has been true since George Adamski in 1952. His alien, from Venus, warned Earthlings about the dire consequences of nuclear war. It was remarkable how closely this fit with the warning from the alien in *The Day the Earth Stood Still*, which featured a completely humanoid alien who tried to teach us a thing or two about peace. For the most part, the aliens have only two notes: They warn us not to kill ourselves, or they try to kill us.

Eventually you realize the aliens get all their information from the same sources that humans do—newspapers, CNN, the Sci-Fi Channel, the Cartoon Network. The Assumption of Mediocrity extends even to advanced civilizations from deep space.

One day I went to a meeting of the Washington chapter of MUFON. The ufologist Bruce Maccabee analyzed the famous videotape of a flying saucer moving through downtown Mexico City. As Maccabee quickly acknowledged, the video does not look real. The flying saucer wobbles as though dangled from a string. Maccabee even showed how the scene could have been faked with computer graphics. I was convinced: It was a hoax. But Maccabee is not the kind of lazy researcher who accepts the easy, obvious explanation. He probed further, making a close analysis of the movement of the saucer, and became convinced that it was mathematically, geometrically impossible for the footage to be fake. Still, he had qualms. Something about the case was eating at him. Maccabee knows that the UFO world is full of deception and nonsense. In this case, the videotape had been shown on national television on the Mexican equivalent of *60 Minutes*. Only then—after this telecast—did people start coming forward to claim they'd seen the saucer. So that was a troubling scenario. You got a flying saucer slowly wobbling through one of the most populated cities on Earth, and yet no one manages to see it—or confesses to seeing it—until after a videotape of the event has been aired on national TV.

Maybe the first impression (what an obvious fake!) was the right one.

It is so very hard to find a UFO story that doesn't have a soft spot. It would be so much easier to believe in alien abductions if they didn't consistently occur when people are drifting in and out of sleep. We might hesitate to say that Whitley Strieber lacks credibility simply because, before he wrote a "nonfiction" account of being abducted by aliens, he wrote science-fiction novels; but our generosity runs out when we learn that he's admitted to a troubled relationship with the truth. He told people, for example, that he was nearly shot by the Texas school-tower gunman, Charles Whitman. Not true. Why does he dissemble? Because, he says, the aliens have messed with his head.

When you hear a story like Philip Corso's *The Day After Roswell*, you aren't faced with many options: Either he saw an alien corpse, and later became engaged in a massive program to reverse-engineer UFO technology, which in turn helped win the Cold War and stave off the full-bore alien invasion—or his tale is a lie. There's not much middle ground there. How do you decide? Lacking direct information, one must go on feel and smell and instinct. You have to ask yourself if there might be a narcissistic impulse behind his book. You have to linger a moment on the wonderful penultimate sentence: "Sometimes, once in a very long while, you get the chance to save your country, your planet, and even your species at the same time." (*And* write a best-seller.)

The UFO debate can't and won't be resolved. The debate could theoretically end through the accumulation of unambiguous, broadly accepted information about the location and activity of intelligent civilizations in the universe—information that does not exist at the moment, and which, for reasons I'll explore further in chapters ahead, isn't likely to be obtained anytime soon (even if such civilizations are out there). The UFO narrative would certainly be resolved if the alleged aliens stopped being covert, but they show no desire to come out of hiding. It's hard to imagine unmysterious, avuncular, discursive aliens. Perhaps Gillian Anderson had it right, more or less: We have created the kind of aliens we think there ought to be. And we think they ought to keep a low profile.

The ufologists argue that mainstream folks (including most journalists) are unable to accept that they've bought into a false world-view. And yet the record is clear: The scientific method has given us an amazing, weird, disconcerting, humbling vision of the universe. We've adapted to that, and we can and will adapt to extraterrestrial visitors, as soon as they find a way to resemble something other than a figment of human imagination.

The UFO movement's strength is not in its evidence but in its overall narrative, its theme. It has an elaborate eschatology, a host of apostles, and a recurring theme of salvation versus doom. It is not the evidence of extraterrestrial creatures but, rather, the idea of the Alien that makes ufology such a powerful faith. The skeptics can dismiss the purported tales of aliens and show the logical flaws in the story, but it will never make any difference. If an idea is sufficiently wonderful, if it springs from deep yearnings, it can easily beat back the yappings of logicians and skeptics and disbelievers. It can overcome a persistent absence of facts.

It can endure forever.

PART THREE

ONWARD
AND
INWARD

Henry Harris and I communicated a lot by e-mail, and he could always be counted on to deliver an angry diatribe about the scandal-obsessed Washington media. I asked him to keep me posted on UFOs. He wrote back:

OK but please don't portray me as a UFO nut. I'm just passing on the statements of astronauts, former Pentagon officials, a former US secretary of defense, a British defense minister, etc., not to speak of the hundreds of thousands of people who saw and video-taped the massive flyover in Mexico City recently. For some reason, the US media is generally ignoring the biggest story of all time (according to these people) while concentrating on trivia like O.J. and Paula Jones.

By the way, have you been following the recent discoveries in China?—the possible discovery of the so-called fifth civilization which was ruled by Huang-ti. According to legend, Huang-ti taught the Chinese language and mathematics 4,500 years ago. Also according to legend, Huang-ti (who was not human in appearance—his head was said to be constructed of metal) constructed large mirrors which tracked astronomical objects and had something called a "tripod" which

23

Henry Reports the News

tracked the star Regulus, a star he claimed was his home. Interesting that evidently the cradle of civilization also invented "hard" science fiction.

—Henry

I hadn't heard of the emperor with a head constructed of metal. There was so much still to learn.

I would meet Henry at the Flintridge Inn in Pasadena whenever I was in town, and we would brainstorm and argue over a couple of drinks. Henry had passionate theories and even more passionate suspicions. (In life, an idea, a belief, or even a revelation is not so powerful and obsessing as a suspicion. A suspicion is classically "nagging" in the manner of a terrible itch.) Henry often seemed to be couching his words, calculating precisely how much he could reveal at this time. He knew the inner workings of NASA, he said, and many things were "political," unbelievably so, even in the exotic field of Advanced Projects. What I noticed was that every time we met he had a new fascination, a new project or screenplay or novel he was working on, and a new theory about what it would be that would rupture the status quo and destroy the paradigm of modern astronomy.

One day he began talking about the interstellar medium, for example. Not many people are focused on the interstellar medium. Most of us don't realize that "empty space" is actually filled with tiny particles of shmutz. The solar system exists in a medium that is not empty and stable but is, rather, a heaving, swirling, dynamic arena. It's one big Roller Derby out there. There are areas of space that are more crowded with particles than other areas. Henry was just tapping into a new paradigm among astronomers. No longer do astronomers view our solar system as some static, inert structure. Rather, we are part of a galactic environment. "We're really a member of the galaxy," Henry said. And we just happen to be in an unusually rarefied part of the galaxy. For the past few million years, our solar system has been zipping along through this relatively empty zone. There are thick clouds of hydrogen lurking out there with thousands of times the density of the region we're currently occupying. We could slam into one of these clouds, with potentially awful consequences. The "solar wind" that helps protect us from cosmic radiation would be diminished, or pushed backward toward Earth. The "heliopause" would no longer be way out there beyond Pluto. Hydrogen from deep space could leak into our atmosphere and combine with oxygen. The free oxygen would be gradually depleted. It would become hard to breathe.

This was Henry's new concern. He was an interesting man.

On this day, as all days, our conversation drifted to UFOs. He reiterated that it is a historical issue, not a scientific one, a distinction that had not grown in persuasiveness in the months since he first offered it. Aliens, he said, are thoroughly documented, and the controversy over their existence reflects a perceptual rigidity on our part. The acceptance of a "massive extraterrestrial presence" will take a different mind-set, he said.

"Human beings are creatures of fixed mental states. And it takes a lot to go to a different mental state," he said.

I suddenly asked Henry, "Are we a doomed species?"

On the TV in the corner, Larry King was filling up the screen with his oversized head. When Larry King leans forward, his head takes on a bulbous immensity normally associated with the Grays.

Henry thought for a moment and said, "We're kind of evolving. We're on a path to raise our consciousness in some ways. I don't have any more details than that. We're still at a primitive level in general. There are individuals who are spiritual people who are trying to raise everybody up, but there are others who are trying to crush everybody."

He said it's like a pot of crabs: If all the crabs worked together, they could easily climb out of the pot. But instead they scuffle, and no one gets anywhere.

He was surely right. There were moments like this when I felt a brief spasm of shame, for it was quite obvious that I had no intention of lifting people up. I was a freelance invalidator, a self-appointed truth squad. I was a force of negativism, of negation. I had become crabby! Meanwhile, Henry, decent Henry, was working on something positive and uplifting—interstellar spaceflight, Goldin's pet project. He would soon hold his big conference, where the leading thinkers would figure out how to get us from Earth to Alpha Centauri.

It was, I felt certain, a ludicrous and unworkable idea.

24

Forward Thinking

Henry Harris had been a useful but underappreciated scientist at the Jet Propulsion Laboratory. He had tremendous smarts and creativity; what he lacked was a Ph.D. His friend and conference co-organizer, Dave Pieri, felt that the culture was too obsessed with paper credentials and that Henry, as a man who never finished his dissertation, never received the respect he deserved. The interstellar workshop was a chance for Henry to shine. If he could pull this off, the event might even register as something historic. The space buffs had their Latin motto: *Ad astra.* To the stars. Henry wanted to turn a subculture's dream into a real NASA program.

Henry called his August 1998 workshop "Robotic Interstellar Exploration in the Next Century." He selected Caltech as the venue, a lovely choice. The campus is serene, elegant, and very quiet. Volleyball games break out on the lawns, but a visitor gets the impression that they aren't the kind of rowdy, drunken, sexually charged games you'd find at a big state school. Instead, they're precision games in which the students may very well be doing field work in geometry or differential equations—testing to see what happens if a ball is redirected from trajectory X to trajectory Y. There are no keg parties here, no bonfires before the Big Game.

Everyone's a genius here, at least by ordinary standards. If the world is suddenly annihilated by a freakish rupture in the space-time continuum, a basement lab at Caltech is likely to be ground zero.

On the opening morning, Henry's hair was carefully combed, and he wore a suit and tie. He kept the jacket on even when he went outside, into the heat. It was a brutal day in which the sun seemed stationary at the highest point in the sky. With a solar cell on top of your head you could have sucked up enough energy to launch a Saturn V to the moon, but Henry was not about to opt for casual clothes. He was in charge.

From the start it was apparent that this workshop was all about vision, with no requirement to do anything, no test of practicality—all ideas welcome. This was precisely the kind of thing Dan Goldin had been talking about. People were even discussing antigravity! One participant, Mark Mellis, reported on his own efforts in a small and controversial NASA program in Cleveland to search for "breakthrough" propulsion techniques. It was controversial because the "breakthrough" techniques were by definition the ones that involved new physics, stuff outside the laws of Newton and Einstein and Planck, and there were a few folks who thought it smelled of shamanism. But that, too, was a Goldin-inspired program. In the Goldin Age, you could talk about things that in the past had been considered kooky. The mainstream had widened.

Hour after hour, the scientists and engineers lobbed fantastic ideas back and forth. One popular idea was to shoot atoms into space. The atoms would serve as pellets of fuel. They'd line up neatly across billions of miles of space in a kind of galactic runway. Then you'd send in a Bussard ramjet—the kind of craft that Sagan had cited in his 1962 paper on space aliens—scooping up the fuel as it went. The runway idea still required a lot of energy, since the fuel would have to be propelled to extremely distant stations. Another engineering challenge, and one that Goldin was particularly interested in, would be to design a spaceship that could function on its own, with something like human-level intelligence, even though it had no pilot and was beyond terrestrial control. When a probe got about one-fourth of the way to Alpha Centauri, it would be a light-year from Earth, and any problems it encountered couldn't be solved with messages sent back and forth between the craft and the scientists at Mission Control (a message would take two years, round-trip). So obviously it was necessary to invent starships that think.

For his conference Henry managed to snare a leading visionary of our time, Bob Forward. Forward had invented the light-sail idea in 1962. For many years Forward had been one of the only people in the world who could talk with expertise and intellectual heft about possible techniques for interstellar travel. He was an odd duck. He wore multicolored vests and white bow ties, and had a giant white mop of hair too bizarre to be a wig. ("Carnival barker" is a term that comes to mind.) Soon after the workshop began, I cornered him and asked what he thought of the Fermi Paradox.

"I believe in the Zoo Hypothesis," he said.

"Why?"

"Why not? It's an answer," he said.

No other answer made any sense, he said. He believed the aliens did not want to interfere with our technological development. They were acting out of selfishness: The aliens are technological creatures, and they view the Earth as a technology proving ground. "The best way to be of value to them is to let us invent our technology without being contaminated by their technology," he said. "If they leave us alone, we might invent something they didn't think of."

Only a true technophile would view all of human history as a laboratory for inventing gadgets on behalf of aliens. (Were they proud of us when we finally got around to inventing the automatic teller machine? What did they think of the Palm Pilot? Aren't they shocked that people in big cities send packages back and forth via "messengers" who are actually insane people on bicycles?)

A couple of hours later I went to lunch with Henry, walking a few blocks to Hamburger Hamlet. The sun was cranked to the red zone— perhaps the sun is literally closer to certain parts of California. I asked Henry if he feels there's good evidence for the Zoo Hypothesis.

"I think it's overwhelming," he said. "It just doesn't fit within our paradigm, that's all."

Over lunch he talked about the novel he's writing.

"It's about a guy who works for NASA, who's also a piano player. And they discover one of these interstellar hydrogen clouds is coming into the solar system and compressing the heliopause."

The hydrogen cloud screws up the atmosphere. It rains a lot, and the oxygen level drops, and people are dying. The NASA guy gets assigned the job of building a giant laser to push the cloud back out into space. Then there are further developments involving an asteroid-mining project and an alien spaceship. The NASA guy becomes an astronaut. I didn't quite follow every twist and turn, but it sounded as if the NASA guy saves civilization and simultaneously shatters several scientific orthodoxies.

The strange thing, Henry said, is that in real life he'd just been assigned the job of managing a space-power system. "That's a huge laser in space whose job it is to save the world by providing energy from space." He smiled. I acknowledged the obvious, that there was a conflation here of fiction and reality. Henry found this pleasing.

"I invented myself," he said. "I wrote a book, and became that character."

The no-aliens idea, this lingering possibility that there just aren't very many out there, is unthinkable even for the most visionary of people. They all do the math. Even if you make the factors in the Drake Equation extremely pessimistic, there still ought to be millions of civilizations in the universe as a whole, and some of them ought to be precocious enough to explore the place, and so they ought to be watching us even now, adhering to a strict, hands-off, no-touching-with-the-antennae policy.

Ben Zuckerman, a student of Sagan's in the 1960s, and later a leading naysayer on the issue of intelligent life in the galaxy, scoffed at Forward's decision to default to the Zoo Hypothesis. Zuckerman is a very thin man with a jumbo head, bald on top, with coils of antigravity hair exploding from either side. If Hollywood needs a body double for a mad-scientist movie, Zuckerman can step right in. (In fairness, he doesn't actually behave like a mad scientist. He's too skeptical. A mad scientist is at core an incorrigible optimist.)

"It doesn't make any sense at all. It makes absolutely no sense," Zuckerman said of the Zoo Hypothesis. He waved his arms, and his eyes widened with the effort of trying to breathe some sense into this silly notion. "The extraterrestrials would have discovered there was a place with something interesting going on as much as *billions* of years ago. . . . All these civilizations would have had a chance to look us over. They all uniformly would have had to decide not to come here."

This objection was an essential part of Michael Hart's argument back in 1975. One alien race might treat us as a zoo, and perhaps even a hundred alien races would treat us as a zoo, but would *all* alien races do that, for billions of years? Wouldn't one of them say, "Screw this zoo business," and just move on in? Do the aliens actually work in concert, like some galactic League of Nations? Surely there are the Borg out there somewhere, disobeying the Federation principles.

Not many people at the conference realized that Henry Harris and Dave Pieri, the two main representatives of NASA, were open to the idea that aliens were visiting the planet. Pieri is a big, good-looking guy who looks more like a tennis pro than a scientist. He's also a former graduate student of Sagan's, and worked for Sagan on the Viking mission. He and Sagan occasionally skirmished about aliens. Pieri teased Sagan about von Daniken, about how the charlatan had clearly stolen a page or two from the respected scientist. Sagan had backed away from his youthful suggestion that archeologists search for signs of alien visitors. He had, in fact, turned somewhat sour on the idea of interstellar travel. He felt that it was awfully hard, and that aliens would rather communicate through radio signals. This was one of Sagan's reasons for backing SETI programs. Pieri told Sagan that there might be undiscovered laws of physics that would make interstellar travel possible. Who were we, an "infantile" civilization, to say that we had discovered all that there was to know? Pieri remembers Sagan's response: "We have to work with what we've got."

Pieri told me he is not as convinced as Henry that there is an alien presence.

"I think, as in most things in nature, the answer is far more complex, far more than black and white," he said. He mentioned the Sturrock Report as an important change, a chance for scientists to look at anomalies without suffering the giggle factor. Pieri said most scientists now believe that there is life elsewhere in the universe and that some of it must be intelligent. Of those civilizations, there could have been some who had to abandon their solar system when their home star began to exhaust its fuel or threatened to blow up. But accepting alien visitors to Earth would require a paradigm shift, Pieri said.

"I think it's entirely analogous to the Aristotelian-Ptolemaic versus Copernican paradigm shift," he said. "There's a huge psychological barrier to accepting that possibility."

At one point during the conference, Pieri approached Zuckerman and asked a pointed question: "How do we know they aren't here?"

It should be obvious if they are here, Zuckerman replied.

To which Pieri had a comeback: "What does my cat think when it sees a 747 take off?"

There was a night reserved for the public, a show of sorts starring Henry Harris, Bob Forward, Lou Friedman, and *Star Trek* adviser André Bormanis. It was a light and jolly evening full of true believers. Forward told the audience that the scoop on a ramjet would ideally be infinitely strong, to withstand collisions with particles, but would also have no mass. Therefore, he said, it would have to be constructed from a special element: "unobtainium."

Lou Friedman played the role of doubter and rationalist. "There are two problems with interstellar flight," he said. "Space is big, and space is empty. It's twenty-five *million million* miles to the nearest star. . . . We'd be there now if we had started at the time of *Homo erectus* and Neanderthal." To go one-tenth of the speed of light would require designing a spaceship that traveled sixty-eight million miles per hour, he said. It would make a lot more sense to try to go to Mars, a tough enough task but not one requiring breakthroughs in physics.

Then came Henry's turn. I was nervous for him, but Henry let the old Vegas entertainer out for a prowl. He was smooth, calm, cracking jokes. He recounted how his teacher had in October 1957 insisted that the moon was too far away to be reached with a rocket. To say the same thing about stars would be to repeat that mistake. He said it's not necessarily true that there's nothing but empty space between here and Alpha Centauri. "Maybe between here and there there's a lot of interesting objects that we can't see," he said. There could be brown-dwarf stars that no telescope has ever noticed. There could be . . . other stuff. Alpha Centauri is the target only because we can see it. We behave, he said, like the drunk searching for his car keys under the streetlight. Someone asks the drunk, "Is that where you dropped them?" And the drunk says, "No, but that's the only place the light is." Henry's message: We have to be willing to look in the dark places.

I was relieved—because I liked him, and was rooting for him—that he didn't say anything about Element 115 or the alien invasion.

One day I took a break from the proceedings and walked over to the office of Bruce Murray, the cofounder, with Sagan and Lou Friedman, of the Planetary Society. Murray said he'd been hearing these interstellar-propulsion ideas since 1961, when he'd talked about it with Freeman Dyson and Bob Forward.

"It's never going to happen," he said. Murray is like Sagan, the kind of smart person so confident of his cerebral skills that he is not shy of making bold declarations. He said he realized, in a "negative epiphany" in 1980, that interstellar travel is impossible. This depressed him. "It meant that we were trapped in this solar system. Not just me and my immediate progeny, but maybe forever." Sagan didn't agree with him. A lot of engineers didn't, either. Murray said that the "technological optimists" tend to assume that everything will go right. They anticipate success.

"It's why you don't put a technologist in charge of a project, normally."

It's not just humans who can't go to the stars. The same applies to the aliens, he said. There will be no magic wands from new physics.

"It's too goddamn difficult to travel between the stars. That's the answer to the Fermi Paradox."

Murray should have heard the talk by the astrophysicist Bernhard Haisch. Haisch not only believes that there are new laws of physics waiting to be found, he thinks he's found one. Haisch is the editor of the *Journal of Scientific Exploration*, the trade publication for the Society for Scientific Exploration, the group that put out the Sturrock Report. Simultaneously he's a scientific editor of *The Astrophysical Journal*, which is as mainstream as it gets. His day job is working for Lockheed Martin at the Solar and Astrophysics Laboratory in Palo Alto. Haisch has received grant money from NASA to work on what is called ZPF, the Zero Point Field, which Haisch believes is a revolutionary breakthrough in physics. He feels that he has discovered where inertia comes from—the secret to what makes matter have "mass." This was the idea I'd heard about from Jesse Marcel, Jr., in Roswell.

Mass, Haisch said, is so deeply embedded in physical reality, most researchers take it for granted. Everyone knows that $F = ma$, that force equals mass times acceleration, but Haisch asked why. Why doesn't force equal mass times velocity? And what exactly is "mass," anyway? It's not the same thing as matter. Matter is stuff, but mass is a property of that stuff. Mass increases when an object is accelerated. One of the problems with making spaceships that go extremely fast is that they become more massive the faster they go. On paper, a spaceship could never travel at the speed of light, because as it neared that speed its mass would approach infinity. It would take an infinite amount of energy to accelerate it just a wee bit more. Light goes the "speed of light" because it would have no mass at all if it were at rest. (My own way of thinking about it—and, kids, don't try this at home—is that if light could be slowed down to a complete halt there'd be nothing there. Thus the speed of light is the maximum speed of a packet of nothingness. Anything that's a *something* must necessarily go slower.)

Haisch argues that mass is not a fundamental property of matter but is, rather, created by an electromagnetic reaction between matter and the energy contained within the vacuum. Inertia, he believes, is an "electromagnetic drag force." When he got up in front of the scientists at Caltech, he laid out this theory in precise, cautious language, knowing that it would be quite dramatic without any extra hand-waving. The audience seemed dazzled. The only catch was that there was no way to know if it was true. Haisch hadn't proposed any way to test it. What he could say, however, was that the implications were tremendous. First, it would rewrite some of the most dearly loved laws of physics. It would be a radical reinterpretation of gravity itself. Einstein's general relativity would be pretty much trashed. (Brashly, Haisch said of his theory, "It's a little like 1905 here." The reference is to the year when Einstein published four brilliant papers that changed modern physics, including the Special Theory of Relativity. Of course, that was also a year when Percival Lowell was making new discoveries about the advanced civilization on Mars.)

If Haisch is right, the energy of the vacuum theoretically could be tapped as a power source for interstellar spacecraft. Why bring fuel when empty space can give you all the zing you need? Even more amazing would be the possibility of canceling the inertia in a spaceship. It could essentially accelerate instantly to near-light velocity with no push whatsoever. The ultimate free ride. Or, as Haisch put it, "the ultimate miracle."

He told a small group of colleagues, "The ultimate miracle is to see if you could do away with the need to have any energy or forces at all." The spaceship would just go. Here I had been making fun of *Starship Troopers* for sending such a big, heavy craft to wipe out the evil bugs on a distant planet, and it turns out that maybe the weight of the ship doesn't matter. If it has no inertia, it can be the size of an aircraft carrier or the size of a VW Beetle. It just goes. *Violà!* We've crossed the galaxy.

This did not quite fly with Lowell Wood, a huge, bearded man from the Lawrence Livermore Lab. Wood had been a major player in Ronald Reagan's Star Wars space-defense initiative. Everyone respected Wood's genius, though he was prone to dismissive comments and arrogant pronouncements. He expressed admiration for Haisch's argument, but had serious reservations. He told me, "There's no way of testing it or refuting it. It's such an exceedingly fundamental thing, we have no idea how to get a grip on it." Haisch, he said, has been floating this idea for four years. "I've been carping at him for a test." Until then, he said, it's not science. "It's a reinterpretation with no consequences. It's not observable. And that is not science." It's not even a theory. "There are assertions that can be made that are objectively refutable. Those are the assertions that science is made of. Everything else is speculation."

As Wood said this, Henry jumped in to defend Haisch.

"The reason I think it could be true," Henry said, "is there is beauty in it."

Wood conceded, "It is beautiful."

Haisch is open about the source of his expansive views. He is a spiritual man.

"The scientific revolution to a certain extent has played itself out. We kind of chose as a civilization to abandon the spiritual world so that we could master the physical world, which we have done. But the problem is, we've mastered it to the point that we could destroy ourselves."

One day over lunch, he said, "I think the spiritual and the scientific perspectives will merge in the future. I think our consciousness will give us evidence of the reality of the spiritual world."

He was born in Stuttgart, Germany, grew up in Indianapolis, and went to seminary with thoughts of the priesthood. He then decided he didn't agree with Catholic dogma on divorce and birth control, though he remained a believer in the divinity of Jesus. He describes Jesus as a "spiritual entity" who may have been incarnated many times. Science has gotten everything backward, he thinks. Science says there is a physical plane, and anything spiritual beyond that is unverified and fundamentally external to reality. Haisch sees the spiritual world as the fundament, with the physical world something created by spiritual beings for their education and evolution. "It's kind of like we've created a Monopoly game on this physical plane, and now we've come to play it."

He spoke of love. The phrase "God is Love" may be the literal truth, he said. "Somehow this thing of love really is the essence of the divine, and on this plane, love really is the driving force, including things that are the polarity of love. You and I really are these little flames of God who are now experiencing what God has made. . . . The only way to play the game is to forget that you built the game. And so we, being little sparks of God—because there's nothing else—conveniently forget who we are, because that's the only way to experience His fullness. I guess that's kind of a Buddhist idea."

In the two and a half years since I'd talked to Sagan, I had been conscious of being, in most situations, an unbeliever, unapologetically so. Talking to Haisch rattled me a little. I found myself worried that I had been too easily unimpressed by the spiritualism of others. I wasn't just checking up on people, I was actively rejecting their beliefs. But who was I to dismiss or even to challenge the spiritual revelations of other people? Could it be that I was truly closed-minded? I couldn't even get totally hypnotized back in that hotel room. What a spiritually wimpy human being I had turned out to be!

Let's face it: The question that really matters, that underlies everything else, is not whether you believe in aliens, or anomalies, or the new physics, or the vastness of the unknown. The important and timeless question is: *Do you believe in God?*

Haisch had embraced a quasi-scientific, quasi-religious, and very New Age conception of a universe shot through with divine purpose and intent, with God a force built into the fabric of nature as surely as gravity or electromagnetism. His words weren't that dissimilar to something I had heard from Mark Williams, the gasping man in the trailer home in Las Vegas who wrote *Ascend!* (the book with the blurb from God). Williams had said that the universe was itself conscious, even at the subatomic level. The particle physicists, Williams said, have smashed protons together and seen them ascend to a higher energy level. This ascension is love at work. "They've concluded that there's a cooperative and loving universe. . . . There's intelligence even at a subatomic basis," Williams said.

Haisch, at least, didn't claim that his ideas had yet been confirmed by particle physics. He had the honesty to say up front that it's just his personal take on things. I was tempted to try to join him in some way, to embrace some element of his "God is Love" universe. But a last-minute conversion wasn't going to happen, because I personally don't believe that the love we experience in our lives has to be sanctioned by a higher authority to be powerful and authentic. His idea of the spirit world was like his idea of the Zero Point Field, something beautiful, something aesthetically pleasing. But science has achieved so much in part because it's not a beauty contest. It's not based on what human beings think the truth *ought* to be. An idea doesn't become scientifically valid because it's pretty. It starts to rise to the level of truth when it passes a few tests.

The spirit world by definition is not subject to confirmation by material instruments, and thus is detected only by the mental instruments of "intuition" or "revelation" or (lapsing into circular logic) "the spirit." When I asked Haisch how he knew all these things to be true, he said simply that he knew it. He *knew* it. Which meant that he felt it. (He said later that this is not the same thing as blind faith, which he opposes. It's an "active, conscious knowledge of my own awareness," he told me, "knowledge based on inner experience.") Haisch did not feel like a purely material creature, the result of billions of years of mindless evolution. That was inconceivable. And thus, with a heave of his gut instinct—or his inner experience, as he put it—he rejected the basic Creation narrative of modern science.

There is a divide here that stretches across all of human society. Bernie Haisch is on the side of countless "spiritual" people, surely the majority of human beings. Over on my side, it says that Grand Design cannot be discerned through feelings. You can attain "truths" through your spiritual tools, and they may be very powerful truths, but they are subjective truths, and your truths might not necessarily be the same truths that I discover using my own spiritual tools. Whereas science offers a technique for finding out things that are true for everyone, of all nations and beliefs and races, true even for the aliens from Alpha Centauri.

Perhaps someday Bernie Haisch will find a way to cancel the inertia of a giant spaceship and send it off to the other side of the universe. But for the moment I was grateful I had a rental car with a combustion engine, and could drive to the airport and go home.

25

On to Mars

It was a perfect day on the eastern flank of the Rocky Mountains, which was pretty much a perfect place even in a cosmic context. The skies over Boulder, Colorado, were blue and clear all morning, and when the clouds rolled in they were gentle and puffy, the kind of clouds that you don't take seriously, that don't really register as being suspended liquid-water droplets. The air was perhaps a bit thin—more than a few millibars under the sea-level average—but otherwise it was clean, snappy, scrubbed by evergreen forests. People wandered the pedestrian mall along Pearl Street, dipping into bookstores, eating ice cream, watching street performers. Bikers pedaled up a winding path into the foothills next to a tumbling stream that is popular with kayakers. Hikers rambled through the meadows of Chautauqua Park, climbing to better views of the knife-edged Front Range and, to the east, the limitless high plains.

But even on this perfect day there were people who did not go outside. Instead they gathered in a ballroom on the campus of the University of Colorado. They stayed inside all day long and into the night. They had a goal: Go to Mars.

They wanted to leave Boulder behind and go to a small, cold planet with no pedestrian malls, no tumbling streams, no forests, no mountain parks, no Mexican restaurants or brewpubs, not even week-day delivery of the *Wall Street Journal*. They wanted to go somewhere with no *air*. Forget kayaking—there's no liquid water on Mars at all, except perhaps in the pores of rocks far below the surface. An unprotected person would more or less instantly die on Mars. Then the remains of the person would be fried by solar radiation, so that even the bacteria would die. The flesh would freeze. The temperature is routinely more than a hundred degrees below zero. Dust storms (as Sagan realized) sometimes envelop much of the planet. It has been observed that Mars is a lot like the moon, only with bad weather.

Not that this would diminish the enthusiasm of the people in the ballroom. They had come here to form a new organization, the Mars Society. They had signed their Founding Declaration. They wore buttons: "Mars or Bust." And they had a leader: the charismatic Robert Zubrin. He was arguably a practical man in terms of futuristic space travel, because he wasn't talking about going to Alpha Centauri, he wasn't claiming to have discovered an inertia-canceling scientific theory, and in fact he was perfectly happy to employ the old-fashioned principle of rocket propulsion. The only impractical part of his program was that he wanted to go to Mars.

Zubrin had figured out the How of the matter. He knew the When. Mars was the Where, we humans were the Who. The toughest question, however—and perhaps it seemed particularly pressing on this sunny day in Boulder—was the Why.

Zubrin took the stage. About five hundred people were in the ballroom, and most had read Zubrin's book, *The Case for Mars*. There were boxes of it stacked by the ballroom entrance. This was clearly Zubrin's show, a Zubrinfest.

Zubrin conceived the Mars Society in early 1998, and appointed himself president, unpaid. He lined up a steering committee that included NASA scientists Chris McKay and Carol Stoker and science-fiction writers Kim Stanley Robinson and Gregory Benford. Hundreds of people sent in abstracts of talks they wanted to give at the convention, and Zubrin weeded out the ones that might tarnish the proceedings, including the proposals from the Face on Mars folks, the Planet X people, the antigovernment paranoiacs, and various people with nothing coherent to say. He knew he needed a good screening process, because advocating human migration to Mars was strange enough.

Now he had his moment, his audience. He looked out upon a vast room of true believers. "I believe this is going to be an historic occasion," he said.

The sound system immediately failed, a squealing mess (was it the same saboteur who struck at the Mars meteorite press conference?). Undaunted, Zubrin tossed his microphone away and shifted to a controlled shout, which seemed to suit him just fine. At first he was a bit stiff, shifting his weight from foot to foot, hands in pockets. But as he warmed up, the hands came out and hacked at the air. Zubrin began roaming the stage, scowling, cracking jokes, extemporizing, speaking at manic velocity, without notes. This was part technology lecture, part sermon. If he had looked at first like a nerdy engineer, he managed, over the course of the hour, to reveal himself as a brilliant evangelist of interplanetary travel. It is time, he said, "to take the steps to make man a multiplanet species."

He explained how NASA had gone wrong. He cited the Space Exploration Initiative of President Bush, that ill-fated proposal in 1989 to build a bunch of orbiting space stations and giant interplanetary vessels that would cost an obscene amount of money—what Zubrin called "the parallel universe." He showed a slide of an enormous spacecraft. "I call it the Death Star," he said. No wonder NASA had failed to get its manned Mars mission past Congress, he said.

Zubrin said he had a better plan. He called it Mars Direct. Zubrin's vision requires no space stations or moon bases or even any advanced propulsion techniques. Instead, he would launch an unmanned craft to Mars using existing rocket technology ("unpiloted" is the preferred term these days, but I find it confusing, since it suggests the craft contains wide-eyed passengers who are wondering who's flying the damn thing). It would land on Mars and set up shop as a manufacturing station of fuel, using just the elements naturally available on the Martian surface. Zubrin said this process is not that difficult; you can make propellant with Martian dirt. Only when this stage of the project had been successful would a crew of astronauts zoom to Mars on a second rocket. The genius of the Zubrin strategy is that the astronauts don't have to bring the fuel they'll need for the return journey—it'll be waiting for them.

This, said Zubrin, is the same strategy explorers have used on our planet for centuries: "Live off the land." (In his book he put it another way: "When on Mars, do as the Martians do.")

Zubrin the engineer speaks with authority. But to be the leader of the Mars Society, of an entire movement, Zubrin has to be something more, a spiritual force, a moral philosopher, a Sagan figure. Zubrin doesn't shy away from that role. Onstage, he outlined the reasons why we should—why we must—go to Mars. First, he said, there's science. If we go to Mars, we can discover whether there was ever any life there. Viking didn't settle that issue. If we find signs of life, present or past, it will indicate that life, including intelligent life, is probably spread across the universe. "It means we're not alone, we're part of something bigger than we suspect right now."

But then comes the grander agenda—a matter that would, in coming days, spark much argument and acrimony. Zubrin said the issue at hand is not whether life used to be on Mars but whether it will be on Mars in the future—human life. He said he believes we should settle Mars, that we should "terraform" the planet to make it habitable for humans and other forms of life. It's our destiny, he believes. Mars, he tells everyone, is not just a planet but a *world*, a place of unlimited potential, a place where good things can happen in the same way that a great nation rose upon the North American frontier. Humans, Zubrin said, are now a "Type I civilization," capable of taking advantage of all the resources of our home planet. But to become a Type II civilization requires that we extend our reach across our solar system. "We have mastered our own environment," he said. "Will we settle for that or will we reach out further and become a Type II civilization?"

He shared his vision: "Two thousand years from now, there will be people not only on Earth and Mars and the asteroids, there will be people in civilizations orbiting hundreds of stars in our region of the galaxy."

He finished to a standing ovation.

We were supposed to be on Mars already. In 1952, the German rocket scientist Wernher von Braun published *Das Marsprojekt*, a plan to send ten spaceships, each with a crew of seventy, to Mars. Going to the moon would be just a preparation for this Martian adventure. In 1962, NASA began preliminary studies on the feasibility of manned voyages to Mars. In 1969, a couple of weeks after the moon landing,

Proposals for a manned expedition to Mars have been circulating since the 1950s. This was a NASA vision in 1970. NASA

von Braun testified in Congress that it would be possible to have a manned fly-by of Mars in 1978 and a landing in 1982.

It never came to pass. NASA gave up on Mars. In the mid-1990s, with Mars having returned to the forefront of scientific interest, there were some blue-sky discussions at NASA of a manned Mars mission in 2014, or maybe in 2018. There were people at the Johnson Space Center who started putting together tentative plans for such an adventure. But Goldin knew that there was not yet the political support for any grandiose initiative. He liked Zubrin's idea, and spoke enthusiastically about Mars Direct, but Goldin seemed to feel that too much talk about Mars would alarm the bean-counters in Congress.

It was easy to see why Zubrin was frustrated. He knew that the space program had achieved wonderful things in the dozen years from 1961 to 1973. Scientists and engineers invented spaceships, orbital rendezvous techniques, space suits, hydrogen-oxygen rocket engines, multistage heavy-lift boosters, life-support systems, interplanetary navigation, soft-landing technology, nuclear rockets, rovers, and on and on. Zubrin can make a Mars mission sound like simple pragmatism. Americans already pay $13 billion a year to NASA for a space program. A lot of that money, Zubrin said, maintains an infrastructure that was originally designed precisely for manned planetary exploration. We paid for it, he says, so why not use it?

But going to Mars is not like taking a Sunday drive to the zoo. To get to Mars, a human crew would need to spend at least 150 days in interplanetary space, cut off from supply lines, locked in a can, and forced to drink water from recycled urine (among other unsavory prospects). If a solar flare erupted, the astronauts would have to cram their bodies into a tiny central compartment for protection against lethal radiation. One speaker, Mike Griffin, a former NASA associate administrator, told the conventioneers that American society has become excessively risk-averse. We are cringing and weak, soft and nervous. We have forgotten the hardships of our ancestors. Griffin told a story about Jebediah Smith's attempt in the early nineteenth century to find a route to California from Colorado across the deserts of Utah and Nevada. He and his party followed the Humboldt River, which ran in the right direction, toward the Pacific. But then the river dried up. It just died out there in the middle of the scorching nowhere. The men ran out of water. To survive, they killed their mules and drank their blood, and two of the men jammed on their wide-brim hats, buried themselves in mud up to their nostrils to stay cool, and waited while Smith hiked sixty miles on foot to the Sierras and replenished their water supply. He returned with the water and found the men not quite dead. "How many people think a trip to Mars will be anything near that grueling?" Griffin asked.

The people in the audience did not have to be sold. They were ready to work night and day for a Mars mission. They were ready to put on jumpsuits and go. They were ready to be buried up to their nostrils in mud.

They all had their reasons. Over a beer after midnight, one scientist said that if we go to Mars, "it means we are immortal": No asteroid could ever wipe out all of humanity. Another man, a libertarian, told me, "You won't be able to have bureaucrats on Mars regulating everything." He wants a Mars mission to be solely a private venture. The idea I found most provocative was the assertion that young people in America need a Mars mission, that they need a cause, a rallying point, something analogous to the Vietnam War protests or the civil-rights movement. A high-school student told me, "The kind of opportunity to challenge yourself and push the limit of human ability, you can't get it here on Earth."

The low point of the convention came Friday night, at the terraforming debate. Terraforming Mars is a popular concept with the Mars enthusiasts. The idea is that Mars can be made habitable through a massive program of introducing organisms, increasing atmospheric pressure, warming the planet, and gradually constructing a more Earth-like environment. There are two problems with terraforming. First, it's exceedingly difficult, because you can't simply plant some seeds. It would be easier to turn Antarctica into a tropical paradise, or to build great cities at the bottom of the ocean. The visionary Zubrin is the first to admit that his young daughter will never splash in Martian lakes.

The other problem is the principle of the thing. Would it even be ethical? This was the subject of the debate. Mars, though apparently lifeless, could have native organisms in the permafrost or deep beneath the ground. Terraforming might wipe out that indigenous life. Some, like Sagan, have said that if life exists on Mars the planet should be left alone.

There were five panelists on stage, plus moderator Chris McKay. The night got interesting when Zubrin took the microphone. He said he believes that some things in the world are simply better than other things, and the societies that believe in science are more successful than the ones that believe in mythology. Humans, he said, were something more elevated than animals, and certainly than microbes. Western civilization, with its humanist orientation, is the only one that people in the room would want to live in. If we alter Mars, he said, we are actually acting on behalf of other forms of Earthlife. We can take fish there, for example. "We eat a lot of fish. But we can return the favor by establishing fish on Mars. They can't get there any other way. It's payback time."

Then came Lowell Wood, the huge Star Wars guy I'd just seen at Henry Harris's workshop. Wood promised to make Zubrin sound like a moderate. Humans, he said, have increased in number one-hundred-thousand-fold in the last hundred thousand years, and the reason is terraforming. By altering our environment, we increase the odds that our children will not starve to death. "At least half of all kids ever born starved," he said. The people who terraformed their environment had a higher percentage of surviving children, and therefore they increased their representation in the gene pool. One of his viewgraphs stated that terraforming Mars is "the sociobiological manifest destiny" of *Homo sapiens,* "the terraforming species." He boasted that the United States has a society that builds new worlds. Americans made America! "We built it into the shining city on the hill on this planet."

No one applauded. Wood had the audience a bit rattled. What was all this talk about starving kids? Manifest Destiny? Shining city on a hill?

In the span of half an hour, the panelists had managed to advertise the soft underbelly of the Mars movement. The technocrats hadn't mastered a jot of the science-meets-religion language of the 1990s; they had no feel for New Age sentiments. They sounded angry. The drive to colonize another world requires an aggressiveness, a grasping nature, that many people find distasteful. Colonization meant tragedy for millions of indigenous people around the world. "Manifest Destiny" is generally associated with the belief among European-Americans in the early nineteenth century that they had a divine duty to spread their society across the continent. One effect of settling the frontier was the genocide of Native Americans. Sagan, for one, understood the historical tragedies of colonization, and he believed that this time we should get it right, that we should be cautious in our interplanetary aspirations, that we should have as our paramount concern the respect for indigenous extraterrestrial life.

McKay asked for questions and comments. A park ranger stood up. He said he objected to the use of the term "Manifest Destiny," and added that the shining city on the hill is sometimes invisible for all the smog.

Zubrin jumped in.

"Denunciation of Manifest Destiny is a fashionable pose," he said, "but by opening the American West we made a place where millions of Mexicans are trying to get in."

A man in the audience leaped to his feet and said, "That's good, because they'll replace all the millions of Indians we killed in the process."

At that point no one in the room could avoid contemplating the demographic situation onstage. All six panelists were white, male, and American. They represented pretty much the entire spectrum of white American male technophiles who want to colonize Mars.

Things got even testier. A schoolteacher, Judy Paulsen, objected to the panel's "arrogant" rhetoric. Another woman, Victoria Friedensen, an officer with the National Academy of Engineering, stood up and complained about the brute-force approach to terraforming Mars. "I find some of the use of technology terrifying. And I find it deeply, deeply arrogant." Zubrin, who appeared increasingly exasperated by the discussion, responded: "What do you find offensive about making Mars into a living world?" Friedensen said, "I would ask that you perhaps modify your language so that the rest of us, who are not as technologically oriented, can join in."

This riled Lowell Wood. He had been misunderstood, he said. There was not a moral imperative to terraform, there was a genetic imperative. Then he said something that brought gasps from the crowd. Apparently he wanted to make the point that it didn't matter in principle if Mars was terraformed gently or brutally, so he cited a famous line by George Bernard Shaw: "Madam, we have established what you are, we are merely haggling over the price."

A whore metaphor? What was going on? Whatever Wood's precise intention may have been, many people heard the comment as an insult to Friedensen. One woman stormed out of the room. She said later, "There were six white males with testosterone flowing, talking about Manifest Destiny. You're transporting me back a hundred years!"

Afterward, many people walked up to offer Friedensen their support. "Frankly, I was scared by some of the rhetoric here tonight," said David Grinspoon, a planetary scientist. Sam Burbank, a documentary filmmaker and musician, said, "I was dying over here. This was a disaster. This was my worst nightmare of what this convention would be like."

The next morning Zubrin felt dispirited. He knew the debate had been awfully acrimonious. He sought out Friedensen and apologized to her. But over the course of the day his spirits lifted. New organizations always wobble at first. What he had seen was ideological diversity. That was a strength!

The steering committee met for many hours and developed a plan. The society would start off with the establishment of an Arctic base to simulate conditions facing Mars explorers. Next would come some kind of unmanned exploratory mission to Mars. The society would earn a solid reputation, gain support, raise money, and eventually put together a mission for humans. This was Chris McKay's dream. The organization would be like the Cousteau Society, doing real things with real people.

By nighttime, with the convention banquet going full-bore, Zubrin was peppy again. About seven hundred people had come to the founding convention of his society. The way he saw it, his organization included people who made A-bombs and people who had gone to jail rather than fight in Vietnam, and they all wanted to go to Mars. "To get to Mars, we're going to need poets *and* pilots," he said.

He felt bad about the fish comment—too flippant. But he couldn't bring himself to change his tune on Manifest Destiny. "There are people here who are more politically correct than I am," he said. Manifest Destiny was a belief system that motivated ordinary people to pull up stakes and set off on a perilous journey to the frontier. The world has benefited from that, he said. "It was tragic for the Indians. But there are no Indians on Mars. And to compare microbes to Indians shows extreme disdain for the Indians."

After dinner it was time to ratify the Founding Declaration of the Mars Society, written by Zubrin and author Kim Stanley Robinson. Zubrin read it aloud, onstage, in his most dramatic, declarative voice:

"The time has come for humanity to journey to Mars!" he began. He gave all the reasons: for knowledge, for the challenge, for the youth who demand adventure, for the opportunity to carry the best of our heritage to a new world.

"Mars is not just a scientific curiosity," he read. ". . . It is a New World, filled with history waiting to be made by a new and youthful branch of human civilization that is waiting to be born. We must go to Mars to make that potential a reality. We must go, not for us, but for the people who are yet to be. We must do it for the Martians."

When he finished, everyone jumped up and applauded. Zubrin called for a voice vote. The ayes had it, unanimously. Zubrin was ecstatic. The crowd was pumped. They had a society. They had a movement. They were going to Mars, a cold little planet millions of miles away—but a world, they believed, that would someday be wonderful.

I said goodbye to Zubrin and went for a hike before heading to the Denver airport. Maybe it was the altitude, but I found myself having irrational, goofy ideas, including the thought that perhaps Zubrin was right.

Mars, after all, is close. Mars is not Alpha Centauri. It's a piece of solid land, some turf that, in cosmic terms, is right next door. Though Zubrin was a zealot, he still appreciated the nuances of the task, both socially and technologically. He had quoted something that John Adams had written to his wife, Abigail, on July 2, 1776, after he had succeeded in getting agreement for the Declaration of Independence. Adams said that centuries hence, July 2 would be celebrated with parades on streets across the North American continent. He had it right, within forty-eight hours. Zubrin allowed that he didn't want to sound maniacally egocentric by comparing his Mars declaration to the Declaration of Independence. The stakes weren't so high, and "I'm not as good a writer as Thomas Jefferson." But he did think that this might be the start of something amazing, that Mars represents an unlimited potential. Mars could be the next America.

Maybe even Lowell Wood, with his chilling talk of the sociobiological Manifest Destiny, would be

26

At the Flatirons

borne out over time. Although he did a fair impersonation of a scary person—the word "fascist" is probably too easily thrown about these days, but it did pop into my head a few times—Wood is espousing ideas that have the power of reflecting the historical record of the species rather than the political ideology of the intelligentsia. He may simply be right, horrible as that is to contemplate. His vision of humanity is that of a virus. As the science writer Oliver Morton said after the terraforming debate, this talk of inevitable sociobiological destiny could be applied to *E. coli*. Humans, Morton said, are supposed to be able to choose their destiny, unlike a bacterium (which just wants to become bacteria).

If their vision is correct, then someday humans will spread across the galaxy. Even if there are only a few civilizations out there, we're bound to meet them. *Then* what? Are we even ready for contact? Are we coherent enough to carry on an intelligent conversation with our peers from outer space? Imagine how hard it would be to craft even a one-page message for the aliens. The first sentence, hammered out by a committee of world leaders picked to represent different viewpoints, would be an utter lie:

We are a peace-loving species on the third planet from the star we call the Sun.

Then the hard part would come. Would we say we worship a god? Many gods? Or that we follow the precepts of the scientific method? How would we address the persistence of our mythologies even as we increasingly use the tools that reveal them to be fictional narratives? This is why there will always be a job for professional writers—world leaders will always need someone to finesse their contradictory ideas.

Should we admit that within six years of discovering that nuclear fission can be controlled we used the discovery to wipe out two cities in Japan? How do we deal with the Holocaust? With all the Kosovos, Northern Irelands, Rwandas?

How do we explain the inner cities of America? What would we tell an alien about slavery and its legacy? Would we say we have fixed the problem? Would we say it's all in the distant past, that we've repaired the racial divide?

How would we explain the smog in Yosemite Valley? How would we explain the destruction of the Florida coral reefs or the ruination of the Everglades or the burning of the Amazon rain forests?

We would say, "Though our species has had problems, we are working them out under the auspices of an institution called the United Nations."

We'd have to lie! We could never tell the aliens the truth! That's why people are so desperate for contact—what we really want is for the aliens to save us from the trouble we've manufactured for ourselves.

The technophiles would hoot at all this and say that it's liberal hand-wringing, and that, moreover, the creation of a frontier would provide the kind of opportunities that would unite people and bring about an improvement of our civilization as a whole. This is certainly questionable as a historical fact, since the opening of the New World specifically led to the expansion of the African slave trade. More to the point, growth and expansion are not cures for human failings.

The Invalidator says no to Mars, for now—maybe someday. For the kids out there who insist that their generation needs a challenge, here's a suggestion: Read some books. Learn. Love. Explore. ("Clean your room," suggests my space-savvy friend Dwayne Day.) There's hardly a young person today who could walk through a forest with a tenth of the capacity to observe that Meriwether Lewis had. A baby is a universe of mystery; so is my back yard. What I lack is not a challenge but, rather, the time to meet the challenges I already face. I want to read, someday, a small but sturdy fraction of the great books that have been written over the last thousand years. I would like to know some of the thoughts that passed through the minds of Leonardo, Galileo, Thomas Jefferson, Martin Luther King. There is so much unfinished work for all of us. My life does not need, on its list of Things To Do, an item saying "Travel to Mars."

Another reason we aren't ready to go is that we do not have, as Chris McKay put it, "a theory of life." McKay was talking about life as a complex chemical system, but the words could be applied to human civilization as a whole. The Lowell Wood theory of life is utterly different from the theory of Jan Bingham, the Starseed person. Wood would expand human civilization through massive application of technology, whereas Bingham would retreat into mysticism as a technique for reaching an elusive higher consciousness. Perhaps we need the best of the opposing ideologies: the willingness to adhere to the methods of science and the aspiration to become better human beings.

I was hiking through a meadow, thinking: Don't want to go to Mars. Want to go on hikes. Perhaps this was, as I said, the altitude messing with my mind. I came upon a chipmunk. It sat upright, manipulating a morsel of food in its tiny paws. Small of brain but nimble of hand, such a creature had potential. The dinosaurs must have thought little of our mammalian ancestors. You never know what unimpressive creature may later emerge to master its environment.

And there were gnats. They were hovering in clouds, and emitting a strange buzzing sound, high-pitched, artificial in tone, the sound we have all heard in countless science-fiction movies when something paranormal is about to happen. How did I know they were really gnats? Why did they go for the head, the eyes? What were they trying to get from me? The thought could not be avoided—they could be those things that Goldin had talked about. Gnats were nanoprobes!

Any and all official government activity with respect to Mars will be
accompanied by a counterreality, a
shadow narrative. After the Mars
Observer disappeared, just as it was
reaching Mars orbit in 1993—the
big space probe may have exploded,
though no one is sure—there were
rumors in the UFO world that the
accident was a hoax. The Mars Observer, some felt, was engaged in a
secret, black-budget program to
photograph the Face on Mars and
other anomalies. As NASA geared
up its new Mars program, the first
step would be to take high-
resolution pictures of the surface of
the planet from an orbiter called
Mars Global Surveyor. Remarkably—incredibly, one might even
say—NASA eventually decided that
among the features it would target
would be the famous Face.

The Face had been discovered
in the region of Cydonia on July 23,
1976, by the Viking 1 orbiter. A
member of the imaging team, Toby
Owen, found it while examining a
photographic mosaic through a
magnifying glass, simply looking for
good sites for the Viking 2 lander.
He was initially amazed. He showed
the "Face" to Jerry Soffen. A few
hours later, Soffen showed it to the
assembled press corps—"Isn't it peculiar what tricks of lighting and
shadow can do," he said. He then
said that a subsequent picture from

27

The Face

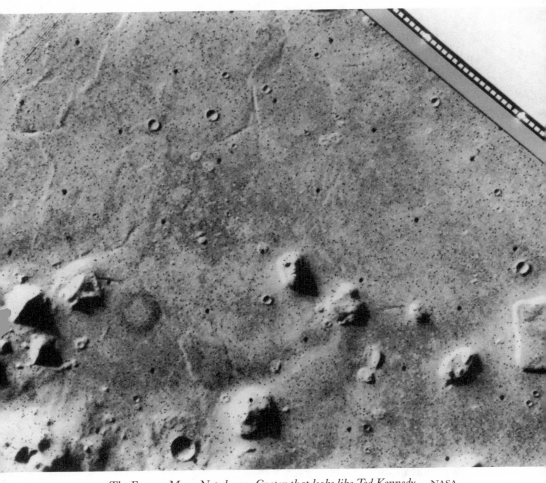

The Face on Mars. Not shown: Crater that looks like Ted Kennedy. NASA

a different angle revealed no face at all. Soffen was actually making things up on the spot—he knew the feature was just a natural formation, and he assumed it would go away if seen at a different angle. NASA itself distributed the picture with the caption "Face on Mars?" As with Roswell twenty-nine years earlier, the government itself initiated the story (before allegedly making a dramatic pivot and initiating a cover-up). In another Roswell parallel, the "serious" ufologists took many years to develop much interest in the Face. Alien stories are organic in their growth; they often have a juvenile period before reaching maturity. The big break for the Face was its appearance in a screaming story in 1984 in the *Weekly World News*, the most overtly ridiculous of the tabloids. By the time Richard Hoagland wrote *The Monuments of Mars* (1987), there was not only a Face, there was a City, a Fortress, a Pyramid, and a circular mound that Hoagland calls the Tholus. If you look very closely at the Face, Hoagland writes, you can see "teeth" in the "mouth." He and his fellow exoarcheologists found signs of battles on the site, remnants of explosions. Most sensationally, Hoagland discerned that there are a number of structures with a relation to one another involving geometric angles of 19.5 degrees. The Pyramid, he writes, also echoes the famous Leonardo da Vinci drawing of a man in a squared circle, both of which are attempts to mathematically reconcile—his phrase—the five-sided symmetry—his phrase again—of living systems, such as man, with the six-sided symmetries—his phrase—found in nonliving systems.

The Face spawned a full-blown cosmology, complete with an exploded planet, a doomed civilization, and a rewritten origin of humankind. The builders may have evolved on the parent planet, according to astronomer and ufologist Tom Van Flandern. The parent planet isn't around anymore—it blew up and became the asteroid belt. Van Flandern thinks the Face can be dated to about 3.2 million years ago—which would mean it was built at roughly the same time as human beings first appeared on the Earth. It could have been that the civilization on Planet X, knowing their world was doomed, migrated to Earth. They wouldn't have been adapted to our planet and would have had to survive by producing a hybrid species. This would explain why we're here. The builders of the Face bred with apes to produce human beings. Van Flandern: "There's a missing-link problem in the evolutionary chain between us and other primates."

Sagan rejected the Face, and as the gatekeeper to credibility he could ensure that the topic remained a fringe passion. An article by Sagan in *Parade*, in which he used the Face as a classic example of misperception, proved particularly devastating to Hoagland and his allies. But they didn't give up. In early 1998, several researchers (though not Hoagland, who was considered radioactive) persuaded NASA official Carl Pilcher to meet with them. Pilcher listened politely. He had no doubt that they were wrong, but he respected their seriousness, and felt they were the kind of people who could have made great contributions to science had they not gotten lost on this bizarre tangent. To their shock, Pilcher endorsed the idea of targeting the Face with the Mars Global Surveyor (MGS).

He was following Goldin's lead. Goldin felt the taxpayers deserved to have their questions answered. He didn't think the Face was artificial, but he believed it was of public interest. Give the customer what he wants. That's the Goldin philosophy.

NASA's decision dismayed one man in particular, Michael Malin, the scientist who had designed and operated the camera on the spacecraft. As a contractor, he was the "principal investigator" for the camera and everything it did. He was furious. Malin is not a man to dress his words in lacy frills. This was stupid, he told everyone. This was silly and irresponsible and a waste of money. The possible payoff to science was so unlikely it couldn't justify the effort, he said. Malin had what he thought were sound technological objections to the imaging of the Face. The camera on MGS didn't actually move. It didn't swivel around on command. Rather, it was fixed to the spacecraft, an unblinking and immobile eye. The entire spacecraft had to move in order to aim the camera. Despite the marvels of modern telemetry, such steering is not easily accomplished. Simply knowing the spacecraft's precise location at any given moment is extremely difficult. This little probe was something like 240 million miles away— almost on the other side of the Sun. Newtonian mechanics had a way of brushing against modern chaos theory when it came to spaceflight. A little unexpected atmospheric drag here and there and within a matter of hours the spacecraft simply wasn't where the trackers thought it should be. There was a serious possibility that even with the best of efforts the camera would miss the Face entirely. Malin knew that would play into the conspiracy theories of Hoagland and the rest of the Face group. He was sure that, no matter what happened, the Face people would never accept the results. Malin had dealt with them for years. He'd even tried to keep the location of his operation a secret, and his company had an unlisted number, but still someone tracked him down and plastered the exterior of his office building with Face on Mars stickers. He advised anyone who would listen that debates with ufologists were a waste of time: "You can't win these things."

To Malin's surprise, Sagan changed his mind about whether the Face should be re-examined. In his book *The Demon-Haunted World*, he wrote that although it was "extremely improbable" that the Face was artificial, it should nevertheless be studied further: "It was probably sculpted by slow geological process over millions of years. But I might be wrong. It's hard to be sure about a world we've seen so little of in extreme close-up. These features merit closer attention with higher resolution. . . . Unlike the UFO phenomenon, we have here the opportunity for a definitive experiment. This kind of hypothesis is falsifiable, a property that brings it well into the scientific arena."

Malin read the passage and was taken aback. Why had Sagan changed his tune? Perhaps Sagan merely wanted to prove the ufologists wrong, conclusively. Or perhaps Sagan couldn't let go of his old dream of finding Martian life. Malin didn't know Bruce Murray's theory about Sagan: that life on Mars was the mushy spot in Sagan's cosmology. Malin never got the chance to question Sagan directly.

And so Malin lost the battle of Cydonia. Indeed, he was directly rebuked by NASA headquarters. Wes Huntress sent him a formal letter instructing him to take the Cydonia pictures. "I would like you to cooperate fully with all aspects of the Mars Global Surveyor mission including efforts to reimage the Cydonia region of Mars," is how Huntress put it. He asked Malin to work "in a spirit of cooperation and teamwork." In case all this didn't sink in, Huntress followed with an e-mail, more sharply worded. "You must know your customer," Huntress wrote.

So Malin took the pictures of Cydonia. He posted a photograph on the Internet for anyone to see. On closer inspection, the Face looked like . . . nothing at all. There was just a hill there, with some varied topography, some ridges and depressions. The Face had been obliterated by the higher-resolution images, in the same way that the canals on Mars had vanished with the coming of larger telescopes.

But Malin was right: You can't win these things. Richard Hoagland went on the counterattack. The pictures, he said, were fakes.

"What in the hell can he be thinking? It's just so blatant and so idiotic!" Hoagland said. Malin, he believed, was clearly perpetrating a hoax. The evidence was so complicated I couldn't quite follow it—something about streaks appearing in the images that indicate they were not taken by the MGS camera. Also, there was something wrong with the shadings.

"This raw picture that Malin has given us has only forty-two shades of gray," Hoagland said. The devil, he said, is in the details. "Malin is lying to us and the American people and the entire world, and I'm going to get him! This is insane!"

I asked him why Malin would want to perpetrate such a thing.

"I haven't a clue!"

But he mentioned the famous Brookings study of 1962. The public would not be able to handle the truth about the alien presence. The suppression is meant to stave off madness and despair. Something like that.

The other researchers on the Face then brought up a new argument. They said the hill still looked like a face if you examined it in the right manner. This was a variation on thinking outside the box: You had to look outside the box. The researchers said there were other pictures taken by MGS that were more facial, as it were. They talked about erosion of the Face over the millions of years since the Martian civilization went extinct—of course it wouldn't *still* look like a face.

One day I got a call from a source in the UFO world who told me about an obscure Web site containing information that could blow the extraterrestrial-life debate wide open. This might be the paradigm-buster, he said. He and another researcher had found something in an MGS picture that others had somehow missed. My source imparted the information as though it were as sensitive as the detonation technology on a fusion bomb. I went to the computer and called up the site and saw it: In the picture of the Martian surface, there was what appeared to be writing. It looked like . . . scribbling. Literally like a child's version of writing. But over to one end there were a couple of markings that looked like letters, including the letter "B."

There had been a "B" seen by the Viking lander in 1976. That "B" had been on a rock. Now here was another "B," only much larger, so huge it could be seen from orbit. The Martians had a thing about "B."

But there was more. The researchers had also found what appeared to be an enormous animal pictogram. To my eye it looked like a fairly random blotch, but the researchers had added their own outline, highlighting the features. If they were correct, the Martians had drawn in the soil the figure of some kind of mooselike creature, with antlers.

It looked alarmingly like Bullwinkle.

28

Silence in Green Bank

Someday we'll find the extraterrestrials—possibly—or some remnant of them, or some echo or "signature" of them, which sounds rather elegant but might merely refer to a spew of radiation blasting out the back of an antimatter-powered garbage incinerator. Finding them remotely, as an extension of our survey of the sky, is the goal of SETI, the Search for Extraterrestrial Intelligence, originally called CETI, for "Communication with" ETI—Sagan's favored construction—until more jaundiced observers suggested that the feat of detecting them is difficult enough without also insisting that we hold a conversation.

For forty years, SETI has been a mandatory security check for anyone on alien patrol. I felt obligated to deal with SETI even though it has received more ink than probably any other experiment in the recent annals of human inquiry, publicity made all the more remarkable given that it hasn't (permission to speak freely?) worked. What redeems SETI is that it might actually succeed at some point, and the payoff would be mind-blowing. SETI is an experiment that by design can never conclusively fail. Absence of evidence may mean only that we haven't looked hard enough, deep enough, broadly enough—or at the right wavelength. It's like the situation with the Mars rock: So what if it doesn't actually have worms in it? That means nothing, or very little, about the overall biology of the universe, and doesn't even prove that there's not life on Mars. It merely shows that there are not any fossils in one particular potato-sized rock.

There are few scientific experiments with as loyal a following among lay people as the radio search for intelligence. Ordinary folks give money to the search; so does Steven Spielberg, among other forward-thinking zillionaires. SETI has the benefit of being the kind of science that everyone understands in an instant. You aim a listening device at the stars and maybe you pick up The Message. It's an experiment that not only satisfies the searching instinct of the public, it also seems morally correct, a sign that we've embraced our true position in the cosmos, our mediocrity. The actual job of being a SETI researcher isn't an easy one. For years and years the search will yield no publishable data. It is potentially on the verge of the greatest breakthrough in the history of science, but in the meantime it yields a lot of static. Worse, the SETI scientists suffer the indignity of being associated in the public mind with the UFO community. The SETI people have to remind everyone, continually, that they aren't looking for flying saucers or little green men. All they're doing is looking for some extremely powerful, communicative alien civilizations that are beaming signals our way from deep space.

In 1960, fifteen years before the Andromeda experiment in Puerto Rico, Frank Drake invented SETI in the remote mountains of West Virginia. You get to the site by driving on narrow, winding roads through a part of the world that would need to add some human population to rise to the level of being merely rural. The mountains corrugate the land, the ridgelines are perfectly horizontal, the landscape a jumble of ancient rocks battered by eons of erosion. In an accelerated world, where everything's disposable, there's something reassuring about mountains that have been hanging in there for two hundred million years.

At the end of the road is the town of Green Bank, and there you find the humble campus of the National Radio Astronomy Observatory. There is usually a herd of deer wandering the grounds. The observatory is here precisely because it is in the boondocks. Green Bank is what is known as "radio-quiet"—a place so remote even radio waves don't go there. The government built this monastic place in the 1950s, and Drake was among the first to check in. He read a paper by Philip Morrison and Giuseppe Cocconi in 1959 arguing that ETs could be detected, potentially, with radio astronomy. A few months later, he tried it, using the eighty-five-foot telescope, the biggest on the site at that time. He called the search Project Ozma, and looked at two stars, Tau Ceti and Epsilon Eridani. He had a brief false alarm with the latter star—he thought he had a signal, but it persisted when he moved the telescope off-target. Other radio-transmitting objects, like satellites and planes, would over the years cause many heart-stopping moments in the SETI business. Drake, to his eternal credit, never tried to argue that he'd found something. Other researchers would talk of amazing near-discoveries, but not Drake. He and his associates kept champagne in the fridge at Green Bank but never popped it open prematurely.

"In a way we're like gold prospectors. We know that over the next hill is El Dorado. Except that it's much bigger than anybody's El Dorado's ever been," Drake said over lunch in the observatory cafeteria.

What I didn't know when I visited Green Bank was that just the night before there'd been an epochal false alarm. Jill Tarter, who ran a SETI program called Project Phoenix—and who was to some extent Sagan's model for Jodie Foster's character in *Contact*—told Ann Druyan by phone that she'd found the best candidate signal she'd ever seen. It turned out to be a satellite. By the time I arrived a few hours later, the astronomers were back into their long-distance-runner mode.

The search takes place inside the 140-foot telescope. This is a monstrous contraption, twenty-five hundred tons, designed in the days when no one was sparing with the steel. The base of the telescope is a building with walls three feet thick. The dish slowly whirs back and forth on a massive sprocket. With this dish, the searchers can pick up the signal from Pioneer 10, the spacecraft with Sagan's plaque on it, six billion miles away and receding. Pioneer 10 has a five-watt radio transmitter, with the same power as a bike lamp.

One night I spent a few hours with Seth Shostak while he stood guard over the computer monitors, waiting for the signal. Also in the base of the telescope was a guy named Nathan, who wore a West Virginia fishing cap and spent most of his time, he freely admitted, loitering in various Internet chat rooms. ("There were five cyberbabes in that chat room," he said when forced to log off.) Nathan ran the telescope—he knew how to move it to and fro. "I'm like a taxi driver for astronomers," Nathan said.

Shostak's job was to watch for candidate signals and decide whether they were worth pursuing. Shostak, like Drake and Tarter, is on the staff of the SETI Institute in Mountain View, California, and had flown in for a two-week run on the 140-foot dish. (He had with him a manuscript of his book *Sharing the Universe*.) Though his hair is turning gray, he retains a boyishness that is rather common among scientists. He giggles. Not for him to be dour and skeptical about the universe. I'd only been with him a few minutes when an alarm went off—a loud beep. It was a signal!

"ET on the line," Shostak said.

But then he had a computer glitch. He'd hit the wrong button. The signal was gone, and he hadn't had a chance to analyze it. He began scrambling. This was terrible! What if that had been the signal— The Message—and he'd blown it by hitting the wrong button? What a goof! By the time he got his computer working again, the signal was gone. Missed it.

"Our ET went away. I bet you next month's paycheck it was a satellite."

The computer is calibrated in such a way as to sound an alarm no more than every couple of hours. The system is analyzing twenty-eight million frequencies, and theoretically could be designed to sound the alarm anytime there's anything even slightly suspicious. But if that happened, the SETI scientists would go crazy dealing with false alarms. About one alarm every two hours, Shostak said, is the limit of the human nervous system. Some false alarms don't get resolved immediately, and Shostak knows what it is like to try to sleep while still coping with a very plausible candidate signal from the Galactic Brotherhood.

"I'm sure they're out there. Of that I have no doubt. I didn't take this job for the money," he said. Was it possible that we could get a hostile message from space? "I think that would be very disquieting," Shostak said. "Presumably that would mean they know we're here. It would also mean they're not far away. I think we'd be toast."

Why would they hurt us? Shostak said they might simply not care much about us. "They're not going to be at our level. So I don't think they're going to give us a lot of consideration."

I was getting the creeps a little. We were in this hulking, old, thick-walled contraption in a remote valley in West Virginia, and a storm was blowing up, and the aliens might not even be nice! We could wind up in a drawer somewhere with pins through our torsos.

Shostak said he believes that extraterrestrial civilizations (the SETI people don't use the word "aliens") are abundant. He said, "I think there's tens of millions, myself. Oh yeah." He meant there are that many in our galaxy alone. In the entire universe there would be, by Shostak's estimate, quadrillions of civilizations. His reasoning is that there are twenty billion Sun-like stars in our galaxy. (Most stars are of a different class than our Sun, and aren't as likely to be orbited by habitable planets.) He figures half of the Sun-like stars have planets. That's ten billion solar systems. How many solar systems have a planet with life? Shostak said that three planets in our own solar system could have life (meaning Earth, Mars, and Jupiter's moon Europa). Again, he skews to the optimistic. All solar systems, he says, probably have life *somewhere*. If only one out of one thousand of those ten billion biospheres manages to develop intelligent life, that's ten million planets with civilizations.

Then he delivers the most important point of all:

"There have to be ten million for us to have a chance at a hit."

This is the deal-breaker. This is the problem with Project Phoenix and other narrowly targeted searches. They will work only if there are aliens all over the place. Phoenix looks at only a thousand nearby Sun-like stars. If Sagan was right, and there are one million civilizations in the galaxy at any given moment, that comes out to one civilization per twenty thousand Sun-like stars. But Phoenix is searching only a thousand Sun-like stars, so the odds are against finding a single civilization, if the Sagan numbers are correct. If Drake's less optimistic numbers are right, then Phoenix would need an incredibly lucky break. Drake didn't think Phoenix would be successful; the numbers were against it. This brought to mind the Viking biology experiments. Even if all the equipment operates perfectly, the most likely result will be the "failure" to find life.

There are so many stars—where do you look? The needle in the haystack is a fairly exact analogy (and no one knows what "the needle" will look like in this case). The aliens may be hidden among an almost infinite number of lifeless places. A careful search takes forever. We throw around these numbers, like "twenty billion Sun-like stars," without pausing to grasp how many that really is. Most stars don't even have names. Who could possibly have named them all? Most galaxies haven't even been looked at. Entire galaxies are waiting for someone to get around to taking a gander at them.

Most of the stars we see with the naked eye are extremely hot, bright supergiants that are several times the size of the Sun and, more important, much younger. The supergiants reach their demise after only a matter of some millions of years, not billions. That's too short a period, probably, for the evolution of intelligent life. Everything we know about life on Earth tells us that it requires a tremendous span of time to evolve into anything like a thinking organism. We don't know for sure, but life probably needs billions of years, not millions, to reach the level of worms and plankton, never mind intelligence. The next time you look at a star-filled night sky and wonder who might be out there, slap yourself upside the head and remember that *those* stars, for the most part, are not friendly to life. Those are just the showy stars, the flamboyant stars, the most extravagant examples of nuclear physics.

The Drake Equation is not universally popular. Joshua Lederberg, one of Sagan's mentors in the 1950s and '60s, chastised him for inserting wild guesses into the equation and acting as though they were reasonable estimates. Lederberg felt the Drake Equation didn't deserve its popularity. "Hocus-pocus," he called it. The first factor in the equation, the number of stars, is the only one that is well known. The number of planets has started to become less of a mystery, thanks to the labor of Marcy and Butler and the other planet-hunters. But after that the equation disappears in the mist. For most of the crucial factors we have only a single data point. Does life always spark up on warm, wet planets? Or is that an exceedingly unlikely event? (Like everything else in this matter, you could argue it round or square.) Does life typically evolve into technological intelligence, or is that a rare evolutionary digression? For the final factor, L, the lifetime of technological civilizations, there isn't even one data point, since no one knows what human destiny might be: We could last for another billion years or wipe ourselves out in a decade. Even proponents of the equation, like Jill Tarter, admit that it's just a way to "organize our ignorance."

In theory, SETI is neutral on the issue of the abundance of intelligent civilizations. What SETI traditionally has not been neutral about is the likelihood that aliens wish to communicate in the radio portion of the spectrum. The word "radio" is just a label for a certain region of the spectrum, in contrast with "X-rays" and "gamma rays" and "light." Fundamentally it's all the same stuff, photons (which, to make matters more confusing, have properties of both waves and particles). Inanimate objects like the Sun produce radio waves in addition to heat and light and every other kind of electromagnetism. Radio has an advantage as a medium for interstellar communication. Its long wavelength allows the photons to pass around interstellar dust particles and even penetrate a planet's atmosphere. We couldn't get a visible light signal from many places in the galaxy because there are dust clouds in the way. We can't get gamma rays or X-rays because they don't make it through the atmosphere to our detectors on the surface. But we can pick up radio from all over.

The radio search, however, has a challenger. It is called "optical SETI." The idea, first proposed in 1961 by Charles Townes, the inventor of the laser, is to search for signals in the visible or infrared portion of the spectrum. A laser beam can contain a tremendous amount of encoded information, and can be more focused than a radio signal, which is rather sloppy. A radio signal is a bit like shouting out the back door for a kid to come home from the neighbor's house, rather than making a direct phone call to the neighbor. But the technology for optical searches was primitive in the early 1990s. Drake and Tarter remained radio chauvinists. The optical strategy hovered on the fringe of a research effort that already didn't seem exactly centered in the mainstream of science.

But then the laser people started ramping up their technology, making more-powerful lasers, and by the late 1990s it was clear that an advanced civilization might find light the way to go. There was a scent of conversion in the air. Even the SETI Institute and the Planetary Society, which had sunk large sums into the radio strategy, gave light a whirl, funding several optical SETI programs. In a sense it was all the same idea: Ransack the electromagnetic spectrum for some scrap of information about intelligent civilizations in outer space.

The debate over the search strategy carried with it all sorts of philosophical issues, such as: Why would the aliens want to be heard? Were we likely to pick up a greeting, or would we detect merely some navigational beacon, the equivalent of a lighthouse on a rocky shore?

"It could be something as trivial as purchase orders," Drake said.

The breakthrough, in other words, may come by eavesdropping on a conversation. To assume that extraterrestrials would broadcast a declaration of themselves would assign to these creatures a motivation that we ourselves lack. Human civilization doesn't announce itself to the heavens. The one exception took place at Arecibo, in 1974, when Drake sent a message toward a globular cluster in the constellation Hercules, about twenty-six thousand light-years away. A British astronomer, Gilbert Ryle, immediately sounded the alarm, saying Drake had imperiled the planet by effectively screaming in a dark forest that might easily abound with hostile organisms. We've been silent ever since, except for the "leakage" from TV and radio stations and military radar instruments. (It would be hard for us to detect leakage from another solar system. We'd have to build a truly enormous radio dish, one that dwarfed even Arecibo.)

Shostak is open to the idea that we've already seen the artifacts of intelligence. He is on the lookout for what is known as "astroengineering." This would be something like a Dyson Sphere—an artificial shell constructed around a star. It is possible, he said, that we have already seen some objects and events in space that are associated with intelligent civilizations but have chosen to interpret them in conventional astrophysical terms. For example, there's a galaxy, M82, that seems to be in the process of exploding. He is suspicious about this galactic event: "It could be that something went wrong with a galactic civilization."

I asked Drake how real the aliens are to him—whether they are just an abstract concept, a statistical probability that requires further research, or are something more tangible, like distant lost cousins, fellow searchers, colleagues in the quest to decode the cosmic riddle. He said they've become more real to him over time. "You talk about something enough times, you begin to believe it. And we sure talk about this a lot."

Drake thinks aliens will stand upright, which would free their limbs for the manipulation of tools. The head will be on top, which is good for hunting and gathering. The eyes and ears will be in the head, so that there can be short nerve connections between the sensory organs and the brain, allowing quick reaction to visual and auditory stimulation.

"You end up with a design," he said, "that is basically us."

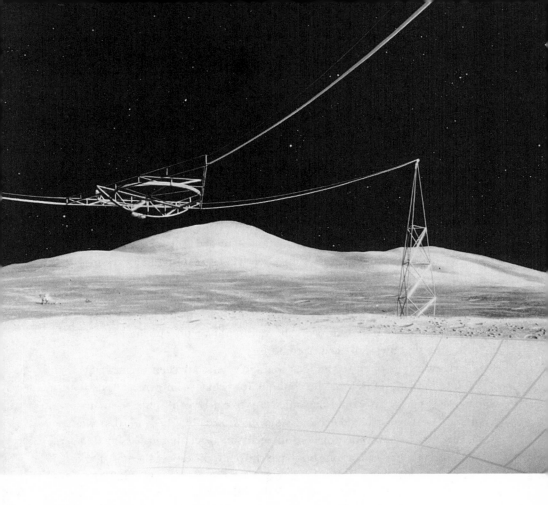

Drake's theory could be viewed as chauvinistic, but, again, any position on the matter is still pure speculation. A case can be made that the forms we see on Earth are not random or accidental, that there's a powerful logic to our design. Animals that hunt have eyes close together; animals that are hunted have eyes on the sides of their heads. Predators tend to be smarter than their prey. At least this is usually true—as soon as you decide it's an absolute rule of nature you remember those dimwitted sharks.

At midnight, Jill Tarter took over from Shostak. She looked rested, fresh from a shower. Facing a long night in the base of this giant old telescope, she remained, as is her nature, optimistic and cheerful. She had lived for fifty-three years, and a great many of them had been dedicated almost exclusively to the long-shot search for ETI. Only

SETI scientists would like to build a huge radio telescope on the far side of the moon, since even rural West Virginia is too noisy these days. NASA

when the movie *Contact* came out did she start to get a piece of the fame that had come the way of Sagan and, to a lesser extent, Drake. Sagan dabbled in SETI, Drake made it a serious line of research, but Tarter gave herself to it completely. It was she, not Drake, who ran the SETI Institute's major search program, Project Phoenix. And she was angry. She felt the sting of SETI's poor treatment by pessimists and congressional opponents. Congress had insulted her career, had spit on the passion of her life, when it suddenly cut funding in 1993 for NASA's SETI program.

But Tarter had discovered that direct government support wasn't necessary. For one thing, NASA could still support other SETI Institute research that did not overtly involve the hunt for aliens. She also lined up some sympathetic tycoons, like David Packard, to contribute money to Project Phoenix and keep the search going. She had a dream of building an even bigger instrument, perhaps an array of radio telescopes a full kilometer in diameter, an ensemble that could pick up the merest leakage from distant planets. And maybe someday we'd build a radio telescope on the far side of the moon, away from all that terrestrial noise. Until then, she was doing her best with this aging contraption in Green Bank.

We had some more false alarms, and then the fire alarm itself went off. This was turning into a jangling night. A massive thunderhead rolled over the mountain and into the valley. Lightning crackled every few seconds.

"I really do have the best job in the world," she said somewhere around one-thirty in the morning.

It seemed like a difficult and frustrating life, to be waiting for the aliens to call. The telescope was cold and exceedingly remote. Perhaps I had grown spoiled, too comfortable a searcher, too unnerved by the distance between me and the nearest Starbucks. But something about SETI didn't make sense. Perhaps the reason we have not made contact with aliens—and will not in the near future—is that the whole process is anthropocentric. We presume that there are species of living organisms out there that possess the same craving for contact, that they are garrulous and extroverted like us. Years pass, experiments are conducted, and still we do not see them. The Copernican revelation looms anew: The universe is not about us. It is not even about the things we are interested in.

Tarter told me that she realizes she may spend the rest of her life waiting for a signal that doesn't arrive, but she is sure it will come eventually, for someone. If nothing else, she can tell herself, "We were able to set in place something that would eventually make that discovery happen."

It rained some more. I went outside and stood under the telescope, which was like a giant mushroom, dripping from the rim. The rainwater splashed a few feet away. It was an astonishingly dark part of the world on a dark and rainy night. I kept a wary eye out for bears. I wasn't sure bears lived around there, but I knew I didn't want my last thought on Earth to be that instead of discovering aliens I had been eaten by a bear.

I went back inside and resumed the watch. The huge dish whirred slowly back and forth, an immensity of heroically engineered steel, targeting stars that are unimaginably distant. The computer monitors showed little dots. Each dot represented a radio frequency. Some dots were brighter than others. There were a lot of bits of information coming in.

It was all just static.

29

The Mystery Constraint

The signal could arrive at any moment. Geoff Marcy, for example, who found all those extrasolar planets, is re-examining his data for some underlying pattern suggestive of intelligence, something overlooked in the search for gravitational disturbances. And there will be other strategies, other nifty ways to parse the information from the night sky. There is a program called SETI @ Home in which participants can use screen savers on their computers to help process data from SETI programs. Some bored kid, taking a break from a gothic death-cult game, could tap into the Galactic MTV.

In all likelihood, the aliens will continue to be elusive. They tend to be farther away, and quieter, than we think they should be. They weren't on the moon, they weren't on Mars, and they don't seem to be sending radio signals at any obvious wavelengths from nearby solar systems. They seem to lack fingerprints entirely. There is something limiting them, a constraint on the rise of intelligent beings, or at least on the kind that send signals and build starships and zip around to visit interesting planets.

We can refer to it as the Mystery Constraint. The Mystery Constraint could be geological, chemical, biological, physical. It could be something psychological. It could be connected to a planetary or stellar phenomenon that we know nothing about. It could be some-thing characteristic of galaxies, limiting their habitability. It could be almost anything from the territory that Henry Bauer called the *un-known* unknown.

For a long time astronomers wondered if the Mystery Con-straint might be something over on the left side of the Drake Equa-tion, in the fraction of stars with planets. For hundreds of years we hadn't found any planets out there, and the pessimists believed that planets might be anomalies. One theory of planet formation required that two stars pass near one another, causing a severe gravitational dis-turbance. Because stars are, in general, so ridiculously far apart, such encounters would be extremely rare, and thus planets would be sparse. That theory has not survived the past few decades. More and more planets are being discovered, and Marcy and Butler have made yet an-other breakthrough, finding three planets—an actual *solar system*—or-biting a Sun-like star named Upsilon Andromedae. So the Mystery Constraint probably isn't the lack of turf.

The Mystery Constraint could be carbon. In late 1998, Mario Livio, a theorist at the Space Telescope Science Institute, published a paper saying that carbon did not become plentiful in the universe un-til the universe reached about one-half its present age. Carbon pro-duction, he said, didn't peak until just a couple of billion years before our own Sun was born. Carbon is the basis of complex life as we know it. If it takes billions of years for intelligent life to emerge, then intel-ligence should be a relatively recent phenomenon everywhere, not just on the Earth. Moreover, there are interactions between a planet's atmosphere and the physics of the parent star that take place over long periods of time and are relevant to the appearance of land creatures. Only when the atmosphere developed ozone, for example, was it safe for animals to emerge from the sea. This is an argument fraught with far more uncertainties than facts, but it suggests where the Fermi Paradox may be headed in the years to come—into examinations of when the universe became habitable and to what extent it remained so even as strange and violent astrophysical phenomena took place, not the least being the potentially lethal gamma-ray bursts that Henry Harris has been so worried about.

Perhaps the moon, our familiar companion object, is something unusual, and a key player in the rise of intelligent life on Earth. The moon has stabilized the Earth's rotation and kept the planet at roughly the same axial tilt, which ensures that seasons are predictable. The tides potentially played a role in the origin of life, as water splashed into tidal pools, which then evaporated, possibly creating the concentrated "soup" of molecules that eventually—somehow—became organized and self-replicating. We look at the solar system and see that such useful companion objects are rare. No other planet has a moon like ours. Mercury and Venus have no moons, and Mars just has a couple of dinky rocks. Jupiter and the other gas giants dwarf their moons. The closest thing to an Earth-moon system is that strange quasi-planet Pluto and its satellite Charon. "Big, close moons could be incredibly rare," says Ben Zuckerman.

The Mystery Constraint could be something cultural. The aliens might have a culture that does not lend itself to discovery. They may have a bad attitude. They could live in caves—the Bat-Men of Betelgeuse. They could be Mole People. They could be like the subterranean Morlocks of H. G. Wells's *The Time Machine*. For that matter, would we ever find a marine civilization? What if the aliens live around hot vents beneath a distant sea?

Civilizations may, over time, become increasingly silent, at least from the distant perspective of a radio astronomer. Frank Drake has watched the world change around him since Project Ozma, and he's noticed something that makes him slightly more pessimistic about detecting aliens. Earth is getting wired. No longer is it common to see a giant aerial antenna on top of a house. When Drake initiated Project Ozma in 1960, it was something of a national pastime to adjust the aerial in order to get the snow to disappear from *The Ed Sullivan Show*. Now we have cable. We also see that optical fiber is the best way to transmit a phone call. Maybe, says Drake, an intelligent civilization goes through only a brief period of detectability.

Perhaps the Mystery Constraint is a kind of Darwinian law that says that the civilizations that survive and prosper are the ones that don't call attention to themselves. Maybe there is a cosmic natural selection that favors these circumspect species. If you do blast signals into deep space, you run the risk that you'll be visited by ravenous space predators with mandibles the size of Volkswagens.

The Space Age dream is built on the notion that it is inevitable that human beings will expand across the cosmos. That is our nature—to expand, conquer, replicate, penetrate, smother, absorb.

That's what life's all about. Unfortunately, it may not be what life is like elsewhere. This assumption that colonization is our best bet for survival may not be a "cosmic truth." With SETI, the best bet is to find long-lasting civilizations, to find the Old Ones, who have been around, say, for a hundred million years (because civilizations that last only a few thousand years are very likely to have vanished long before we built our radio telescopes). The secret to survival on multimillion-year time scales may be *sustainability*, not expansion. The Old Ones may be members of the Green Party. They may be into Deep Ecology—a bunch of hardcore tree-huggers. Or they may be extremely kicked-back, nonexpansive, barely able to cross the room to get a beer. (They may be stoned!) It's true that societies that have ceased to explore and expand have often gone into decline, but those declines are measured in years and decades, which is the scale of human history, not cosmic history. You could argue that the expansiveness and consumerism of Western civilization (which for the most part has been the leader in space travel) are a long-term threat to survival, not a requirement of it. The industrialized nations burn through finite resources as though they couldn't care less about what the world will be like in a hundred years. The "developed" world may be a good model for the kind of civilization that's ephemeral in cosmic time scales, that simply doesn't stick around long enough to make contact with other civilizations.

There are those who believe the Mystery Constraint is nothing other than God. We can entertain the idea that a Creator has decided that there should be but a single species in His image. This is a notion that has not weathered well even in some religious circles in the centuries since the use of telescopes has revealed the incomprehensible dimensions of the universe. Sagan's characters in *Contact* say that if we're alone that means there's an awful lot of empty space—a slim but elegant argument. In any case, it's not an explanation that lends itself to research. God isn't subject to scrutiny by human machines. God can be cited, praised, worshiped, but not so easily investigated, interrogated, or corroborated.

There is ultimately only one place to go when you want to study life in the universe and get some handle on how it comes into existence, and how it evolves. You have to go home, to Earth, to terrestrial biology. And here is where the big surprise comes. It turns out that we do not really understand what life is, or why it is here. We don't understand ourselves. *We* are the mystery! You don't have to look into the night sky to find an enigma. Life on Earth is a phenomenon begging for an explanation. We don't know very much about its origin or how it evolved. Over the course of doodling around in alien country, I began to realize that the crucial questions were not necessarily astrophysical ones but, rather, biological—and terrestrial. The truth is not "out there," in other words, but somewhere here, within the cells of our bodies, within the biochemistry of living things. It's possible that the Mystery Constraint is somewhere in the long path from nonliving matter to intelligent being—that somewhere along the line we made a leap of cosmic significance. Intelligent creatures might not be unique, but they might be extraordinarily rare, extraordinarily unusual.

It's worth checking out: *We* might be the anomaly we've been searching for all this time.

In Sagan's novel *Contact* there's a moment when Ellie Arroway is face to face with an extraterrestrial entity. This presented a tricky problem for the astronomer-novelist: Any description of the alien's morphology could be ridiculed. If the alien looked too humanoid—with bilateral symmetry, arms, legs, and a dominant processor encased in what is clearly the alien's "head"—Sagan would be guilty of the same narrow-mindedness and disrespect for cosmic diversity, the same untamed anthropomorphism, seen in the typical Hollywood movie. To come to terms with our true station in the universe, our total insignificance, our complete and utter mediocrity, our almost infinitesimal relevance, it is important to avoid projecting our biological assumptions onto others. But then what could Sagan do with his alien? If his alien was just some kind of blob, some green viscous slime, or if it had fourteen heads or ninety tentacles, or if it looked like a giant moth with a cockroach head and a prehensile tongue—or whatever—it might come off as silly, or distractingly weird, and would be a more difficult role to cast when the novel became a movie.

Sagan had a solution. It was scientifically defensible and, more important, it fit the larger theme of the book. Sagan wanted his aliens to represent something grand and even

30

The Rules of Life

reassuring about the cosmos, to suggest that we have a deep connection, as humans, with all the intelligent life forms scattered among the stars. And so when Arroway meets the alien, he (it?) looks exactly like her father. The alien is a perfect simulacrum of her dad, who died of a heart attack when she was a child. This alien has somehow read Arroway's mind, figured out that her father is the most important influence in her life, and taken on that image. (How did he manage this trick? He used that fabulous, mysterious, inexplicable alien technology.) A wonderful aspect of this alien is that a human could play the role without any bizarre makeup. The only bad thing you could possibly say about this solution was that it may have merely reinforced the idea that, even as we say we're open to cosmic diversity, we're really just craving contact with ourselves. We don't want to find aliens, we want to find Dad.

Arroway can ask this alien anything that comes to mind. After traveling through wormholes across most of the galaxy, a bone-rattling trip that has left her a tad dazed, she comes up with:

"I want to know what you think of us, what you really think."

The savvy reader would be excused at this point for screaming, "Wrong question!" What kind of self-absorbed, narcissistic human being would travel across the universe to get some verbal strokes? This reflects the first rule of alien lore (which Sagan himself had noted in his nonfiction): The main job of any alien is to comment upon, lecture, warn, study, and otherwise obsess over the human race. Aliens are *crazy* about us.

Arroway does manage a few other queries, wondering, for example, about the wormhole transportation system and the general history of the aliens—amazingly, her alien contact doesn't seem to have many answers—but there is still one important thing that she forgets to ask. She fails to ask what the aliens are made of. She doesn't ask the alien about "his" biology—or even if he's a he or a she. Do the aliens reproduce sexually? Are they predominantly asexual, like the inhabitants of the planet Winter in Ursula Le Guin's *The Left Hand of Darkness*? Is their biochemistry based on DNA? If not, what is the mechanism for storing and replicating the instructions for creating proteins—or do they not use proteins, either? Are they even truly matter-based, or have they evolved to some kind of energy state?

Arroway's catastrophic dropping of the ball here is not really a flaw in the book. Sagan understood the major questions about the nature of life as well as anyone alive, but he surely didn't want to gum up his narrative with a lot of boring biology. He knew that people don't think about biology except when they're losing their hair, gaining too much weight, or suffering from severe constipation. No one pauses to reflect upon the incredible fact that we are able to experience our world subjectively through an elaborate biochemical system over which we have little control. No one asked to be human. No one designed his or her DNA, RNA, proteins, cells, tissues, brain. We just glide along, oblivious of the machinery that makes our existence so vivid an experience.

Occasionally we try to connect with some larger truth about our "selves," but if anything we tend to ignore the fact that the self is a biological entity, preferring to focus on the self as a spiritual being, or at least a personality, an experiencer of the world. No one thinks of the "self" as an agglomeration of a few trillion cells. When we get in touch with ourselves, it is our transcendent spirit we are seeking. We go to church, pray, meditate. We write poetry. We read the classics. We stare at sunsets. But what we all get from these moments is a vague sense of wonder and awe about the world itself, not a keen understanding of the true complexity of our own personal biology. Who could possibly grasp such a thing and still put one foot in front of the other?

This general ignorance of biology—this ability to live our lives while paying no heed to the biochemistry and neurobiology that make it possible—is at the root, I think, of our tendency to overestimate how many aliens are buzzing around the galaxy. When we take ourselves for granted, we make it that much easier for life on other worlds to reach our own level of technological intelligence. We are so generous—we assign the aliens all the gifts of biological complexity that we enjoy ourselves.

It is noble to aspire to improve. It is right and proper that we not assume we are the zenith of biological complexity in nature writ large. But neither should we assume that there is something modest and inevitable about the journey of life on Earth from simplicity—from prebiotic molecules, most likely—to an intelligence capable of guiding a tiny rover to check out rocks on Mars.

When we imagine aliens, we focus on their anatomy and their

hardware. They have antennae on their heads but for some reason they don't need them on their flying saucers. In the better brands of science fiction, aliens are allowed to have a history, a back-story, often involving their own experience with war, conflict, hubris, and other terrestrial problems that we discover are universal. But rarely do aliens have an evolutionary history. This is the sort of technical detail that's too boring and inconsequential to belabor, yet it's also the crucial question of life in the universe. Does it take four billion years, roughly speaking, for life to evolve into intelligence? (Not to mention lots of liquid water, an atmosphere, a dynamic geological environment, and so on?) What ingredients are absolutely essential? We're stuck, as always, with a single data point: life on Earth. All we know for sure is that a human being is the distillate of four billion years of genetic change. We embody the magnificent patience of chemistry, its ability to plod ahead, atom by atom, molecule by molecule, and gradually bumble its way toward something remarkable. *And we take it all for granted.*

In Sagan's case, he wasn't ignorant about biology, he simply was optimistic. He felt that life was likely to arise on many worlds, and that it would tend to evolve into intelligence. Unfortunately, that supposition remains nothing more than a guess. No one knows what the cosmic rules of biology might be. All we can do for the moment is look at ourselves, our dogs, our cats, the ants in the yard, the flowers, the microbes in the petri dish, the tube worms around the hot vents at the bottom of the sea—and extrapolate like mad.

Aristotle, who banged out theories on just about everything, was in some ways the first biologist. He studied the development of embryos, examined the roe of fish, and mulled the design of an octopus tentacle. Still, he had scant observational data to support his theories, and he could not shake his conceits and prejudices. He wrote things like:

"We should look upon the female state as being as it were a deformity."

And:

"A deep voice seems to be the mark of a nobler nature."

And:

"It is clear both that semen possesses Soul, and that it is Soul, potentially."

In other words, he was just winging it, tossing out ideas while getting drunk with the boys.

Over the centuries, knowledge slowly percolated through the hard sediments of superstition and misapprehension. One idea, slow to vanish, was that life arose through spontaneous generation, that species simply appeared—as, for example, when maggots emerge without provocation in manure and garbage. In 1668, an Italian physician named Francesco Redi finally showed that flies lay the eggs from which maggots hatch. Nothing spontaneous about it.

In 1859 Charles Darwin provided the world with a compelling and rather unnerving answer to why we exist. The depth of his argument in *The Origin of Species* made the theory of evolution impossible to wave away with the magic wand of religious doctrine. Evolution today remains highly inflammatory among large segments of the public, but among scientists it is no more controversial than the heliocentric solar system. One could argue (as has, for example, the journalist John Horgan) that there will never again be a discovery about life on Earth as revolutionary, as paradigm-busting, as Darwin's.

In the nineteenth century it also became increasingly clear that life was not the result of a "life force." It cannot be reduced to a simple type of energy or vital substance. The principle of "vitalism" did not survive the coming of modern biology. Long ago, it seemed plausible that there was some kind of juice, some current, some crackling energy, that lights up cells and gets them buzzing and humming. But the life force just isn't there, according to modern science. This is a huge fork in the road for science and spirituality. New Agers and Eastern mystics and acupuncturists may tap into various biological-energy fields, but a biologist sees only cells, tissues, fluids, chemical interactions. A biologist sees structures. Structures again! It's the universe, the Eagle Nebula, the galaxies and galactic superclusters, reconfigured in the human body.

The secret of everything is geometry. The structures have properties. A protein molecule performs its function not because of any intent or "power," but because of how it is folded in three dimensions, and how, in that shape, it fits into various receptors in the body, like a key in a lock. If enough of these properties are in place, and are dynamic in certain ways, we can venture that we're dealing with something that meets our definition of being alive. Life, the exobiologist

David Des Marais told me, is "an emergent characteristic that is fundamentally structural in nature." By "emergent" he means that life has no singular cause, like a life force, but is, rather, an overall trait of the system. A good analogy is that, when you get a bunch of water molecules together, you have something that has the property of being wet. No single molecule is even damp. Nor is a single molecule alive, at least no molecule we have yet discovered. It takes teamwork to get this thing called life to operate. When Des Marais says life is "fundamentally structural in nature," he means that the carbon and hydrogen and oxygen and other ingredients of life are coordinated in an organized system. They're not randomly configured. They are orderly.

A living thing is made up of the same stuff as a nonliving thing—it just tends to be a lot more complicated, a lot more organized, and in that incredible organization it takes on certain characteristics. Life is sort of like an overall attitude. A living thing has an inside and an outside, it uses energy, it emits waste, it replicates, and it evolves according to Darwinian natural selection. The definition of life, used by many exobiologists, states: "Life is a self-sustaining chemical system capable of undergoing Darwinian evolution."

Biologists have trouble even imagining a nonevolving life form. Quartz, for example, crystallizes in a fixed pattern in such a way that one could say that one crystal is making a copy of itself, and "reproducing." But all those exact copies result only in a rock. Nothing about that rock seems alive. By having the power to mutate, a life form can adapt to different environments, become more complex, and become more interesting than rock or mud. It can get organized. It can rise to the level of . . . well . . . scum. And then go from there.

Here's what we see on Earth: A living thing has a genetic code made of DNA. This is true of everything alive on this planet. There's no particularly obvious reason why there couldn't be some animal or microbe with a different kind of biochemistry, but no one has found any such creature. The biochemistry of the world is straight out of a Bill Gates fantasy—there's only one operating system for everything.

The DNA is a string of four nucleic acids. They are adenine, guanine, cytosine, and thymine. The DNA copies pieces of information onto a similar molecule called RNA, which is the "messenger" for that information. The RNA carries the information to something called a ribosome. The ribosome makes proteins. The proteins make the creature. The central dogma of molecular biology is that DNA makes RNA, which makes protein. This process is how you get from a "genotype" (a code made up of the nucleic acids) to the "phenotype" (a squishy animal with eight tentacles and a tendency to squirt black ink as an escape mechanism). The infinite elements of our culture—religion, music, art, sports, space probes, the Arecibo dish, the little cars that Shriners drive in parades—are what might be considered the "extended phenotype" of the human species.

A gene, generally speaking, is a segment of DNA, sometimes many hundreds of thousands of nucleic acids (sometimes called just "letters," in homage to A,G,T,C). A single gene typically provides the instructions for the production of a single protein. In a human being there are something like sixty thousand genes, perhaps as many as a hundred thousand—a number that will be nailed down by the Human Genome Project and its private competitors within a few years of this writing. A simple *E. coli* bacterium has twenty-five hundred genes. That translates into something like six million "bits" of data. This is the conundrum for origin-of-life researchers. Even if one imagines an Earth shot through with interesting molecules floating around a dense primordial broth, there's almost no chance that they will arrange themselves into anything like a bacterium. Obviously there have to be some intermediate steps—many of them—between the soup and the microbe. An interesting element of this is that it implies that Darwinian evolution not only is necessary for life, but has to play a role in prebiotic chemistry—that you need dirt that evolves, so to speak. You need a "molecular community" that can transmit information over time, and give rise to new, different molecular communities, even before you have anything that resembles a living thing. Evolution has to go back in time to a point *previous* to the beginning of life.

No one has made life in the lab. It's been a tremendous disappointment, and a surprise. The synthesis of life wasn't supposed to be this hard. In 1953, a graduate student named Stanley Miller, working in the lab of Harold Urey, showed that by sparking a mixture of gases he could produce two amino acids and some simple organic molecules that are among the building blocks of life. The Miller-Urey experiment was interesting chemistry, but chemistry is not quite the same thing as biology. Stanley Miller presumed at the time that it would be only a few years before the final leap was made. Everyone thought the same thing: Just tweak a few inputs here and there, and soon enough the test tubes would begin frothing over with newly created life. But years went by, and no one could ever seem to generate anything that could be defined as alive. The conclusion sank in: Life is not just some funky carbon.

"Looking at the single known biology on Earth, it is clear that this biology could not have simply sprung forth from the primordial soup. The biological system that is the basis for all known life is far too complicated to have arisen spontaneously," biologist Gerald Joyce has written. Chemist and author Robert Shapiro has written that the mixture of chemicals manufactured in a Miller-Urey type experiment "no more resembles a bacterium than a small pile of real and nonsense words, each written on an individual scrap of paper, resembles the complete works of Shakespeare."

This doesn't mean that life required magic, a divine touch, or some alien seeding. It simply means that the experiments we've done so far don't come close to matching the genius of nature. We just haven't figured it out. No one even knows how the first organism managed to form a protective layer so that it could have an "inside" distinct from the outside world. Another hairy question: How could the first organism have survived without the existence of a food chain? Why didn't it consume everything consumable and then starve to death?

Microbes could have arrived on meteorites from another planet. A few eminent scientists raise the possibility of what they call Directed Panspermia. The idea is that an intelligent civilization could have sent little packets of microbes, such as blue-green algae, to Earth about four billion years ago. This is, alas, a daffy theory. It has zero evidentiary basis, for starters. Worse, it doesn't make any sense on its face, since it presumes that aliens have a powerful desire to spurt microbes into deep space. Such an action certainly makes little scientific sense, because if the universe is already amenable to the emergence of life (as it appears to be) there's no compelling reason to prime the pump. Moreover, it might be destructive. We would never do it, for example, because our microbes might destroy any existing microbes on another world. That's why we sterilized the Viking probes before they went to Mars. One person's panspermia is another person's contamination. Panspermia also doesn't really answer the question of how life started but, rather, removes it to an alien world that we haven't discovered. It passes the buck in a big way. It declares that we can't figure out the origin of life until Henry Harris gets us to the stars (or the alien biologists land with some charts and a helpful rack of slides).

The crazy theories probably aren't necessary. At some point people have to be patient and wait for the answer to come, rather than demanding that nature deliver the secret of life. The biggest advantage that the Earth has over a scientist working in a lab is an abundance of time. Humans are insanely rushed, never more so than in today's accelerated, googly-eyed society, where we can't bear being delayed forty-five minutes on our flight to the other side of the continent. We want to invent life in the lab *right now*. A planet can wait for happy accidents and doesn't have to schedule them. Great discoveries tend to be serendipitous, and so, too, presumably, was the appearance of our first oozing ancestor.

An unresolved issue with ramifications for the ET debate is whether life must be "carbon-based." (Was Ellie Arroway's "dad" a carbon creature?) The expression refers to carbon's role as the structural foundation of organic molecules. "Organic" is often used as a way of saying that a molecule has a carbon superstructure. The carbon atom is amazingly useful in forming long chains of molecules. It has four electrons in its outer shell that it can share with other atoms, almost as though it had limbs extended with grasping hands. Not only can carbon form large molecular structures, but these structures are stable at many different temperatures, a useful feature in a universe that is alternately freezing and broiling. Silicon is sometimes suggested as an alternative to carbon in the foundation of life, because it, too, forms large molecular chains, but silicon in many forms isn't stable in water or at high temperatures. You might say that silicon is the poor man's carbon.

Sagan ultimately declared himself a carbon chauvinist. He also thought liquid water was probably essential to life. Water is the perfect medium for biochemistry; there's nothing else in the universe that compares, so far as we know. The molecule is an excellent solvent. It's made of hydrogen and oxygen, the two most common elements that are chemically active. (We know that hydrogen is the most abundant element in the cosmos, followed by helium, oxygen, carbon, neon, and nitrogen. Helium and neon are chemically inert. Therefore, life uses the other four elements, almost exclusively.) Life doesn't require exotic material—it's nailed together with the material that is lying around everywhere. As Michael Papagiannis has noted, the universe is "predisposed" to the emergence of life on suitable planets.

The problem with liquid water is that it so easily becomes something else: ice or vapor. On frozen or broiling worlds, biochemistry probably can't happen. Where else in the universe have we found liquid water? That's the strange part. We have found none anywhere. We have found ice. There was great excitement about the discovery of water on the moon, because it could be used to sustain a human colony someday, but the stuff is quite thoroughly frozen and is not likely to be inhabited by a single tadpole.

"Liquids," said David Des Marais, "are not a very common phenomenon in the universe."

And so the search for life in the universe is really at the moment the search for liquid water. This is why there's such excitement about Europa. This moon may have an ocean under all its ice, and even if there's nothing alive down in that frigid dark environment, at least we've made contact with something wet. There is also evidence that Callisto, another of Jupiter's large satellites (they're each roughly the size of the planet Mercury), could have a subsurface ocean.

All of this makes us appreciate anew our own little world. Earth had many potential spawning grounds for life, from tidal pools to the hot vents at the bottom of the sea. Earth is that rare place where water stays liquid and splashes around at the surface. It is not some lousy, desolate planet like Mars or Venus. We live on an amazing world. There may be other planets that are equally amazing—maybe billions of them spread across the universe—but they remain, as with so many other marvels, in the realm of the unknown.

To appreciate the distance between ufology, with its gallivanting aliens, and mainstream exobiology, which would settle for a drop of water, it's necessary to do some conference-hopping. There's a Gordon Research Conference every other year that attracts the extraterrestrial-life crowd, though the official title of the meeting is "Origin of Life," since that's a more credible topic than "Space Aliens." The meeting in Henniker, New Hampshire, in July 1997 drew the stars of the exobiology community, people like Chris McKay, Chris Chyba, and Mars rock naysayer Bill Schopf. For five days the scientists talked about Mars, Europa, and chemistry, lots of chemistry. It became readily apparent that the origin of life is a subject that has relatively little to do with life and much to do with the way molecules interact. At the beginning, there was nothing gooey and squirmy; there were just polymers that could perform an interesting trick of replication.

After a few days at an origin-of-life conference, a lay person begins to crave a discussion about something as beguiling, as prepossessing, as cuddly as a paramecium. (From the perspective of the origin of life, a paramecium is so highly evolved, so incredibly complex, as to be essentially no different from Katharine Hepburn.) The scientists

31

The Brain of an Alien

have a schedule: lectures in the mornings and evenings, goof off and canoe and send e-mails in the afternoons. Everyone in Henniker ate breakfast, lunch, and dinner in the college cafeteria, then drank late at night in a pub conveniently located about fifteen feet from the lecture hall.

These scientists were jazzed by their work. They'd had some good years. The Mars rock was a tremendous boost, at least in terms of funding (most of the scientists were skeptical of the conclusions of the Houston team). More generally, there had been a series of discoveries that had made life in the universe seem more plausible. For example, it had become clear that complex carbon molecules saturated the galaxy. Billions of star systems have the "vital dust" that could provide the foundation for life. And just in the last few years, life had been found in bizarre places on Earth, from hot vents at the bottom of the sea to the pores of rocks many miles below the surface. The scientists had picked the right field: Life was hot stuff. Life was proving itself to be more adaptable, penetrative, and promiscuous than anyone had realized. Another important discovery was the existence of microfossils of organisms that lived 3.5 billion years ago. Others had found evidence of past biology in rocks nearly 3.9 billion years old. It was almost precisely that long ago that the planet stopped getting bombarded by rocks from space. It appears that, as soon as Earth became potentially habitable, it became in fact inhabited. If life appeared quickly, then it might be argued that life requires no great miracle. This was the Sagan dream. Start with carbon, hydrogen, nitrogen, oxygen, phosphorus, and a few odds and ends, and add heat and water. Let simmer. Presto.

I found myself talking one day to a Tufts University professor named Eric Chaisson, who has been going around for a couple of decades talking about "cosmic evolution." He looks a bit like Gene Wilder, and has some of the same manic energy. He can quickly reel off the seven stages of cosmic evolution:

Particulate
Galactic
Stellar
Planetary
Chemical

Biological

Cultural

These seven stages stretch across three separate "eras." They are the Energy Era, the Matter Era, and the Life Era.

The Energy Era lasted for a few thousand years after the Big Bang. Energy was all there was—hot, dense, furious. Energy shaped everything in the universe. Then matter appeared, launching the Matter Era, which was marked by the formation of galaxies, stars, planets. Matter dominated the structure of nature. Gradually, the Matter Era is giving way to the Life Era. We're in the thick of the transition. As we go along it will be life, particularly intelligent life, that starts to alter, shape, and create the reality of the cosmos. Talk about empowerment! This is a kind of remedial anthropocentrism: We may not be in the center of the universe, literally—we may be, for the moment, grunting beasts on a rock orbiting an ordinary yellow star—but just watch what we do next. Stand back! We're going cosmic.

"Nature would not have built this table," Chaisson told me over lunch, slapping the picnic table on the deck outside the cafeteria. Then he pointed. "Would not have built that wall. Would not have built that automobile. We have become the agents of change. We are beginning, for better or worse, to manipulate matter more than matter manipulates us."

At the conference one night, everyone mingled around a new exhibit called A Walk Through Time, which graphically showed the evolution of life on Earth. The exhibit had a serious flaw: Life has been, for most of its history, extremely boring. It's sad but true. Life just sat around for about three billion years. Bacteria, and more bacteria. The illustrations for A Walk Through Time were spaced in hundred-million-year increments, but some of those hundred-million-year intervals were real yawners.

Here is the highlight reel for the planet Earth 2.7 billion years ago: "Special joint ventures occur in communities of mixed populations. A sluggish, ancient fermenting bacterium and a small, swimming, spirochete-like bacterium may have formed a particularly brilliant partnership. . . ." It goes on to describe how these sticky microbes manage to glom onto one another. Another hundred million years pass on by. Life is slow, viscous, gooey.

Finally, about 2.2 billion years ago—several of these hundred-million increments down the line from the sluggish, fermenting bacteria—came the huge innovation that may even today separate the Earth from your average habitable planet. At about that time, life invented the trick of oxic (or "aerobic") respiration. The early Earth didn't have much oxygen in the atmosphere. In fact, oxygen destroyed life; it was a poison. That meant that life on Earth couldn't use oxygen in its metabolism, an unfortunate situation given that oxygen is a chemically powerful molecule. But about two billion–plus years ago, oxygen levels soared, pumped out by photosynthetic organisms that could tolerate an oxygenated atmosphere. At roughly the same time, a new type of organism—larger, more complex—appeared on the scene. These "eukaryotes" exploited the oxygen. Oxic respiration was more efficient, by leaps and bounds, as a way of driving a metabolism. Life had suddenly thrown itself into a higher gear. "It took billions of years for the Earth to figure out how to make that happen," Des Marais said.

On another world we might find life that hadn't yet figured out oxic respiration, much less how to fix a jammed Xerox machine. Add "oxic respiration" to the constraints on the presence of intelligent aliens in the universe.

Another great biological leap occurred about 544 million years ago, the beginning of the Cambrian Period, when multicellular animals spread throughout the oceans. It's hard to know if the "Cambrian explosion" has echoes on other planets in the universe. Each of these biological leaps could follow a general pattern, a cosmic tendency, or they could all be anomalies. We don't really know if Darwinian evolution is the norm, or if there is some other mechanism for variation and survival on other worlds. Absent another example of life, we are left with dinner-table arguments, no more substantial than the vapor from the brandy or the smoke wafting from the cigars.

Stephen Jay Gould has argued for years that there's no "arrow" in evolution, that it's not necessarily progressive, that it has not been an inevitable march toward *Homo sapiens*. His position on aliens is sensible, if a bit unadventurous. He signed a SETI petition asking Congress to fund the radio search for aliens. His argument against the inevitability of humans is not against the possibility of intelligence. When I asked him if he thought there are aliens out there, he said simply, "No data." I rephrased the question and he said it again: "No data." Thus concluded my Gould interview.

The debate over the origin and evolution of life is one that relies on tiny scraps of evidence, and these scraps can mean different things to different investigators. Consider the organism called Ediacara. The Ediacarans swarmed the oceans of the planet for a brief and glorious period before the explosive emergence of marine animals. Ediacarans may well have been animals themselves, but some researchers claim they were something entirely different, a kingdom of their own. They were large, soft, squishy, without heads or mouths or guts. One researcher, Mark McMenamin of Mount Holyoke College, has argued that the fossils of these organisms provide evidence that there are intelligent beings throughout the universe. This is exobiology at its finest—a wondrous feat of extrapolation.

McMenamin's argument is that the Ediacarans were developing elaborate structures, by the standards of their day. They had momentum. This kind of momentum could have eventually—someday—somehow—given rise to the even more complex structure of the sentient brain. But that didn't happen for the Ediacarans, because their hopes and dreams were dashed with the arrival of the animals. The animals were also swimming creatures, and they also had developed complex structures. For example, animals had invented the nifty aperture we call the mouth. The mouth gave them a great advantage in their encounters with the Ediacarans. The Ediacarans went into the mouths of the animals and didn't come out. In other words, they invented the act of eating, only they were the losers in the deal. They never had a chance to become intelligent.

The animals, the creatures that ate the Ediacarans, eventually evolved into humans who began wondering whether there were aliens in space. What's important, according to McMenamin, is that, at the dawn of the Cambrian, complex anatomical structures emerged not once but twice, a sign that there is a natural tendency for life to get complicated. McMenamin told the magazine *New Scientist*, "My work changes everything in the debate. It shows vitalism in which the forces and structures of the Universe evoke life, which evokes complex life, which evokes intelligence."

All this from an indescribably ancient fossil. Those of us who are not paleobiologists are put in a difficult position. We don't know what an Ediacaran is, or how to interpret a fossil, and for that matter we don't know much about McMenamin as a researcher. So, although we may wish to ground our belief in ETI in scientific knowledge, the evidence is so thin we might as well rely on the water-cooler paradigm. In fact, the record of life on Earth can be manipulated to make the counterargument that we're alone. For this we turn to the woodpecker problem.

Jared Diamond, a distinguished science writer, came up with the woodpecker argument in the context of extraterrestrial intelligence. Woodpecking, wrote Diamond in 1990, is a great adaptation for obtaining insects, sap, and other helpful things underneath the bark of trees. It's also a way of building a fresh, clean nest almost anywhere. And yet woodpecking as an adaptation has occurred only once in the biological record, and all woodpeckers on Earth are closely related. Other useful traits have occurred many times (eyeballs, wings, etc.), but woodpecking has happened only once. There are places on the planet, specifically Australia and New Guinea, where there are no creatures at all that fill the gaping biological niche of woodpecking. Diamond says this is but one example of a perfectly logical adaptation that for some reason is quite rare. Where are all the woodpeckers? It's the Diamond Paradox. *Where are they?* Diamond extrapolates the situation to the cosmos as a whole, and says this is evidence that there might not be many radio telescopes being built in the universe. Radio telescopes are useful, a great adaptation to the environment, just like a beak that can peck wood. But that doesn't mean that a distant species will decide to build such a contraption. It might simply never get around to it. Sometimes things happen, sometimes they don't.

Diamond makes an even more provocative assertion: It is not clear that intelligence and dexterity, our treasured traits, are particularly useful for long-term survival.

> On the basis of the very recent evolutionary experience of our
> own species, we arrogantly assume intelligence and dexterity to

be the best way of taking over the world. In fact, few animals have bothered with much of either; none has acquired remotely as much of either as we have. Those that have acquired a little of one (smart dolphins, dextrous spiders) have acquired none of the other, and the only other species to acquire a little of both (chimpanzees) has been rather unsuccessful. Earth's really successful species have been dumb and clumsy rats and beetles, which found better routes to their current dominance.

The Diamond Paradox amounts to another constraint on our ever-growing list of constraints on communicative civilizations. You need a planet, liquid water, life, billions of years to waste while the bacteria decide to evolve into something, intelligence, technology, and finally you need a civilization that actually decides it's a good idea to build a radio transmitter and send us a signal (a factor the Drake Equation wisely considers). You start adding these things together and you can get a bit of a chill, an omen of present and future loneliness.

The most exotic creature in the field of extraterrestrial life is the genuine nonbeliever, the unapologetic naysayer, the person who feels compelled to shout to the heavens that we are alone after all. In fact, the only person I met who fit that description well was Ernst Mayr. He found the notion of aliens an affront to his scientific knowledge.

When I saw him, he was ninety-three years old. He lived in a retirement home near Boston, a widower, surrounded by books, many of which he had written. He was a living archive of twentieth-century biology. Mayr never became famous like Sagan, but his name will forever be part of the history of science. When he got into the game in the 1920s, the theory of evolution was still squishy on a lot of its details. No one knew how genes were passed along, no one knew that the Earth was more than four billion years old, and no one knew that life had existed prior to the Cambrian Period. Biology lagged far behind physics as a mature enterprise. In the 1920s, Mayr, a German, roamed the highlands and valleys of New Guinea, working on a dissertation on birds. In 1931, he came to America. He set to work on a massive and difficult project, the reconciliation of genetics with the Darwinian theory of natural selection. That we understand them today to be part of the same process is due to the work of Mayr and Theodosius Dobzhansky, known as the "evolutionary synthesis." Mayr went on to write twenty-one books. All that, however, has been overshadowed in recent years by his vocal opposition to SETI. Mayr says the radio search is a waste of time: "They're not out there. Life is almost a certainty, but it would be something like bacteria."

The German accent remains a heavy syrup on his words.

"How many species have existed on the Earth since the origin of life, which was three billion eight hundred and fifty million years ago? A good answer is one billion. How many of those species have acquired advanced intelligence? You know the answer," he said.

Mayr had become anchored on that single point. He even gave the odds against the appearance of intelligence: a billion to one. We're the one. Pump that into the Drake Equation and you could easily come up with a Milky Way galaxy in which we're the only sentient observers.

It wasn't the most persuasive of arguments. Sagan and Mayr had skirmished in the pages of the Planetary Society's magazine, and Sagan, skilled at verbal skirmishing, had made a compelling counterargument. Sagan said that if Earth is our single data point, then we could argue that intelligence arises in 100 percent of the planets with biospheres. There's one planet that we know of that is habitable; that planet has intelligent life. That's one for one! A perfect score.

This debate seems to be something of a semantic argument. They are using the same database. The dispute is once again over what we see happening in the evolutionary record. In Mayr's view, there have been all these species of animals that didn't evolve into intelligent creatures; in Sagan's view, they *did*, as a collective if not individually. It is not an argument that has a winner or a loser, but merely highlights two different ways of looking at Earthlife. Sagan's is the more generous approach—it pats those little microbes on the head and says, Great job, you took your time but eventually you got a big brain.

I asked Mayr about the dinosaurs. Is it possible, as the SETI optimists think, that dinosaurs were on the verge of evolving some intelligence? The idea had been floating around at science conventions, an assertion of nature's beneficence. Our universe is possibly so primed for intelligence that even a stupid lizard has a chance to become smart someday. (One imagines the descendants of the tyrannosaurs sitting around the dinner table, arguing about convergence versus divergence in evolution. Would they settle their disagreements in a civil manner, or would arguments end with someone getting his head literally bitten off?)

Mayr scoffed. Dinosaurs, he said, were almost certainly cold-blooded, contrary to some fashionable latter-day reports. There are no intelligent coldblooded creatures. Hot blood is necessary to operate a big, intelligent brain. "High intelligence is just absolutely a fluke of history." It took a meteor, he said. Two decades ago, Luis and Walter Alvarez presented their astonishing evidence that the dinosaurs had been wiped out by an asteroid or a comet sixty-five million years ago. "If that Alvarez asteroid hadn't hit the Earth at the end of the Cretaceous, the dinosaurs would still have flourished and the mammals wouldn't have done anything," Mayr said.

Another contingency . . . another constraint. You had to have all these perfect things, this metaphorical alignment of the planets, and then you might still need a huge stroke of luck, like this rock from space. Only through luck, chance, a freak asteroid impact, could intelligent humans appear on the Earth. That's the Mayr view. There is life in Mayr's universe, but it is not our kind of life. Perhaps there are dinosaurs out there, stomping around on Planet X, unperturbed, indomitable. The reptilians don't build spaceships because they do just fine eating small, cowering mammals.

Dogging the issue of intelligent life is the uncertainty about what intelligence is. Is it the same thing as consciousness? Can a creature be conscious but not intelligent? Intelligent but not conscious? We struggle, likewise, to figure out whether language is essential to intelligence. It's hard to separate our consciousness from our manipulation of language, our ability to think in words. It is language, to a large extent, that defines us as a species. Would aliens also develop language? Many animals on Earth are essentially intelligent, but don't seem to have much need for language, which would lead us to think that aliens might be intelligent but nonverbal. Nonverbal aliens might be like dolphins, superior creatures in their own environment but not prone to philosophy, long-term planning, and the creation of narratives about the universe such that space travel would seem a nifty idea. No language, no star trekking.

Naturally, this sounds a bit anthropocentric. According to some professional linguists and primate researchers, there are apes that use language. If language is not proprietary to humans, then language may be common in the universe as a whole, which would increase the chances that we will someday communicate with other civilizations. Primate research, unfortunately, is so heavily wrapped in ethical and political concerns that the skeptical outsider may find the expert testimony about apes using language not entirely credible. There seems to be a passion to apply to other animals the adaptations of humans. No ape has ever tried to teach a human to brachiate, but we are determined to get apes to talk to us. For millions of years, apes have survived just fine without needing to hold a prolonged conversation. Animals don't need language, in the same way that humans don't need sharp spines to ward off predators, or a prehensile tail to swing from branches. Language is merely one of nature's many brilliant creations, and it is humans who use it and benefit from it. Only a truly feverish desire to apply the Assumption of Mediocrity to all things human would provoke someone to argue that humans aren't specifically adept at using words and symbols. I bow here to Steven Pinker, author of *The Language Instinct*. Pinker says apes don't use language—they are merely accused of doing so by overly earnest researchers. The assertion, for example, that certain gorillas have mastered American Sign Language is a kind of libel against that language, suggesting that ASL is "a crude system of pantomimes and gestures rather than a full language with complex phonology, morphology, and syntax." Pinker claims the apes didn't learn any ASL signs. Every time a chimp named Washoe put his finger in his mouth, the researchers decided that he was making the sign for "drink." It took a deaf researcher for whom ASL was a native language to see that the ape was just moving his arms and hands naturally.

Pinker argues that Sagan and Druyan romanticized primates in their book *Shadows of Forgotten Ancestors*. They contended that there was no sharp distinction between humans and—they made sure to put the word in quotes—"animals." If Sagan and Druyan are right, there's no significant firebreak between large animals and "intelligent" creatures. By their logic, if there are planets with animals out there, it is merely a matter of time before one of those animals becomes an astronomer.

That's not a truth, however—it's just a narrative.

Here's another possibility: To be an intelligent creature is a near miracle. There may not be many of us. We could, in fact, be alone, for all intents and purposes. This is the model of extraterrestrial life that is in disfavor in contemporary society, because it's such a downer.

I am not advancing this as a closely held position or belief. The only safe approach to the topic of extraterrestrial intelligence is to be agnostic. Without data, all we can do is make guesses and wild-eyed estimates. Dressing them up with the label of "plausibility arguments" won't make them any more accurate. But it's clear to me, as a reporter scanning the horizon for aliens, that people tend to see mirages, that they bring to this particular issue a set of biases and hopes and preconceptions. They project. They want to see themselves out there, and so, squinting hard, they do. The topic of extraterrestrial life is thoroughly contaminated by wishful thinking. This problem is found among scientists as well as among flying saucer–chasing citizens. We have a natural desire to fill the heavens with love, to people the universe, to warm up this vast cold space and rescue ourselves from existential loneliness.

To toy with the idea of being alone seems a bit pre-Copernican. But the Copernican revolution is not complete—we aren't entirely demoted from our special standing in the known universe. One does not want to suggest that it was correct to send Bruno to the stake, but a modified, *metaphorical* anthropocentrism may be defensible, at least until we make contact with another species that can hold up its end of the conversation.

The Assumption of Mediocrity may need to be retired for a while. We could file it away in a drawer, lose track of it for a while, but not throw it away, since there may come a moment when we need to scramble and fetch it quickly. Although there's no reason to believe we are unique in the universe in possessing intelligence (liberally defined, we're not even unique on this planet), the term "mediocrity" seems to imply that human beings are in every sense ordinary. The scientific record at the moment doesn't support that conclusion. Barring evidence of superior civilizations, we are the singular marvel of biology in our corner of the universe. We are specially blessed and highly evolved. We are an immensity of time translated into organized matter.

We could be, as Shklovskii put it, "functionally alone." Our space brothers may be too scattered and too indifferent and simply too weird (too inhuman! too alien!) to become a part of our own reality, *ever.* Any civilizations that are extremely far away, in other galaxies, are removed from us not only in space but in time. If there's a cosmic speed limit, as Einstein declared, then there's no realistic hope of making contact with any creatures that are separated from us by thousands or millions or billions of light-years. Bernie Haisch offers hope, but he has his work cut out for him.

"Functionally alone" may sound like an oddly hedged concept—the wiggle room is generous—but in a universe this large there is nothing simple about two civilizations' getting together. The Kantian conception of galaxies as "island universes" is appropriate. Our home turf, the Milky Way, is already so enormous that it is hard to imagine how, in the foreseeable future, we could explore anything but the tiniest fraction of it. Unless there are aliens zooming all over our galactic ZIP code, we will probably never meet them.

Perhaps the question we should ponder is not whether we are alone in the universe but whether in our part of this one galaxy there are any examples of life that meet our definition of intelligence and with whom we might potentially communicate. Everything we know about the evolution of such intelligent life indicates that it is a rare event. Earth had no creature capable of interstellar communication until the past hundred years—after four billion years of horsing around. Our sector of the Milky Way galaxy may have a fair amount of life in it, but the bulk of that life could be microbial, perhaps similar to our own prokaryotes, simple things that lack cell nuclei. When we venture forth to explore the galaxy, we might find, at first, nothing as dramatic as an amoeba, much less something as flamboyant as a flower.

The Earth is quite possibly a paradise, even in cosmic terms—a great world with a stunning biological history. It is hard, at this particular moment, to support the theory that Earth is an ordinary rock. Until there is reason to believe otherwise, we have the right to view our planet and its inhabitants, and especially its sentient creatures, as the treasures of the known universe. (Who you callin' mediocre?)

Even as I was investigating the alien question I could see that it was rapidly being overtaken, if not entirely superseded, by a new and equally millennial fear. Suddenly, the world was obsessed with killer rocks from space. Who cares about those alien breeding experiments when the horrible truth is that any day now we might be obliterated? The experts assured us that the destruction of civilization could occur at any second, that there would be no warning, that these rogue planetary assassins were out there by the thousands. Perhaps life itself would survive. Even some pockets of humanity might endure. But billions would perish in a massive die-off. Pending this disaster there would be some excellent Hollywood movies.

This was, like the alien business, the perfect joint venture of hard science and popular culture. In exobiology circles (an appropriate term—there's so little information, everyone seems to be going in circles) the catastrophic impact of sixty-five million years ago was more than a trivial matter. It was, as Ernst Mayr argued, a crucial and perhaps essential event in the rise of intelligence on Earth. All this led me to leap at the chance to join an expedition to Central America, sponsored by the Planetary Society—the group Sagan cofounded—which planned to search for debris

32

Death from Above

(ejecta, as they call it) from the Crater of Doom. I could justify the trip as fieldwork in the Drake Equation—an attempt to understand, through terrestrial geology, the role of contingency in the longevity of dominant fauna. On planets throughout the universe there is probably an impact hazard from debris left over from the formation of the planets' solar systems. That debris may play a role in mass extinctions on a cosmic scale. It's all part of a bigger picture. Besides, going to Central America couldn't be any more of a lark than going to Roswell.

Sagan, the guiding spirit (looking all the more calm, sober, and rational over time), would have approved. In the final years of his life, Sagan began worrying about catastrophic impacts. He'd seen Comet Shoemaker-Levy pelt Jupiter and leave a series of Earth-sized bruises on the giant planet. It was an incredible show, and gave pause to anyone who could think in deep time. These rocks came not just from space, but from the larger dimension of space-time, the smaller ones coming every few hundred years, the big ones coming in million-year increments, and the ones that cause mass extinctions, the real planet-whoppers, every hundred million years. There had even been a big impact in the twentieth century. In 1908, an object about two hundred feet across is thought to have exploded over the Tunguska region of Siberia, leveling hundreds of square miles of trees. (People point out that the Tunguska object could have wiped out New York City, but the fact remains that it wiped out Tunguska—a place so remote and uninhabited that its only chance of being mentioned in history books was to get hit with an asteroid.)

For Sagan these catastrophes had a significance above and beyond the dinosaurs. He wondered if they constituted a cosmic rule of life. Perhaps this is how it always goes: A lucky species is blessed with the gift of intelligence, has its moment as the dominant animal, and then is erased, obliterated, by a piece of space junk. The solution is to go into space and colonize other worlds, he felt. One of his final papers was titled "On the Rarity of Long-Lived, Non-Spacefaring Galactic Civilizations."

Chris Chyba, an astronomer and one of Sagan's protégés—and one of a handful of people who might plausibly fill Sagan's role as the leading explainer of all things exobiological—had taken a similar view. He said that only civilizations that protect themselves from catastrophic impacts are likely to last long enough to be detectable through SETI. Sagan and Chyba both felt that this represented a cosmic selection effect, that it put pressure on organisms to develop technology. The galaxy, Chyba says, isn't likely to be home to many civilizations of pure poets, pure philosophers—if they can't build widgets and sprockets, they simply won't be around long enough; they'll get nailed by a Death Rock. The universe, therefore, favors technologically advanced creatures (particularly those who study space and build anti–Death Rock systems) and selects against whales, chimpanzees, pigs, horses, and other smart animals, not to mention dumb ol' dinosaurs. As Chyba puts it, "Dolphins can't protect the planet."

There is only one road north from Belize City. The pavement has no center dividing line, no shoulders, no railings, and there are hardly any signs. It is just a ribbon of asphalt, winding through marshes and jungles and pastures and great expanses of sugar cane and occasionally blasting through a village. The cars on the highway are outnumbered by trucks hauling mounds of wobbling cane. Sometimes there is ash in the air and the scent of caramel from the burning of the fields.

The villagers have few cars, and instead walk along the side of the road. Children stroll to school in neat uniforms, long pants for the boys and pleated skirts for the girls. Some of them live in thatched huts, others in wooden homes raised on pilings, with dogs sleeping and chickens scratching underneath, but there are also modern homes of concrete block, a few with naked columns shooting skyward in aspiration of a second floor. Belize has defied the stereotype that a developing country must be chaotic and crowded. The whole country is as quiet and peaceful as Caltech on a summer's day. There aren't many people—fewer than a quarter-million, so few that a single phone book easily handles every number. Hit the seek button on a car radio and it might well circumnavigate the spectrum and finally return to the same lonely station. The Information Age hasn't arrived here, and there are only tenuous, fleeting signs of the Industrial Revolution.

The past is layered upon the present. Queen Elizabeth II stares from the currency, a reminder that this was once called British Honduras. In the cities there are many blacks, descendants of slaves. There are Creoles and mestizos. There are Garifunas, people of African-Caribbean ancestry. The great majority speak English, but at the borders the inhabitants speak the language of the long-gone Spanish conquistadors. In the mountains there are villages of Maya Indians. Their history is still being scraped from the jungle, temple by temple, stone by stone.

Eventually the Northern Highway reaches Orange Walk. At that point you can veer west, toward Mexico, and navigate the back roads, dodging car-swallowing potholes, until you reach a dark, calm river. A bridge is made of planks stacked on floating barrels. On the other side is Albion Island. There is a famous quarry there, a quarry that is not so much a hole in the ground as a great gouge through the heart of a hill rising high above the surrounding coastal plain. In broiling heat the quarry workers drive bulldozers and massive groaning trucks, hauling rocks toward a conveyor belt to be crushed and ground and turned to gravel. It is a loud, uncomplicated operation, an exercise in raw power applied to stone.

Other people also come here to handle the rocks—geologists. They are struck by something high on the quarry wall. There is an undulating line in the rock face, a distinct boundary between one geological era and another. This is the K/T boundary. Below the line, the geologists believe, are normal sedimentary rocks formed in shallow seas over tens of millions of years. Above the line is a tremendous jumble of stuff, a matrix of rock and grit and gravel and stones and, here and there, boulders the size of houses. This thick layer of material, the geologists say, did not take millions of years to form, only a few minutes. It is splatter from the great catastrophe that ended the Cretaceous. All this rocky debris apparently rained from the sky in a matter of moments. That day a large object from space—a comet or an asteroid—smashed into the planet. It was roughly seven miles in diameter, and moving at hypersonic speed. This cosmic cannonball blasted a hole roughly 120 miles wide and thirty miles deep near what is now the Mexican town of Chicxulub. The Chicxulub impact and its aftermath killed about two out of three species on the planet.

The dinosaurs had ruled the world for 150 million years, a heroic span. They were, by any measure, a blue-chip family of animals, real crackerjacks. But they didn't become astronomers or geologists, didn't scan the sky for its timeless story, didn't learn the language of the rocks and fossils. They didn't know where they were in the universe. That was their doom. To survive in this world, you have to be more than a great eater.

• • •

I went to Belize just before Hollywood delivered two rocks-from-space blockbusters, *Deep Impact* and *Armageddon*. Paramount Pictures' *Deep Impact* was about a Chicxulub-size comet on a collision course with Earth. *Armageddon*, the product of the cheerful minds at Disney, told the tale of an asteroid "the size of Texas" (as big as Henry Harris's hypothetical lens-in-space!). The official movie synopsis:

> Its impact would mean the end of mankind. With only 18 days to spare, NASA discovers the Global Killer bearing down on Earth. NASA's director, Dan Truman (Billy Bob Thornton), is left with only one option—send a crew up to destroy the asteroid. Truman enlists the help of Harry S. Stamper (Bruce Willis), the world's foremost deep core oil driller. . . . Stamper and his Dirty Dozen crew must land on the asteroid, drill into its surface and drop a nuclear device into its core. . . .

So it's a bit tense out there. (One thing's for sure: No matter how much technology and teamwork go into a caper like this, ultimately the job of saving the planet always falls to a single individual, a lone, ornery, rugged, orthodoxy-ignoring, bureaucrat-disliking person with a bad habit of waiting until the last possible second.)

Hollywood had also gotten a real-life boost that was better than any paid advertisement. The International Astronomical Union announced that astronomers had discovered a potential impactor. A rock the size of—well—not the size of Texas, but definitely the size of Santa Monica, was heading our way and would possibly strike the planet at 1:30 p.m. Eastern Time on Thursday, October 26, 2028. It seemed so appropriate that the world would end on a Thursday—a day that has always been sort of lame. (The only worse day is Tuesday—an assertion that I can't prove but feel in my bones.) The rock, named 1997 XF11, seemed on course to pass as near as four thousand miles from the center of Earth—a serious problem, because at that distance it would strike the surface of the planet. The *London Times* wrote in its lead story, "Apocalypse could be just 30 years away."

Fortunately, the Apocalypse lasted only one news cycle. Astronomers double-checked the rock's orbit and announced that it would actually miss by six hundred thousand miles. NBC's popular Thursday night lineup would not have to be pre-empted after all.

But the threat was still there. Once again we had a numbers game on our hands, a Drake Equation–type situation in which no one knows the truth but everyone has an estimate. There might be, for example, as many as two thousand asteroids a kilometer or more in diameter that cross or come near Earth's orbit. A rock that size would be large enough to inflict global damage if it hit the planet. There could be several hundred thousand smaller objects capable of wiping out entire cities (with a bull's-eye impact). Most of these rocks, large and small, remain undetected, trajectories unknown.

The most common risk assessment in the small community of people who study these things is that a person—any person—has a one-in-five-thousand chance over the course of a normal lifetime of being killed by an asteroid or comet impact. Obviously this is a startling number. This is a much greater risk than that of an airplane crash, for example. How could any rational person offer the public so dire an estimate? The secret is that the experts presuppose that a big impact would kill a billion people. Even if the impacts are extremely infrequent, the risk is still relatively significant because of this presumption of death on a massive scale. Duncan Steel, an Australian who has searched for asteroids, argues that a person living in San Francisco has a greater chance of dying from a catastrophic impact (because a big rock in the Pacific would generate a horrific tidal wave) than of being killed in an earthquake along the San Andreas Fault. Planetary scientist and asteroid-finder Tom Gehrels says, "This is the most serious environmental danger facing humanity. It can come on you all at once. Kaboom! And you're out."

The likely warning time for the planet, says NASA scientist and former Sagan graduate student David Morrison, would be zero. "Feel the ground shake," Morrison told me dramatically. "See the fireball over the horizon."

In the town of Corozal, not far from the Mexican border in the north of Belize, I caught up with the scientists and volunteers from the Planetary Society. They all felt exultant—they were sailing a new paradigm into an exotic location. In a world that seemed already totally mapped, the Chicxulub crater and its outwash are new, large, dramatic features. Voyages of discovery aren't completely over; you may simply need to look at old places with a new vision.

There was no distinct division between the scientists and the volunteers. They all had gear—sunglasses, water bottles, bandannas, ponchos, gloves, flashlights, binoculars, digital cameras, compasses, knives, hand lenses, hand picks, scrapers, jabbers, old toothbrushes, and bottles of acid. They had boots and wide-brimmed hats and vests with lots of pockets. They had lotions, Chap Stick, moleskin, repellents, malaria pills. They were geared up for serious science.

Joe Breeden came from Santa Fe, New Mexico, where he was a business consultant who dealt in chaos theory and genetic algorithms—stuff that I could never quite understand. Sandy Miarecki came from the Mojave Desert, where she was a test pilot at Edwards Air Force Base, dreaming of becoming an astronaut. Roger Nordin flew in from the village of Skellefteå, in Sweden, where he searched for gold and silver and other precious metals in deep mines. Among the things I came to admire about Roger was that he knew that, at his normal pace, his step measured precisely 0.7 meter. From upstate New York came Paul and Jutta Dudley, whose relationship was solidified by a mutual love of geology.

"Our first date was fossiling. In the Devonian," said Jutta.

The most colorful character was Bruce Hartel, a big-game hunter from Colorado who had thirty-five animal heads mounted at home. He'd been to Belize several times, trying to add a jaguar to his collection, back before jaguars were made off-limits. "Man is a predator," he said when someone challenged his love of the kill. Hartel told stories of near-death encounters with raging beasts. "The second cape buffalo I took, I shot him one, two, three times. Every time, he went down and got right back up and charged. I finally dropped him with a frontal heart shot. That put the spice in the dish, that's for sure!"

Adriana Ocampo and Kevin Pope, from Los Angeles, were the expedition leaders. In 1988, Pope was in Mexico at a scientific conference, discussing a ring of sinkholes, called cenotes, that had been recently detected in northern Yucatán by satellites. He was interested in them because they were important freshwater sources for Maya civilization. Ocampo heard the talk and approached him afterward. Is it possible, she said, that this ring of sinkholes marks the outline of an impact crater?

She was right. There had been fracturing of bedrock at the crater's edge that allowed groundwater to seep to the surface. The revelation set them off on a scientific adventure, and at one point they paused long enough to get married.

The crater from the Chicxulub impact is detectable only with high-tech imaging equipment. That's why the expedition came to Belize. Models of the impact show that Belize should be full of ejecta. The best site found so far is the Albion Island quarry, about two hundred miles from the crater. We all drove there one morning, bouncing over the rural roads in rented vans. Pope discovered Albion in the mid-1980s as a good place to swim. He would come and dive into the pond at the edge of the quarry. He had no idea that above him were these rocks from the disaster that had wiped out the dinosaurs.

At the quarry Pope briefed the troops. He pointed out the K/T boundary on the far wall. It included a distinctly orange line about one meter thick. He and Ocampo believe everything on top is the ejecta blanket. Three lines of evidence indicate that this is the spew of the Chicxulub crater. First, it's a chaotic mix of material, from giant boulders to tiny blobs of clay. Second, the rocks have markings, various lines and grooves, that suggest they had been banged up and scuffed in countless midair collisions after being launched from the crater. Third, there are distinct traces of weathered glass, apparently from rock melted by the intense heat of the impact.

None of this has hardened yet into scientific fact. Pope's dream was to find a shatter cone, a peculiarly pyramidical fragment of rock that has been found at other crater sites. He thought he had a shatter cone already, but Ocampo wasn't convinced it was the real thing. There also remained a serious question as to how long it took for the ejecta to get here—seconds? minutes?—and why parts of the ejecta blanket had a layered look, as though neatly laid down.

Here was a geological mystery, a challenge from the remote past. It said: Look at me. Figure me out. The Planetary Society scientists and adventurers got on their hands and knees and scraped at the earth, digging for knowledge as the tropical sun beat down upon them.

For hundreds of years, it has been apparent that the rocks of the Earth tell a tale of life and death. In 1815, geologists recognized that there was an abrupt change in marine fossils at a certain stratigraphic point that later became known as the K/T boundary, the end of the Cretaceous Period and beginning of the Tertiary (the "K" is from the German equivalent of "Cretaceous"). In the decades that followed, scientists became interested in the large bones that were being found around the world. In 1841, the world learned the bizarre truth: The bones belonged to giant reptiles, "terrible lizards," the dinosaurs. These huge creatures disappeared at the end of the Cretaceous. Did they die because pesky little nocturnal mammals ate their eggs? Did they freeze in an ice age? Did their marshlands turn into mountain ranges? Did they get zapped by radiation from a nearby exploding star? There were many outlandish theories. Constipation from eating flowering plants was one inventive explanation. So, too, was the idea, floated periodically, that an object from space had smashed into the planet. In the early 1970s, Nobel Prize–winner Harold Urey—he of the Miller-Urey experiment—pushed the impact hypothesis. Hardly anyone paid attention. There was a problem with Urey's idea: It involved a catastrophe. Most scientists didn't like catastrophes. They were just a bit too . . . Old Testament. The prevailing wisdom was that everything we see—the land, the organisms—came into existence through unimaginably gradual processes. Gradualism ruled.

Less than two decades ago, the strange truth began to emerge: Both sides were right. Things happen slowly—except when they happen suddenly.

During the nineteenth century, many scientists doubted that meteorites were, in fact, of extraterrestrial origin (the favorite history-of-science story among ufologists). Rocks fall from the sky? Preposterous! But slowly the evidence mounted. Geologists discovered that Meteor Crater in Arizona, once thought to be volcanic, had been formed by an impact as recently as fifty thousand years ago. The legendary geologist Gene Shoemaker convinced his colleagues that the craters on the moon were the scars of impacts (he was, tragically, killed in a car accident in the Australian outback in 1997). The Chesapeake Bay basin turns out to be the remnant of an impact crater.

The Old Testament had it right: Things fall from the sky like thunderbolts from God. Small nervous children can be reassured—they needn't worry about being eaten by a dinosaur, because sometimes most of the life on Earth is wiped out by a giant rock from space. Sweet dreams.

In 1973, Walter Alvarez began studying the K/T boundary near the medieval town of Gubbio, Italy. The Gubbio site is striking because the boundary is clearly delineated, with white rocks below and pink rocks above and, right at the boundary, a thin layer of clay. Alvarez called his father, Luis, and wondered aloud how long it had taken for this layer of clay to be deposited. Luis Alvarez realized that it would be possible to study the deposition rate of the K/T clay layer by measuring the abundance of the radioactive element iridium. He knew that iridium is scarce on Earth but comes to the surface from space in predictable quantities from the infall of small meteors and cosmic dust.

Walter Alvarez gathered samples of the K/T clay and passed it on to two researchers in a laboratory at Berkeley. Eight months later, the results came in. The iridium in the boundary was about three hundred times as abundant as in adjacent rocks. The father-son team began to suspect that all this iridium came from a meteorite and had been spread around the globe in a catastrophic impact. In 1979, Walter Alvarez presented his theory at a conference. He was scorned. What was he, some kind of old-time fire-breathing catastrophist?

What no one at the conference knew was that, just a year earlier, Glen Penfield and Antonio Camargo, two geologists working for the Mexican oil company Pemex, had been making gravity maps and magnetic maps of the region around the Yucatán Peninsula. They found a huge ringlike structure about 120 miles across. They thought it might be the remnant of a crater. Penfield and Camargo announced their finding at a conference in 1981. Carlos Byars, a reporter for the *Houston Chronicle*, knew of Alvarez's impact theory and quickly knocked out a story saying that perhaps this Yucatán structure was the hypothesized K/T crater. It was quite possibly one of the greatest journalistic scoops of all time—and it was completely ignored. It just whooshed through the atmosphere unheard and unseen. Great work, no impact.

Years passed.

Geologists started noticing things. They noticed that there was evidence of a massive tidal wave far up the Brazos River in Texas. They noticed that sites around the Caribbean basin had heavy amounts of shocked quartz and glass fragments of the type produced by impacts. They were homing in on the Crater of Doom. The reporter Byars mentioned the suspected Yucatán crater to an Arizona graduate student named Alan Hildebrand, who in turn went to Penfield. Penfield and Hildebrand re-examined the Chicxulub data. Everything seemed to fit, and they published the first scientific paper saying this was indeed the long-sought K/T crater. Pope and Ocampo followed shortly with their research showing that the sinkholes reveal the crater's edge.

Not everyone went along with the story. There were strenuous objections and alternative theories. There is much debate about when, precisely, the dinosaurs died out. No dinosaur fossils have ever been found right at the K/T boundary. Many dinosaur species went extinct long before the Chicxulub event. One theory is that volcanism put the final kibosh on the dinosaurs. At about the time the last dinosaurs went extinct, the planet was in a period of heavy volcanic eruption, particularly from a hot spot in India called the Deccan Traps. Geologist Charles Officer of Dartmouth has argued that the volcanoes could have been the source of the iridium in the K/T boundary. They could have caused global acid rain, destroying marine life at the base of the food chain. Officer has raised the possibility that Chicxulub may itself be the remnants of a volcano.

In Belize, the competing theories provoked only huffing and snorting. "Completely discredited," said one scientist. But there is clearly much work to be done. Just because no one believes a theory doesn't mean it won't wind up a few years later as the unassailable orthodoxy.

We left Albion Island and drove south for half a day, toward the Maya Mountains. We reached Pook's Hill, and a rain-forest lodge run by two South African expatriates, Ray and Vicki Snaddon. Toucans flitted among the towering palms. Iguanas climbed all over one another in a cage near the edge of the forest. We hiked down a trail under the gloomy canopy. When we reached a river, everyone jumped in, leaping from a boulder into a deep pool and dodging swarms of bats emerging from roots along the bank. After dark we ate beans and tortillas, and then expedition member Tim Herman found a guitar and serenaded the troops.

One night Kevin Pope ran through his theories for why this Chicxulub impact caused so many extinctions. The dinosaurs and all those other creatures didn't simply get knocked over or burned up or drowned from the direct impact of the object. Wiping out an entire species is different from a mere mass killing. "An extinction is only true if you remove the *last one*," said Jan Smit, one of the scientists. Pope asked, "Why couldn't you have pockets that survived?" To have such a vast and conclusive effect, the impact had to have triggered some kind of lingering global disaster.

The impact sent dust into the atmosphere that, combined with soot from continent-size wildfires, plunged the planet into darkness. And there may have been a second problem: Sulfuric rock vaporized at the impact site would have filled the stratosphere and formed sulfurous clouds, and these could have lingered for as long as a decade—reflecting much of the sunlight back into space. Earth may have become a planet of vast ice sheets and glaciers. And then, when the clouds and dust were finally gone, it may have gotten hot. Really hot. This is because of another liberated material at the impact site, carbon dioxide. CO_2 is a powerful greenhouse gas.

Freeze them and then roast them: a recipe for extinction.

But now comes another quirky fact of the Chicxulub case. If that rock or comet had hit almost anywhere else on the planet, it wouldn't have caused so much damage, because it wouldn't have ejected so much CO_2. Most of the world has a relatively thin layer of carbonate rock. But at the tip of the Yucatán is what is known as a carbonate platform, with carbon-rich limestone a couple of miles thick. The carbonate platforms are found in clear waters along tropical shorelines with no nearby rivers to muddy things up. Those tourists who go to Belize to dive along the great coral reef know how perfectly clear the water is—just right for the kind of creatures who create carbonate platforms, which, in turn, are capable of spewing massive amounts of carbon dioxide into the atmosphere in the extremely unlikely event that an asteroid or comet comes plowing into them.

Which is what happened! A contingency within a contingency. A bull's-eye within a bull's-eye.

It's not clear what this means in the cosmic scheme of things, whether this is another constraint on intelligent life. The only thing that is obvious is that once again randomness—bane of the spiritually inclined—seemed to be looming large in the narrative of life on Earth. There were no conspiracies here. The aliens didn't do it. It's just . . . geometry, physics, thermodynamics, Newtonian motion, kinetic energy. Shit happens.

One night at Pook's Hill, everyone gathered in the large thatched open-air hut for a lecture by Al Fischer, one of the scientists leading the expedition. Fischer and his wife, Winnie, had driven all the way to

Belize from their home in Los Angeles. They're elderly but unretiring, the kind of people who search and explore and have adventures until they drop. Fischer, in fact, had just seized upon a stunning idea. He planned to discuss it at a scientific conference in Rome but would first give us a preview, here in the hut. He wasn't going to make a fancy show of it. He had some chalk, a chalkboard, but was quite visibly barefooted, wearing shorts and a T-shirt. A lecture delivered barefooted in the jungle carries an extra gloss of brilliance.

He drew a coiled line on the chalkboard. The line circled round and round like the shell of a nautilus. This was our planet, moving through time. Earth, he said, goes through a long, distinct cycle of heating and cooling. The cycle—one circular trip around the chalkboard—lasts 148 million years. We're hot, seventy-four million years later we're cold, and seventy-four million years after that we're hot again. Simple so far.

We are currently in the coldest part of the cycle, he said, so we're down near the bottom of his circle. Earth is in an icy era—the last ice age, after all, was only ten thousand years ago.

What does this have to do with catastrophes? Fischer pointed out that the last time the planet was at the hottest point, up at the very top of the circle, the planet got whopped by the Chicxulub meteor. In fact, he said, more than half of the giant impacts known so far—events deduced from the various large craters around the world—apparently occurred when Earth was near the peak of the heating cycle. There are a few other impacts that are either at the bottom of the cycle (the coldest part) or halfway in between. The Chesapeake Bay impact, for example, occurred about thirty-five million years ago, halfway between the hottest and coldest phases.

His point was astonishing: These random, sudden, catastrophic, totally unforeseeable impacts are almost regular. They happen like clockwork.

We were all gaping at this point, trying to figure out what it meant, where he was going. There was no obvious reason why the heating of the planet would be associated with giant impacts, and why this would be such a regular process. So he gave the answer: the galaxy.

This is connected, he said—standing there barefoot in the lantern light—to the solar system's long orbit around the center of the Milky Way. In the universe, everything moves. The Earth moves around the Sun, but the Sun moves, too, in an orbit around the galactic center, a complete trip taking something like three hundred million years, he said. And what's so special about three hundred million years? It's almost exactly twice the duration of the hot/cold cycle. So there could be an additional pattern here. Earth goes around the galaxy, gets hotter and colder and hotter and colder, and at the hottest and coldest points, and the points halfway between, finds itself bombarded with giant space objects.

Fischer conceded that his revelation stopped at that point—he didn't know why the pattern exists, didn't know the agent causing these periodic impacts. He ventured that they might be related to the passage of the solar system through the two star-dense spiral arms of the galaxy. If the Sun passes close to another star, there could be a gravitational disturbance in the Oort Cloud, the community of comets that orbit the Sun far beyond Pluto. Nudged from their home turf, comets could come flying in toward the inner solar system. Occasionally one would hit Earth.

By Fischer's calculation, we are about eight million years from the next shower of comets. Eight million years until we bottom out on the hot/cold cycle and find ourselves in a catastrophic meteor shower.

"The whole thing has just developed over the last month," he said.

When he put down his chalk, everyone sat in awe. If Fischer was right, these impacts were not simply a planetary phenomenon. This thing was *galactic*. We were going into ultradeep time here. That ghostly pale strip of light overhead on the darkest of nights—that whirlpool of stars seen on edge—is not some remote astronomical structure. It's our planetary environment, our *ecosystem*.

We wandered around for the next hour, amazed, murmuring, a few of our number actually on the verge of a religious experience. It is the dream of science to find out how everything relates, to unify our understanding of nature. On this night everyone wondered if we were getting closer. Maybe it was all somehow connected—the galaxy, the climate, the craters, the bats flitting to and fro, the crawling iguanas, the freakish moths, the riotous trees in the dark moaning jungle. Air. Stars. A planet full of life.

In the morning we would go out and look at the world again.

Al Fischer's barefoot lecture was rocket fuel for the brain. We were part of a whole, part of an integrated system, a bioastrogalactic continuum. Right? (Even I, even the Invalidator, could groove on interconnectedness and secret wisdom and the magic of the unknown. It was almost like being spiritual.)

But of course . . . this hadn't been double-checked and peer-reviewed. Fischer might have been wrong. This hadn't even been published. (I later heard he was backing off a bit. Too many uncertainties.) Wasn't his estimate of the orbital period of the solar system around the galactic center pretty darn approximate? He was saying three hundred million years, but I'd heard 250 million from other sources. Most worrisome was the giant gap in causation, the fact that at the core of the narrative he had a big ugly unknown, this mystery of why a galactic orbit would be associated with the heating and cooling of a planet and—also—mass extinctions. Airtight this wasn't.

But at the very least he was on an interesting track. He was engaging the brawny issues of why we're here and how we fit into the cosmic scheme of things. We were making progress. We were getting somewhere. By the millennium, maybe we'd have it all figured out.

33

Goldin Sees God

• • •

One Saturday morning I had breakfast with Dan Goldin. We met near his house on Capitol Hill and walked to Eastern Market, a village bazaar where the power elite of Washington intermingle with the ordinary folks just looking for some decent tomatoes. We ate breakfast on a picnic table, and Goldin picked at his food. It was not apparent whether he lacked appetite or simply preferred to talk. Goldin is not really someone who can be interviewed, because an interview implies a series of questions, an interactive experience, and Goldin pretty much takes the first question and runs with it for the next hour. You have to interrupt him at times or he'll still be busy answering last week's question.

I asked if he would leave NASA soon. He'd had his run, John Glenn was about to fly on the shuttle, the space station would soon have its first section in orbit, he'd put the Mars program in place and started Origins and Astrobiology, and if he was smart he'd get out before the space station had a chance to get slammed by a meteor. The station had disaster written all over it. It was going to cost something like $60 billion, would have parts from fifteen countries (with the American pieces not using the metric system), and its construction would require a thousand hours of spacewalks. The Russians were always behind schedule with their components. NASA had come up with a list of reasons for building the station, and Goldin claimed that under his leadership the station had acquired a genuine scientific purpose, but it still seemed to make everyone nervous. Bad things happen in every construction project, and this one was huge, and it was in space—a hostile medium. The main reason to go forward seemed to be the benefits of global cooperation, and perhaps the need to transport huge cargoes of money to Boeing. The ads on television argued that we were building this because we owed it to Copernicus and Galileo and so forth, though if Sagan had been around he might have noted that Galileo, for one, would probably have preferred to see the money go to studies of Europa.

In any case, I assumed that Goldin would find the perfect moment, perhaps soon after the first space station components were launched, to make his exit. But he ducked the question. He wasn't sure what he'd do.

I asked if he was spiritual. After all this time, all these months stuck in the Pluto orbit around the truth of extraterrestrial life, I had wound up in the do-you-believe-in-God business.

"I believe in God," he said. He told a story. About twenty years ago, he had gone backpacking with some buddies on Mount Whitney, on the eastern slope of the Sierra Nevada. His foolish friends ate beans at seventy-nine hundred feet and suffered a massive case of high-altitude flatulence effect (HAFE). They were also swigging brandy. When they camped at twelve thousand feet, Goldin felt compelled to steer clear of his increasingly sick and inebriated friends, and he set up his sleeping bag several hundred yards away. As it grew dark, he watched the stars come out. And they kept coming out. They exploded from the firmament. Goldin started getting emotional. He started getting scared. It was overwhelming. So many stars . . . And who was he? What did it mean?

"I came in contact with the concept of infinity in time and space. I had these feelings about, holy mackerel, why am I here?"

And then he had a revelation.

"I realized why I believe in God. I said, If I'm lucky I'll be here a hundred years. I've got good lineage. One hundred years I'll be here, maybe. There's got to be a reason I'm here."

He didn't completely finish his line of thought. It kind of hung there. But I thought I knew what he meant. God is implicit in the universe. We are small creatures living in a tiny compartment of space-time, and we sense that there must be a larger intelligence, a larger force that has created all that we see, and that has given us this moment to witness that creation. We are blessed to be allowed entrance into the cosmic theater.

I asked if he believed that there are intelligent aliens out there, and if so, how would he solve the Fermi Paradox? He said he didn't know: "My rational side does not allow me to propose a solution."

He added, "Anyway, it is irrelevant without information."

He was echoing Gould's line, *No data*. It's the smart, safe approach in a time when many people prefer the flamboyant leaps of faith.

I pointed to all the people wandering around the market, and suggested that most people don't know what they are, have little sense of their evolutionary history, and take for granted their intelligence. Goldin lit up again.

"We live in a fog, man. We live in a fog! And we can't see beyond the nose on our face. We've got to extend the arm of knowledge to blow away this fog!"

Before we would blow away the fog, we would recapitulate the glory days of the Space Age. Two days before Halloween, 1998, John Glenn went back into orbit. It was a mission clearly designed as a spiritual adventure more than as a quest for knowledge. No one with a heart could have protested the move. Glenn symbolized a noble aspiration. He'd sat on a bomb back in 1962—two of the five previous Atlas rockets had blown up. Bravery doesn't get any purer. The technicians watching Glenn's first launch were inside a concrete bunker, the blockhouse, peering at Freedom 7 through periscopes. He was alone out there, sitting on that big, nasty rocket. To recapture some of that spirit was a good idea. There was only one catch. We live in the age of science. You can't do anything simply as a spiritual exercise. And so Glenn refused to say that his mission was largely symbolic. No, he and the NASA people had an elaborate explanation worked up, in which this was an attempt to study the effects of spaceflight on the aging process. There would be eighty-one experiments. Glenn mentioned that in almost every sentence: not just one but eighty-one experiments, *real science*, probes and gauges and blood samples, charts and graphs. You can't launch a seventy-seven-year-old hero into space simply because it would be cool, or because the guy wanted a victory lap. All missions must be legitimized by information rendered in binary code.

STS-95 was one of the four biggest launches, in terms of media attention and public interest, in the history of the Cape. (The others were Apollo 11; the first shuttle flight; and STS-26, the "Return to Space" after the Challenger disaster.) There were more than thirty-seven hundred requests for media credentials. Walter Cronkite came, working for CNN, as did all the anchor monsters and lots of celebrity commentators. Jimmy Buffett covered the event for *Rolling Stone*. Ted Williams showed up to talk about heroism (he'd been a buddy of Glenn's during World War II, in addition to being the last man to bat .400). President and Mrs. Clinton would be on hand, as well as seventy-seven members of Congress, Tom Hanks, Sean Connery, and Leonardo DiCaprio. It is fair to assume that none of these people had any interest in gravitational factors in the aging process. Glenn's return to space was more of a collective reaffirmation in the whole dream of the Space Age.

Before the launch, I spent a few days wandering around the Cape, scrounging for feature stories to file to *The Washington Post*. If any news had broken, any actual unanticipated events, it would have represented a tremendous failure on the part of the NASA planners and media-affairs staffers. This was a planned event down to the camera angles. At about five o'clock every afternoon, the buses left to take the photographers to the beach for the sunset shot of the shuttle as it sat on the pad. I went on the public bus tour—feeling like an intrepid reporter for escaping the media scrum and doing what any tourist can do for about 15 bucks. I was on a bus with a bunch of Swiss people who were particularly thrilled to see the alligators lurking in a ditch. An American sometimes looks at people like the Swiss and thinks: How can they face themselves in the morning, knowing they don't have a space program? Doesn't it bug them that we were the ones who went to the moon? Don't they feel small? (What is it like being *neutral* all the time?) Coming from loudspeakers on the bus were tapes of astronauts talking about the Kennedy Space Center and, more generally, about how great the space program has been for terrestrial society. We learned, for example, that a spinoff of the space program is the Dustbuster.

A highlight is the drive-by of the Vehicle Assembly Building, a structure so cavernous it is said to have its own indoor weather. The taped message said that inside the VAB, the NASA engineers can pick up the entire Space Shuttle with a crane and then lower it so gently it can come to rest on an egg without breaking the shell. (You have to wonder if they've actually tried that with an egg.)

Another highlight is the Saturn V Center. After some Disneyesque maneuvering through two rooms, we reached an enormous hangar with a Saturn V rocket stretching horizontally from one end to the other. It was a bit like a whale in a natural-history museum, hovering above the gaping crowd, only this rocket was far bigger than any cetacean. The rocket engines are each as cavernous as a studio apartment, and they sprout a bewildering array of wires, hoses, tubes, cables, sensors, gauges, toggles, switches, joints, clamps, deeleyboppers and doohickeys. Anyone can see that technology in the good old days was not only larger, but also more transparently complicated and fabulous. As in a well-made Hollywood movie, they showed you where all that money went. You could really see the technology, gasp at its incomprehensible wonderfulness. All around me the tourists used their digital cameras and camcorders to capture the digital image of this great rocket. Today the zenith of technology is something tiny, an etched piece of silicon that operates on quantum-physical principles, a little marvel embedded and hidden inside our modern contraptions. Inside the cameras, the rocket was turning into zeros and ones.

The second Glenn flight wasn't noticeably inspiring much fear and nervousness. NASA's engineering prowess had once again become an article of faith. The Glenn flight was starting to feel ceremonial—not quite an adventure in space.

Everything associated with a shuttle launch goes according to a plan, which follows an established formula. For example, there's the Arrival. The astronauts don't fly to the Cape on a standard aircraft but, rather, in sleek two-seat T-38 jets, a ritual right out of Tom Wolfe's *The Right Stuff.* Glenn flew in as the passenger on a jet piloted by the mission commander, Curt Brown. Before landing, the NASA jets flew in perfect formation over the launch pad, another sacred custom. The spry senator climbed out of the jet with no sign of creaking or groaning. He raked his hand over his head in case there might be a stray hair. It required about 150 journalists to document the moment.

At night there were parties for the visiting celebrities and aerospace folks. Two nights before the launch, Wally Schirra, one of the "Original Seven" Mercury astronauts, was cracking jokes on a stage at a swank reception, the crowd sipping champagne and feeling no pain. Schirra said he had fallen asleep on the way to the pad. He noted that his fellow Mercury astronaut Gordon Cooper fell asleep, amazingly, while on top of the rocket and waiting to blast off. And John Glenn? He fell asleep onstage at Alan Shepard's funeral, Schirra said. So why, Schirra asked, was everyone worried about whether Glenn could be used for a sleep experiment with melatonin? Glenn can definitely sleep, he said. "Join the Original Seven, we'll teach you how to sleep!"

The crowd nibbled on roast beef, crabcakes, asparagus wrapped in prosciutto, new potatoes with dollops of caviar. Someone wandered the room in a pressurized spacesuit. The mirrored visor was closed, so it was impossible to tell who was inside. The spaceman did not speak, communicating only with the universally understood thumbs-up signal. Maybe it was Neil Armstrong. (By the way, it has not escaped the notice of the UFO community that Neil A. spelled backward is "alien.")

Schirra told me, "NASA's never been able to sell itself. They finally picked an old pro to do a sales job. John Glenn by no means is doing anything of scientific value."

I assumed he meant that as a criticism. He corrected me: A sales job is just dandy. "It's gorgeous!" he said.

There were signs that the back-to-space idea was getting infectious. "I'm ready to go on the Mars mission," said the thin, frail-looking Cooper in a moment of exuberance. He was completely serious. He said he's talked to Dan Goldin. If NASA decides to send humans to Mars, Cooper thinks he has a chance of joining Glenn as another old-timer in space. "I'd like to be the commander of the mission."

Also on hand: Buzz Aldrin, Apollo 11 moonwalker. Aldrin, looking pleased to be in the spotlight in his bright-red blazer, put in a word on behalf of an organization called ShareSpace. He said there has to be something more dramatic to come along after the International Space Station. He envisions private space shuttles holding eighty to a hundred people, essentially tour buses in low Earth orbit. He said there could be a lottery to allow ordinary folks to go along for free: "We need to open this up to the ordinary citizen." He personally doesn't want to go into space again, but recently he'd been to the North Pole, to the wreck of the *Titanic*, and he had plans to visit the South Pole. (The moon is a tough act to follow.)

These old astronauts reminded me of something that Freeman Dyson had said. I had asked Dyson if the Space Age dream would ever come true—if people really would go off and live on other planets and travel to the stars. "Yes. No doubt at all. It's just a question of how long it takes. There'll always be enough crazy people to make this thing go."

There are different strategies for taking in a shuttle launch. The best strategy is to be a famous person, a president, a NASA launch-control person, a member of the press corps, or someone who is impersonating a member of the press corps. Most people watch from at least five miles away, a distance from which the shuttle looks rather toylike. The NASA Causeway across the Banana River is on Kennedy Space Center property and draws a lot of NASA families, bused-in schoolkids, and hardcore launch fans who requested causeway passes six months earlier. There's a slightly different feel in Titusville, west of the space center, where you're more likely to find the old folks in earth-flattening RVs, the hucksters who feed off crowds ("Limited-edition John Glenn coins, only ten thousand made!"), and the beer drinkers.

On the causeway, *Post* photographer Bob Reeder and I wandered around in a funk, worried that so few people seemed to be drunk or crazy or really all that interesting. Finally, we found Robert Gass, and he saved the day. A Fort Lauderdale accountant, Gass had been to fourteen previous launches and may have been the most excited person on the Cape who was not actually inside the shuttle. His father had dragged him out of bed as a kid to make him watch rocket launches. This time he brought about twenty friends. He kept a running patter, anticipating each word from Launch Control, announcing his emotional state as it varied from minute to minute, waxing philosophical about the dangers. "Human beings are flawed. Mistakes happen. I worry until the boosters fall off."

Gass gave brief seminars on the Russian space program, the tale of astronaut Deke Slayton's heart murmur, Alan Shepard's refusal to let a technical glitch ruin his moon landing, and the G-forces inside the Atlas rocket that had first put Glenn in orbit.

"The Atlas was designed as an ICBM. When it went up, you got crushed."

The loudspeaker said: "We are continuing to hold. . . ."

"It's taking too long. I don't like this," Gass said, pacing.

The loudspeaker said: "At this time we are go."

Elated again, Gass gave another seminar, this time on what the launch would look and sound like.

"All of a sudden the sound comes, and it starts to rumble, and it starts to rumble louder, then it crackles, and then your stomach starts to shake."

The poll of the launch team began, his favorite part. This was where the launch director asks everyone for a "go" or "no go" assessment.

"ATC."

"Go."

"TVC."

"Go."

"LTS."

"LTS is go."

On down the line. Discovery was a go.

"This is gonna be great!" Gass said.

The final countdown began.

"Ten, nine, eight . . ."

First came a white cloud. From where I stood, the cloud totally enveloped the shuttle—we had a different angle from the shot seen on television around the world. For a moment there was nothing to see but the billowing vapor. Slowly the shuttle lifted up through the cloud, powered by an orange flame. The shuttle moves so slowly at first that it is easy to imagine it might fall back down. (The Saturn V was even slower; it just sort of levitated a bit at the start.) But then the shuttle accelerated upward. *Vwoosh!* I had a dull-witted thought: My gosh, it's a rocket. It goes faster and faster and just explodes into space! On television you don't see the acceleration—the shuttle appears to be moving at a constant velocity. In real life you see this big rocket get suddenly small and then turn into a little white dot out over the Atlantic Ocean, already puncturing the upper limit of the atmosphere just a couple of minutes into the flight. Left behind is the vapor trail, white and gleaming in the sunshine, a serpentine diagram of what just happened.

"Within thirty seconds," lectured Robert Gass, "it was already twenty-five miles up and it was going thirty-nine hundred miles an hour."

The loudspeaker said: "An uneventful climb into orbit today for the seven-member crew of Discovery."

It was a good show. It was literally thrilling. There was, however, a nagging thought from the rational module of the brain. As nifty as this thing was, it wasn't the future. The shuttle is full of modern wizardry, but its essential design is old technology. It's a *rocket*. It's the technology of Robert Goddard and Wernher von Braun. The shuttle still relied on the same old thrusting energy, the same principle of propulsion, as the V-2. Rockets, as the science-fiction writer Thomas Disch has pointed out, seem like wonderful inventions to thirteen-year-old boys because they are basically souped-up cars. They are vehicles for liberation and entertainment.

NASA had a new spaceship in the offing, the Venture Star, which promised to be the first single-stage rocket into low Earth orbit. If the next generation of shuttle could reduce the cost of flying into space tenfold, we might be able to make space a more useful and accessible environment. But this wasn't what the dreamers were thinking about when they invented the idea of space travel. They had grander plans for larger, faster spaceships that would go deeper into the heavens. Watching Glenn's launch, the technophile might easily have become wistful. Where was Henry Harris and his laser-beam light-sail? Where was the antigravity technology, the Zero Point Field inertia nullifier, the Bussard ramjet? For that matter, where were the genetically re-engineered astronauts capable of living in nonterrestrial environments? Where was the astrochicken? Where were the von Neumann machines and the nanoprobes?

Dan Goldin was there, of course, watching from a rooftop above the launch team at Kennedy Space Center. He had a difficult job: He had to shepherd President and Mrs. Clinton through the space center, say all the right things, represent his agency with dignity and intelligence, and all the while pray that nothing went wrong. When the shuttle lifted off, he said later, he got weepy. In the final minutes, he had barely been able to say a word.

"When it goes off, I can hardly breathe. I'm at one with that rocket," he said.

I asked what was so stirring about the Glenn flight.

"You're asking me a rational question about an emotional experience," he said.

A familiar problem.

The rational objection to the flight (it's not futuristic enough) is ultimately outweighed by our innate emotional support for it. Goldin had said it to me more than two years earlier, when I first met him: The human life is more than survival. Life can't just be about going to the grocery store and the shopping mall and watching TV. You have to go do something! You have to search for truth. The Glenn flight had a political subtext and a flimsy science subtext, but ultimately it was about doing something other than going to the mall.

And thus it was a good day for the species.

The Earth hurtles onward. We scurry over the surface of the planet, dip into its waters, and sometimes build machines that rocket us into space. We are an energetic species. We relentlessly apply technology to our environment. The genotype of the species will soon be published on the Internet for everyone to see in its entirety; the phenotype (culture, etc.) could someday be galactic. Freeman Dyson has a vision of human biology re-engineered for other habitats, so that we could live on the billions of comets in the Oort Cloud, where there's all that unused, if icy, turf. (The Oort Cloud makes Mars look like Miami Beach.) Eventually life will find a way to expand at near-light speed. It will race across the universe. There are people thinking deep in the night who envision a descendant human intelligence on a cosmic scale, merged with other intelligences, reaching a point of omnipresence and virtual omnipotence. Finally, we'll be able to set the clock on the VCR.

When I think about these futures, I'm convinced only of one thing, that they are wrong. We always get the future wrong. It is our signature. All we know is that something is going to happen with this species and this planet. The future will be unlike the past. The Space Age story has us going into space, colonizing other worlds, trekking

34

Strange Futures

across the stars, and somewhere along the line making contact with aliens, who lecture us unmercifully. (What a bunch of schoolmarms those aliens always are.) But, Glenn aside, that version of the future has dimmed of late, suffering from heavy competition with other futures. Space may turn out to be irrelevant, except as a database. Space may turn two-dimensional on us, becoming a vast computer screen that relays various interesting factoids. We'll download the cosmic files, and continue with our latest obsession, an internal voyage, a descent into intensified experiences, mind-melding, collective consciousness, transcendence, meditation, enlightenment. Our goal is to know the universe without feeling so constantly *hassled.*

Contact with aliens is still a possibility, though probably less likely than some people think. There is a chance that Sagan had it right in *Contact,* that we live in a privileged moment of sorts, that our radiating television broadcasts are steadily announcing our existence to whatever's out there. Aliens might pick up our signal and decide to investigate. But on a galactic scale the TV broadcasts are exceedingly weak. If there are aliens out there with finely tuned instruments capable of detecting our electromagnetic leakage, they would presumably have had other sorts of detectors for millions of years, perhaps the equivalent of Dan Goldin's Planet Mapper. They would have known, long ago, that Earth had an interesting biosphere, that it had complex life in the form of vegetation. Certain atmospheric conditions, including the rise of methane and carbon dioxide over the centuries, would have given the aliens a hint that intelligent life had appeared on the planet and begun to alter the environment. If we accept the idea that life intrinsically exudes information, then the hypothetical aliens should have known about us long before they received our broadcasts of *I Love Lucy.* There is no solid evidence that they have ever been motivated to visit or to contact us. For the moment we can think like a federal agency: The aliens will stay silent, because that's what they've always done.

In fact, the argument can be made that we'll never voyage across the galaxy ourselves. We won't become galactic colonizers, goes this argument, because that would put those of us on the "home planet" in a privileged position, in total violation of the Copernican Principle. This notion springs from the mind of Gott.

J. Richard Gott III is a theoretical astrophysicist at Princeton. I learned that I could always turn to him for an instant, confident explanation of such topics as Why Is There Something Rather Than Nothing? As a high-school science-fair champion, he came up with new models for sodium polyhedrons with a spongelike quality—and then asked himself if the universe on the whole might be spongelike. He has continued ever since to invent nifty solutions to mind-bending problems, such as his proof that time travel is possible. Gott's time-travel mechanism, unfortunately, involves the highly implausible manipulation of two adjacent ultra-high-density "cosmic strings" moving in opposite directions—exotic structures in exotic circumstances, and not exactly practical for someone wanting to go back in time and beat up Lee Harvey Oswald.

In 1993, Gott published a paper in *Nature* arguing that we can say, with 95-percent certainty, that humans will survive as a species somewhere between 5,100 and 7.8 million years. He said we are unlikely (per his statistical analysis) to colonize the galaxy. However, we are not the dummies of the universe by any means, because most alien civilizations don't engage in space travel.

He has a system for calculating these things. He starts with the Copernican Principle, which states that we are not in any privileged position. This applies to more than just the location of the Earth. It means that none of us exist in a privileged moment in time. What we call "now" should have no special position between the "beginning" and the "end" of anything. Gott first applied it to the Berlin Wall. On a trip to Berlin in 1969, he decided that, as a random observer, he should expect to be located randomly in time between the construction of the wall (1961) and its destruction (as it turned out, 1989). He has a formula in which he says that we can be 95-percent certain that the future existence of something (a wall, a civilization, a species) will be no less than one-thirty-ninth of, and no greater than thirty-nine times, its past existence. This seems fairly generous and hardly radical, at first glance. It becomes controversial only in certain applications. Gott thinks, for example, that the numbers foretell a massive die-off in human population. He points out that the human population has increased exponentially, and is expected to reach something like ten billion during the twenty-first century. The most optimistic scenario is that humans will somehow sustain their population for millions of years, but Gott says that's mathematically unlikely. If, eons from now, some observing entity were to make a list of every human being who ever lived, we should expect that our names would be randomly located more or less in the middle of the list. But if that list has quadrillions of names on it, our own names, today, would be over near the beginning of the list—in violation of Gott's rule.

He also says we shouldn't think of ourselves as being at the beginning of a long period of spaceflight, because this makes us "lucky" to be alive in a special moment. If humans will someday colonize other worlds, then how come we find ourselves on Earth, the home planet, and not, say, on a planet orbiting Alpha Centauri? Why weren't we born on one of our colonies?

His theory was criticized by Carl Sagan in *Pale Blue Dot* as being "vaporous," and I have to confess that I've never found a soul who is convinced by it. But it does stick in the brain, a kind of nagging thought. It suggests yet another constraint on intelligent life in the universe, a constraint imposed by our very existence at this moment, in this form. It says that the fact of being a human being on a planet with about six billion people, and being infantile in one's space program, and being relatively new to advanced technology in general, constrains the advancement of all other life in the universe. It's a weird, slightly creepy constraint. By locating ourselves in space and time, and doing certain things, and having certain conversations, thinking certain thoughts—reading certain books?—we collapse the range of cosmic possibilities.

A statistical argument like this is never convincing. But perhaps the Gott version of the Copernican Principle is a useful tool in thinking about why we're alive at this particular moment. It's something we've all wondered. We do not live in the Stone Age. Nor do we live in a galactic empire. What does that imply about human destiny? The Drake Equation has that big L at the end, the Longevity of an intelligent species, and it's still the greatest unknown.

What is going to happen to us?

Forecasting is a risky business. The atomic era promised energy too cheap to meter. It also promised, as a kind of backup plan, the total annihilation of human civilization. It was a good-news, bad-news thing. So far there is no sign that either scenario will happen. No one builds nuclear-power plants in America anymore, because they're too expensive to operate safely and within government regulations. The superpowers, meanwhile, held off on World War III, and then one of the superpowers folded its tent, and now we're more worried about extremely incremental annihilation through global warming.

Artificial intelligence was a hot idea in the 1970s, but the smart machines seem kind of dumb lately. The IBM computer system called Deep Blue managed to beat Garry Kasparov at chess, but only through "brute-force" computing. No one has yet invented a computer that can take cognitive shortcuts, much less give a hoot whether it wins. There are still predictions floating around that soon the most intelligent entities on the planet will be nonbiological, but for the moment the human brain is a far more complex device than anything made of silicon.

Which brings up a possible future that may supplant, or heavily amend, the *Star Trek* scenario that had been the favored future of our civilization for the last century. Biotechnology promises to alter the very essence of being a human being. It promises to tinker with human beings at the root of their existence. It could change the way people fundamentally are, cell by cell. Michio Kaku, in his book *Visions*, says we are making a fundamental transition here, from "unraveling the secrets of Nature to becoming masters of Nature." Barbara Marx Hubbard, a futurist and author of *Conscious Evolution*, believes this is a moment as important as the origin of life itself four billion years ago. "Human intention is entering matter," she says. "It means that instead of evolution working by natural and unconscious selection, evolution is beginning to happen through conscious choice." The biologist Lee Silver, author of *Remaking Eden*, predicts that biotechnology will give rise to an elite class of genetically altered individuals. In several centuries, the human race may be divided into two distinct species, the "Naturals" and what he calls the "GenRich." And then, of course, there's immortality. The antiaging researcher Michael Rose of the University of California at Irvine has figured out how to double the life spans of fruit flies. He thinks his techniques could someday be applied to humans. His conclusion: "Life spans are not inherently finite." The buzz is out there: The first immortals may already be walking around.

Ed Cornish of the World Future Society mentions, just offhandedly, the possibility that someday people will be able to have Elvis's baby. Elvis, though presumed dead, still exists in physical form in a grave at Graceland. That means his genetic code, his DNA, can be sampled and, presumably (okay, here's the big leap), used to create sperm or somehow generate a genetic donor cell. For that matter, he could be cloned. "How would you like a child by Abraham Lincoln?" Cornish says, obviously in a playful mood. "I think that may be possible. There may be remains of Lincoln that could be used." Eventually, Cornish says, it could be possible to decide what kind of body you want to have. People might want wings. They might want six arms, like a Hindu goddess. You can't help running the scenario to its logical conclusion. A craze could break out among teenagers—everyone has to have six arms or else he'll feel left out, like some kind of freak. And if the Russians start having eight or twelve or twenty arms, then the whole world could break out—it must be said—in an arms race.

There have been, over the past century, two competing futures, which can be simply categorized as utopian and dystopian. The intriguing thing about biotech is the way it splices them together.

The utopian future was extremely futuristic. It was mechanized, electrified, atomicized. It had a lot of nifty machines. Rooms were so clean they looked like laboratories. Humans wore uniforms, with bright colors like red and yellow. Some had V-shaped vests with shoulder fins so wide they could barely fit through a door. No one had any reason to break a sweat. To get almost anywhere, you could stand on a moving sidewalk. Your house would be made of waterproof plastic, and on cleaning day you'd simply hose everything down. Clothes would never get wrinkled . . . because they'd be made of the same stuff as your sofa.

This future could be seen at theme parks. Disney called it Tomorrowland.

The other future, the dystopian, was hideously dark. Human freedom would be crushed by totalitarian governments and soul-destroying technologies. George Orwell in *1984* gave us a vision of the future as a boot heel in the face . . . forever. There was a spell there in the 1950s and 1960s when it seemed as though every science-fiction story was premised on the Earth's turning into an irradiated wasteland. Of the various dystopian predictions of the future, the one that stands out is Aldous Huxley's *Brave New World*. Published in

1932, the novel is set six hundred years in the future, in a world in which humans have become an assembly-line product, grown from sperm and egg in factories and carefully programmed to believe social dogmas. Society is divided into castes that are biologically engineered, with Alphas as the elites and Epsilons as the low-intelligence drones. The elites take a feel-good drug called "soma."

Huxley's vision is recognizable in the contemporary debate over biotech. There is much soma on the horizon. People will be increasingly able to affect their minds with carefully designed chemicals, above and beyond the remarkable antidepressants and stimulants already on the market. But this is not a Brave New World as yet. To the extent that there is a caste system, it is driven by economics, discrimination, educational inequities, language barriers, and other factors, and not by scientists in lab coats manipulating embryos.

Why doesn't the future ever turn out the way it's supposed to? Because (and this is Dyson's argument) we aren't sufficiently imaginative. We extrapolate from the known, but nature is far more inventive. *Star Trek* extrapolated from 1960s technology, and so did not countenance genetically engineered astronauts. The totalitarian vision of Orwell was simply an extension of what already existed in the late 1940s in the communist world. The truly unimaginable thing, as late as the 1980s, was that the mighty Soviet Union would simply go poof.

It may turn out that genetically altered humans will be added to the list of things that were supposed to happen but didn't. The technology is a long way off. There is no human gene marked "intelligence." It's one thing to make a test for a defect in a single gene that partly controls tumor suppression, but it's another to make a test for what gives someone a sense of humor. The Human Genome Project and various private efforts will reveal the entirety of the genetic code, but it will still be coded information, a string of "letters," representing various sequences of nucleic acids, and the interpretation of these sequences will take decades at least. The biological system that creates a human being is fantastically complex. It took four billion years for the system to come together. This isn't like playing with Tinkertoys.

There is something that almost everyone forgets about in making forecasts: human choice. People might not want to have six arms. They didn't even want to buy the biotech tomato called the Flavr Savr. People are not yet robots who can be programmed by advertisers to consume the latest hot product. They might not want to do what a recent electronics advertisement said they'd be able to do: send e-mail while watching their children play soccer. Maybe they'd rather pay attention to the game!

At some point people may decide to make a stand against any truly radical biological engineering. Whether from a religious standpoint or a secular one, most of us understand that we should be grateful for who and what we are. We possess something invaluable. To alter ourselves in any radical way would reveal a raging biological greediness. We all want to live long, healthy lives, and we want our children to prosper, and few people can honestly reject the benefits of medicine and technology. If we could play tennis at the age of a hundred, we'd jump at the chance. But the goal should be to remain in every meaningful sense a normal human being.

Working with what we have, and appreciating it, seems like a more reasonable approach to life.

In the movie *The Arrival*, Charlie Sheen, looking even more beer-faced and burned out than normal, thinks he's detected an alien signal. He goes to a NASA official, his boss. This official (played by Ron Silver) is not Goldinesque—he's not expansive and enthusiastic and ready to jump on a starship—but, rather, is a narrow-minded, persnickety, pencil-pushing bureaucrat. The NASA official explains to Sheen that the suspicious signal does not meet the test of science, because it was not repeated. Without a repetition, it's like a nonevent—essentially, it never happened. The official then tells Sheen not only that his discovery is meaningless, but that the entire search for signals is being terminated, and Sheen needs to find something else to do. Naturally, Sheen is outraged. He can't believe that he has to make a second detection to make his discovery valid. Who made up that rule? he asks. Can we fire *him?*

The NASA official who enforces the hateful rules of science turns out to have a secret agenda. You see, he's actually an *alien in disguise*, part of a subterranean hive of buglike extraterrestrials. The aliens are causing global warming to make Earth more hospitable to their species. The moral of the story: When someone starts talking about what's "scientific" and what's not,

35

There Must Be More

he's probably really worried about something else, like his plan to conquer the world.

Hollywood knows how to press buttons in the public psyche. Hollywood knows that science is not universally loved. Almost everyone has some level of reverence for what science has wrought, in terms of improved medicine, geographic discoveries, the moon landing, and so forth. But as an industry, science is also resented. It is considered elitist and cynical. It questions the truth of people's deepest beliefs. It is incomprehensible, using a language that we do not speak. It is full of exclusions, prohibitions, and negations. It's no fun.

My mystical friends say things like: "When we make contact, it won't be with machines. We're not going to use the tools of science. It's going to be spiritual. We are going to travel through internal portals." They will say this while tapping their chests. (My own inner portal, I'm sad to report, is covered over with plywood.)

Many people want to achieve what they call "higher consciousness." What's remarkable is that, for them, a higher consciousness is not a metaphorical state but, rather, a literal, physical, vibrational condition. It has a reality to it that could theoretically be measured by careful use of a consciousness-ometer. Currently we lack such devices, but that doesn't mean (they would argue) that the higher level is imaginary. People want their cherished narratives of reality to be viewed not as fictional constructions or metaphor-laden imaginings but as literal truths every bit as "real" as the narratives of physicists and chemists and cosmologists. They want, in short, to have it both ways, to be free of the restrictions of science even while demanding that their beliefs be accorded the same level of respect as a scientific principle.

No one alive today can escape the fact that science is a foundation of our civilization. The most spiritual people still repair at some point to their computers to send e-mails. But they reserve the right to reject certain aspects of science, some of the annoying particulars—such as the fussier requirements of the scientific method. They certainly don't think of science as a comprehensive take on reality. It's more limited than that—even a physicist or a chemist would say so. Science describes the world to the extent that it is subject to the empirical investigation of science. For many people, this means that science provides only the *technical* version of reality. There has to be more! There must surely be phenomena that science can't quite handle, because they're too exotic, too . . . vibrational. And so some people who accept the basic tenets of astronomy do not want to exclude from the realm of possibility the basic principles of astrology. All the achievements of reason, from Aristotle through Copernicus to Einstein, have been lovely, but people still want more. The temptation is to lean one's head out the car window, to try to peek around the corner, stretch the neck, catch a whiff of the unseen world.

It is not enough that scientists announced that we live in a galaxy with a hundred billion–odd stars in a universe with tens of billions of such galaxies. People think there may be some other realms, some alternate spaces, parallel realities, hidden dimensions. How can scientists argue that the universe they've discovered is *all there is?*

People are not impressed by the scientific explanation of the mind. Science claims that a chunk of matter in the skull can manufacture the astonishing sensation of consciousness and give a person the ability to perceive the world subjectively, to love, to marvel, to look at a painting or a child's face and notice something beautiful. But that's insufficiently amazing. People want more than that—they want to have brains that can shape the world around them, bend spoons at a distance, change the universe with the amazing power of higher consciousness.

It isn't enough that science has discovered that humans are genetically related to every other living thing on the planet, from apes to squirrels to grasses to plankton to long-dead dinosaurs. No, people want to have something even more "holistic" than that, a connection among all life forms that is spiritual. They want the ability to channel the thoughts of dead people and aliens (though one never hears about anyone channeling a dead alien).

More. It's a wonderful concept. More of everything.

Channeling the thoughts of aliens isn't something that contravenes science, it's merely an extension of it (goes the theory). People want some science in the mix, hence the invention of "pseudoscience." The worst thing, for them, would be to lapse into mere "religion." That's not their game. The wonderful thing about contemporary ufology, astrology, New Age consciousness-raising, alternative medicine, and other fresh perspectives is that these belief systems can be defended as secular truths, fully modern, and nothing like the supernatural stuff taught in Sunday school. Best of all, ufology, astrology, etc., are advocated by a groovier set of people than is traditional religion or traditional science. Religion is pushed by grandmas and deacons and camp counselors; science is pushed by bowel-blocked, middle-aged Caucasians who can't relax.

We live in a world of machines. We do not understand the machines, don't know what's inside, and sense that the machines don't understand us. The computer is a magic box. We yell at it. It has no soul. Naturally, the last thing that anyone wants to be is a machine, to be a soulless entity, and that is the ultimate perceived slander of science, with its talk of neurons, brain chemistry, DNA, proteins, receptor cells, the geometry of life, a common origin in a warm tidal pool, everything descended from slime. That is a blood libel. Where's the spiritual dimension? People are offended. The mortal sin of science is that it does not take seriously the idea of mortal sins. The universe as described by astronomers does not contain good or evil. In the scientific view, no one made us, no one cares about us, and when we die we disappear for all eternity and our bodies are devoured by worms and all that is left is a skeleton that eventually turns to dust and then the Sun blows up and everything is destroyed and then other bad stuff happens. The driving force in the universe is not a Creator, but just some equations—it's just mathematics and geometry gone completely out of control. The universe is a machine.

(All this is my impression—thoughts of a generalizing nature, emerging from my brain in what feels like a random pattern, a synthesis after many interviews and much reading, and it doesn't feel like the product of neurotransmitters. If a machine had written those paragraphs, they would have been better organized.)

Bryan Appleyard's *Understanding the Present* is as articulate and thoughtful a review of the rise of science as anyone could wish to read, but he is also rather furious. He writes that science is widely promoted as "a triumphant human progress toward real knowledge of the real world. This is the official view of the schoolroom and the television spectacular. For me it is nonsensical propaganda which conceals all the important issues. In my version the story is a sad one, a long tale of decline and defeat, of a struggle to hold back the cruel pessimism of science."

He adds, "The struggle is to find a new basis for goodness, purpose and meaning."

What might that new basis be? He doesn't say. His book is essentially a complaint, not a prescription.

A kindred spirit of Appleyard is the historian John Broomfield: "Classical science bore the seeds of its own destruction. A consciousness that conceives an indifferent, spiritless machine universe of which the basic constituents are senseless chunks of matter locked in interaction by random, purposeless energies is profoundly unhealthy for humans, critters, and the very Earth itself."

This is a serious indictment, and requires a spirited defense. It's probably inadequate to go down the list of wonderful practical achievements that have emerged from the world of technology. Citing the eradication of polio or the tremendous decline in infant mortality is not really on point. The accusation, simply put, is that science has devalued the universe by robbing it of its soul. Those of us who would defend science have to make the case that we are not spiritually poorer for our gains in knowledge. I would suggest that this is a universe in which goodness, decency, beauty, love, and timeless truth remain viable notions. I believe humans can find in their own nature—the virtual miracle of being alive and conscious—of being thinking creatures!—a great deal of motivation to create a better and more beautiful and more ethical planet.

For many people, an ethical and loving life makes sense only in the context of a universe created by, and infused with the spirit of, a Supreme Being. There are others who do not consider gods or spirits (or aliens, for that matter) to be necessary elements of an enchanting universe. There are those of us for whom "being spiritual" is a metaphorical idea. We continue to explore that metaphor even as we assume that our loftiest and most "spiritual" feelings are firmly rooted in the material world, in our brains, our neural networks. A secular explorer of the world has to decide when to reduce an issue to its material constituents and when to remain metaphorical. Reductionism isn't all that useful, most of the time. Love, for example, may be a function of brain chemistry, and of animal behavior shaped by millions of years of evolution, but if you wanted to know more about love you'd have better luck talking to a poet than to a biologist.

Science is just a tool for understanding certain fundamental qualities of the world, and it is an extremely efficient and useful tool. What makes it so efficient and useful is that, at least in theory, it has no biases. It has an implicit oath to seek the truth, regardless of what the truth may be. It's irrelevant whether the truth in question makes people feel good. Scientists personally may have powerful social consciences (and form organizations such as the Union of Concerned Scientists), and they certainly make errors and inject their biases into experiments, but, abstractly speaking, science is cold and calculating. That's why it usually works so well as an investigative technique, and is so poor, most of the time, as a spiritual energizer.

There are periodic, noble attempts to reconcile science and religion. Not long ago, *Newsweek* ran a cover story declaring "Science Finds God." Such stories have the flaw of being untrue. God has so far remained undiscovered by the methods and machines of science. That may speak unfavorably of science, or well of God. Science has very little to say about God at all. Certainly there is nothing that disproves the existence of a Supreme Being. To ascertain the existence of divine entities, it is traditional to employ tools other than the scientific method. People discover God through faith, revelation, prayer, and this is as it should be. The assertion that science has found God is often based on the apparently remarkable fact that the universe is perfectly designed for the emergence of intelligent life. A few minor changes in the "constants" of nature would make our existence impossible. (If gravity were slightly stronger, the universe would never have expanded; if weaker, planets and stars would never have formed. A slight change in the weight or the charge of subatomic particles would make fusion impossible in the interior of a star—and the universe would be dark, cold, and lifeless.) The counterargument is usually some variation of Brandon Carter's "Anthropic Principle," which essentially states that only universes compatible with intelligent life will have anyone in them who stands around and wonders why the universe is compatible with intelligent life. There could be other universes, perhaps an infinite number, most of them less "successful" by our standards. Neither side could win the argument with the existing information. What could be said is that science has found that the universe is very much as a Supreme Being might design it.

Let's ask ourselves: What would happen if science really did find God? What would happen if someone found an unambiguous signal from God hidden in pi (which is something like what happens in Sagan's novel *Contact*), or if the Hubble really did find Heaven? It's possible that the response would be mixed. There would be an initial explosion of told-you-so's, of course, and much rapture and right-eousness. But some people would remain unsatisfied. The scientists would point out that there remain a host of unanswered questions as to the precise methodology and sequence of the Creation. "We will need a great deal more funding to study this," they would say. Meanwhile, the New Age people would absorb the discovery of God into their already elaborate agenda for achieving higher consciousness. God would be a great validation of the program. For some people, discoveries are interesting only to the extent that they can be directly applied to their personal lives, thoughts, moods, goals, relationships, and psychological transformations. Some people might even say: "You mean there's only one God? That's it???" And the paranoiacs would raise the possibility that this was an elaborate smoke screen, a ruse perpetrated by the men in black. How do we know He's God and not just the ultimate cosmic disinformation agent?

One afternoon in Estes Park, high in the Colorado Rockies, I took a break from a NASA meeting about the Origins program and climbed a hill near the camp, reaching the top of what is called Bible Point. There was an old metal mailbox at the crest, stuck into the rocks, tilted by the steady winds off the Continental Divide. Inside the mailbox was a notebook, with messages from hikers.

"Dear Heavenly Father, Thank you so much for this Creation. . . ."

"Savior, Trying to fathom your awesomeness and Your will is a feeble attempt at understanding God's hugeness. . . ."

"Dear God, As I sit here, I realize just how small I am. I beg of you to teach me. . . ."

The writers were expressing a sentiment that everyone in the world has surely felt. It's the same thing that Dan Goldin felt the night the sky exploded in stars on Mount Whitney. It's what John Glenn felt when he drifted as an old man above the Hawaiian islands. It's what the people in Montana felt when Goldin showed them the picture of the Eagle Nebula. It's what Bernie Haisch felt when he developed his theory of the Zero Point Field. It's what we all feel at one time or another as we perceive ourselves in the context of an amazing planet. What we feel is that this is not a *random* world.

We do not feel like accidental tourists in a reality created by mindless physical laws. We are certain that if we look hard enough we will see the pattern of meaning, the purpose of all this structure. A mountain cannot be simply a protrusion of rock. A mountain has a soul. A planet has a purpose. An intelligent species has a special mission. That's what you think, staring at the great spine of the Rockies, the jagged peaks layered in snow, dropping into forested valleys.

I could hear a tumbling stream below. Gray storm clouds drifted slowly across the Divide. So much structure! Could it possibly be that all this structure had no meaning? Or was that the challenge set before us? Perhaps that was the real mission for us all. We are given the opportunity of inserting into the picture the significance, the meaning, the goodness, the beauty, and the love. We are blessed to have that assignment.

At the end of the road there's a bottomless pit, a drop-off into nothingness, just mist and vapor. As a kid, I saw an episode of *Lost in Space* in which the Robinson family encountered a bottomless pit, and the pit scared the daylights out of me, just the thought of it. I couldn't conceive of how a hole could be infinitely deep, and this unknown dimension, this endless depth, became the very essence of fear—or so I say now, trying to reconstruct a childish concern. Part of getting older, growing up, is you don't get scared anymore by the pits. You accept them. You peek over the edge and trust your footing. You find mental tricks for thinking about complex problems, even if that requires spasms of illogic. (How far would something fall in a bottomless pit? All the way.)

The alien question is bottomless for today, and probably for years to come. There's no firm ground here. This is likely to remain a matter of infinite possibilities and zero certainties.

The Mars rock is still widely regarded as a misinterpretation rather than a great discovery. David McKay and Everett Gibson haven't given up; nor have their critics. The science is so abstruse, no one is likely ever to win that fight. Even McKay says that the answer is not in the rock. We'll have to go to Mars for the truth.

36

Personal Journeys

In March 1999, I had to make a final dash down to Houston to hear McKay and Gibson make yet another startling and controversial announcement. They'd found nanofossils in a different rock—maybe. There were bacterialike shapes, they said, in a meteorite that fell in Egypt, near the town of Nahkla, in 1911. They chose to tell their fellow meteoriticists at the annual meeting of the Lunar and Planetary Institute at the Johnson Space Center. McKay's talk lasted only about fifteen minutes, crammed into a full schedule of other presentations, and, to some grousing from the audience, the moderator declared that there was no time for questions. (Kathie Thomas-Keprta also had new evidence to present that she said supported the contention that certain elements in ALH84001 were biogenic and not mineral.) The controversy this time didn't reach escape velocity. And McKay and Gibson and their colleagues weren't pressing on the accelerator so hard this time, and in fact were not yet submitting the Nahkla research for publication in a scientific journal. McKay called the new finding a "hypothesis." This was more of a heads-up. We're still at it, they were saying.

Once again they didn't seem to persuade many people. The camps were more clearly delineated than ever. As always there was no way for an outsider (a reporter, for example) to know which side was right. It did seem strange that the Houston team had a knack for finding life where others didn't. Scientists had been furiously slicing up these meteorites, looking for fossils, since the August 1996 announcement about ALH84001, and in that time no one else had found anything. The Houston team was either very good or very wrong. It didn't help their cause that one of their own scientists, a postdoc visiting from England, had discovered that all these meteorites are full of terrestrial bacterial contamination. McKay admitted that the contamination was a red flag.

Someday we may know everything, and there will be no more Fermi Paradox, no more Mystery Constraint. Recently a brilliant physicist, Geoffrey Landis, published a paper suggesting an answer to Fermi's question, what he called the "percolation effect," in which a given galaxy will have, simultaneously, colonized regions and uncolonized regions. For a civilization to colonize other solar systems, Landis points out, there must be a cultural desire to expand, the technology to do so, and a candidate world within a reasonable distance. Some colonies will turn insular and will cease to expand in a given direction. Others will hit an empty region in the galaxy (sort of like the one we live in, perhaps) and decide that it's not profitable to go in that direction. The result is a galaxy teeming with intelligent life, but with much of that intelligent life isolated in uncolonized pockets.

Or maybe intelligence is constrained by those gamma-ray bursts that Henry Harris was so worried about. James Annis, an astrophysicist, recently argued that a galaxy-sterilizing gamma-ray burst occurs once per galaxy every several hundred million years. In the past, however, they may have happened more often, perhaps once every few million years, wiping out any life on land. The aliens, he figures, are out there, but they're likely to be young, like us, and haven't had time to master the technology of interstellar travel.

The quest to understand life in the universe will remain, for the moment, largely a search by analogy. There is much bottom-up astrobiology going on (much of it funded by one of Dan Goldin's creations, the Astrobiology Institute), and the research has essentially nothing to do with aliens. The scientists look at the organisms thriving around the boiling springs of Yellowstone, or in salt ponds in Baja California. They pound around Antarctica and the Atacama desert of Peru. One of these scientists is Penny Boston, who once dreamed of going to Mars. She was a member of the Mars Underground in Boulder back in the early 1980s. She realizes now she is not likely to go into space, so she goes into caves, searching for exotic life forms. She puts on a gas mask and plunges into a sulfurous, bat-filled hole in southern Mexico, *la cuerva de la Villa Luz*, where the shallow stream flows white, the air is lethal to humans, and bacteria drip from the ceiling in formations that Boston calls snottites. This is exobiology at the millennium: intrepid exploration of the alien environments within our own planet.

There is going to be, in coming months and years, more or less continuous breaking news on the extraterrestrial-life beat. There will be stories galore. The only thing these stories will lack, I'm guessing, is any actual evidence of life beyond Earth. Instead there will be evidence of potential habitats. Our own solar system may not be as dead as it looked after Viking and Voyager. The Mars program is aggressive and may yield the first Martian microbe. Europa looks more and more as though it has a liquid ocean. There are other moons in the outer solar system that are exobiologically intriguing—Callisto, Titan, Triton. And most fabulously, there are new planets, the extrasolar worlds found by Geoff Marcy and Paul Butler and their fellow planet-hunters, including the solar system discovered around the star Upsilon Andromedae. True, it didn't look like a terribly pleasant solar system. The planets were massive, Jupiter-like things and two were in tight orbits where they would have swept away any Earth-like planet. But that's a selection effect of the technology. The sensitivity of the telescopes is such that they can detect these bestial worlds only in nasty, close-in orbits that cause serious wobbles in the parent star. In the years to come, the tools of detection will become more sensitive. Inevitably, if there are Earths out there, they will be found.

Dan Goldin remains, at last report, determined to send a spaceship to Alpha Centauri. He said in the spring of 1999 that he hopes to launch a mission in thirty years, using a craft he calls Yoda. It would have both mechanical and biological components. This would not be your ordinary tin can winging off into space. This thing would contain DNA. It would have artificial intelligence, so that it could think its way through sticky situations. It would start out the size of a Coke can—not a nanoprobe but darn small—and would race to the asteroid belt, where it would mine carbon and iron. There, amazingly, it would grow larger, more powerful. It would evolve! Like those frontier explorers, Yoda would live off the land. And then it would zoom to Alpha Centauri, carrying with it our dreams, our hopes, and scraps of our genetic code.

When I last checked in with Henry Harris, he was still feeling gung-ho about laser-powered light-sails, and had just published an article on the topic in *Scientific American*. He complained that the magazine editors did not appreciate his literary style and insisted that the piece read more like standard scientific prose. He'd also written up a road map for NASA, and persuaded the bureaucracy to fund further research on beamed energy propulsion. He felt like a winner in the scientific competition to decide which way interstellar propulsion would go. His way, beamed energy, looked as if it would replace the traditional approach of rocketry. That was his assessment of the situation. Meanwhile, he had more personal passions. The artist within him kept stirring. He'd finished his novel, the one in which the piano-playing NASA scientist saves the world from a potentially lethal cloud of hydrogen that comes wandering in from interstellar space. He dished a few more details. His hero's name is Michael Star. Michael Star figures out how to destroy the cloud with a giant laser built from a nickel-iron asteroid, but while engineers are mining the asteroid they find billions of dollars' worth of gold and platinum, which triggers corruption involving the CIA and the Chinese Mafia, operating out of Paraguay. The novel is called "The Silent Dying." He hadn't managed to sell it and was looking for an agent. He said he was very tired, and I remembered our first meeting, how exhausted he'd been from staying up so late, thinking so hard.

The UFO movement will march onward, indestructible. In late 1998, a brilliant Silicon Valley entrepreneur, Joe Firmage, gave up his job as a CEO to start his own movement in pursuit of the alien mystery and the promise of the Zero Point Field. Firmage had read the dense papers of Bernie Haisch and the others who felt they had figured out the secret of mass, and he concluded that we are on the verge of an amazing moment in human history. When I interviewed Firmage, over dinner at a steak house in Los Gatos, I could see how he might become the next big figure in the UFO world. He was extremely smart (tremendous bandwidth, as they say in computerland), intense, and confident. He had lived in an accelerated world, where companies rise and fall in a matter of a few short years, where his own start-up company, USWeb, could zoom to a $2.5 billion market capitalization. For Firmage, it was perfectly reasonable to think that massive changes in society—formal contact with aliens, for example—could happen in the next couple of years. He wanted to be a part of that massive transition in human civilization; he wanted to be a player. He'd written a book and published it on the Internet. It was called *The Truth.* (He does not emanate a great deal of intellectual humility.) He recounted his own vision of an alienlike being, an entity who had appeared to him one morning in his bedroom, moments after he'd hit the snooze button on his alarm clock. He remembers telling the entity: "I want to travel in space."

So Firmage was fresh blood for the movement. He was only twenty-eight years old (many of the big names in the UFO world were getting geriatric, and the aged Colonel Corso, who had saved the world with alien technology, had passed away the previous summer). His great asset was his roaming intellect. He had a mind like an Internet search engine, a brain that could hyperlink through the universe of UFO information, science, religion, spirituality, even environmental issues. He was also a child of *Cosmos*. He revered Sagan. His father had insisted that he watch *Cosmos*, and the program transformed him. And yet he departed from Sagan dramatically on the issue of UFOs. I asked him why. Firmage answered that he knew things that Sagan didn't know. Sagan never learned, Firmage said, about the Zero Point Field. Firmage had come across Haisch's research—such as Haisch's theory about the Zero Point Field—while surfing on the Net.

Firmage's synthesis of all the incoming data was that the human species is on the verge of discovering the secret of manipulating gravity, of being able to harness energy from the vacuum. This technological ability has made us the target of scrutiny from advanced civilizations. They might decide to become more overt in their encounters with humans. They might, Firmage said, hover over Times Square, or show up at the Super Bowl. And it might happen soon. In the Firmage universe, everything happens fast. Either you adapt quickly or you become utterly obsolete.

Another development: In October 1998, the word spread in the press and on the Internet that an intelligent signal had been detected from an alien civilization in or around the star EQ Pegasi. The "hit" allegedly was made by an amateur SETI researcher, who would not identify himself or herself, but who had gained access to a private e-mail list designed for verification of SETI signals. Soon it became apparent that it was just a hoax, but some researchers—hyperalert to disinformation—suggested that the hoax itself was a hoax, that there really was a signal detected from EQ Pegasi and the cover-up had been diabolically clever. The word spread on the Internet, among ufology sites and New Age spiritual groups, that a spaceship would land in Arizona on December 7. One report stated that the spaceship was hostile. Richard Hoagland's Web site noted a great deal of government activity in and around one mountain in particular. Unfortunately, the arrival was literally a washout, as a heavy storm moved through the area. Hoagland raised the possibility of foul play, pointing out that David John Oates, the man credited with discovering "reversed speech," had divined that a comment by Secretary of Defense William Cohen, when played backward, said "Plan evil weather." Hoagland entertained the possibility that the bizarre weather was induced by a secret government facility as part of a "hyperdimensional physics" experiment, and that this masked the arrival of the spaceship.

The point being that the intellectual flexibility and creativity of the anomalistic community have no limit. The narrative of the aliens will always grow, mutate, and somehow lurch forward.

Again and again, as I tromped through alien country, I noticed that the most powerful battles in our lives are deeply personal ones. I was always struck that so many people who dealt with the alien question had their own struggles to stay healthy, or stay alive, or stay sane, or merely pay the rent.

I thought of how Jill Tarter, the SETI researcher, had to overcome not only the political and intellectual objections to the radio search but also a serious bout with cancer. She survived, having endured the irony of searching for advanced civilizations in the universe while trying to maintain a single, fragile human life on Earth. When I last saw her, in April 1999 at a "Cosmic Questions" conference in Washington (among the exceedingly cosmic questions with zero chance of being answered during the three-day session were "Are we alone?" and "Was the universe designed?"), Tarter seemed every bit as determined and optimistic and forward-thinking about SETI as she had been that stormy night at Green Bank. It's her job to remind people, constantly, how big the galaxy is (not to mention the universe), and how feeble our searching abilities are at this moment in time. Even something as grand (and unbuilt) as Goldin's Planet Finder, she said, would only be able to scan for planets around the nearest two dozen stars, approximately. Even with a radio search like her own Project Phoenix, she said, we only look at the nearest region of the galaxy. "We're going out now a hundred and fifty-five light-years in a galaxy that's one hundred thousand light-years in diameter. It's way too early to get discouraged."

Jerry Soffen, the man behind the Viking mission, confided about his own experience with pain and tragedy. After Viking, he had thought of writing a book on the mission. He had all the documents, the memories. He had lived every emotional high and low of the great

Martian adventure. And then, one day, his wife drowned in the family's swimming pool. She hadn't been a good swimmer. He was shattered. The Viking papers, boxes and boxes of personal records and correspondence, were consigned to an attic, and Soffen never wanted to see them again. He cannot easily disentangle his greatest achievement from his worst experience.

I will always remember my meeting with Dick Joslyn, who told me about Heaven's Gate. It was not some abstract tale to him. We were talking in a Starbucks on Sunset Boulevard in Los Angeles. He seemed rather dispirited, even a bit dismayed, as though he couldn't quite believe that he had wound up in such a sorry situation. He had no illusions about his status in the world. He was nearly fifty, losing his hair, HIV-positive. He was still handsome, well built, but he hadn't made anything of his looks, hadn't become the actor he could have been: "I could have been a star," he said. He didn't have a regular job and could not imagine, for example, working in a coffee shop. "I've dealt with the cosmos for twenty years. How could I work at Starbucks?" Some nights he hung out in a café in West Hollywood, reading tarot cards. Sometimes he worked the phones for the National Psychic Line. There were always people desperate to find out who they were, what would happen to them. Joslyn was a counselor, in effect, but he was still struggling to figure out what exactly had happened to his own life. The one thing he could cling to was that, unlike his fellow cultists, he'd escaped. But where would he go from there?

After Heaven's Gate, people stopped showing up at Miesha Johnston's Starseed meetings in Las Vegas. The first defections were the kids, the members of her teen group. Their parents pulled the plug. "They didn't want their kids involved in anything that might be a cult," she told me. The adults were also leaving. Miesha was distraught. She had labored for five years to put together a solid network of abductees and experiencers, and Heaven's Gate had destroyed everything.

Then she and Jan Bingham had a falling-out. They had never really been on the same page. Jan had a more radical view of what it meant to be a Starseed: She literally felt she was an extraterrestrial spirit in a human body. Miesha had never wanted to commit herself to that idea. Miesha also found herself breaking up with her boyfriend. Finally, she decided to hit the road. "People were too skittish, too frightened." Miesha and Deborah, the woman who had tried to hyp-notize me, and a third friend moved to San Diego. In their group of believers there was a new emphasis on Satan, and Miesha didn't want to go down that path. She didn't think the aliens were all a bunch of agents of the devil. Deborah moved again, and Miesha lost touch with her. When I last spoke to Miesha, she was a week away from moving yet again, unsure where she would find an apartment. She worried that she wouldn't find a roommate who would understand her beliefs. She had two small support groups a month, a core of twelve people, and by day worked as a receptionist.

Her fundamental beliefs hadn't changed, but she'd become more wary. "There is a major amount of disinformation," she said. People were lying. Either that or they were simply too credulous: "They're being gullible." She'd come to the conclusion that the only sure truth about aliens is that nobody knows the truth: "There are no absolutes in this."

I paid a final visit to Jan Bingham in Las Vegas. Jan had gotten even deeper into her Starseed cosmology. We tried to sort out our dif-ferences. She was angry about an article I'd written in the *Post* that she felt had insulted the Pleiadians as a group, conflating them with the Grays. She quizzed me: "Tell me, Joel, what's the difference between a Gray and a Pleiadian?"

I did my best to convince her that I was now fully educated on the difference, though I couldn't help thinking that the Pleiadians are what a newspaper reporter would call "libel-proof." Normally some-one is "libel-proof" on account of being a documented criminal or fa-mously despicable person—Ted Bundy, for example—whereas the Pleiadians fit the description on account of being, as far as I knew, nonexistent. Jan was also annoyed that I had written that humans and aliens would, according to her, someday get along nicely, whereas in fact we'll never get along with the reptilians. The reptilians, she said, are extremely malevolent. "Two of the highest leaders in our military are reptilians," she said.

Suddenly Jan said something that echoed Dan Goldin's rule, that human life is more than survival. "We know that there's more to life than getting up every day and going to work and going to the gas station," she said. "Our immortal life force, and you can laugh at this all you want, and this can be *scientifically proven*, is who we are." She was enthused about the movement to raise our consciousness. "It's a spiritual movement. And that movement is about getting back to what we are. We're not our bodies, we're our life forces."

I wanted to shout: Wrong! There is no life force! I have an entire chapter on that exact topic! We're just the distillate of billions of years of evolution that have turned simple structures into highly organized structures! But by this point I knew it was not possible to argue with Jan. In a way, she was always going to be the dominant one in the room, because all I had were factoids and semieducated notions from my various conversations with scientists, whereas she had *beliefs*. She could dismiss twenty-five hundred years of scientific inquiry with a few deadly sentences.

"Scientists are in their brain. They're in their body. Human bodies are an experience of only the first four dimensions of consciousness. They're just vehicles for getting us someplace. . . . We're children in the universe. We're infants, okay? So these human bodies with these limited, mortal human brains get us through these experiences that help us learn and grow. We're like third-graders in the universe."

The Assumption of Mediocrity.

She said there are eight dimensions of consciousness. We have a long way to go. The Earth is actually a living entity who has reached one of the higher levels of consciousness. She said she's had conversations with the Earth. I asked if she was sure about that. "I'm very sure," she said. But doesn't she have doubts? "Remember, I have a background in academia," she said. She said she sometimes leaves her body behind and travels interdimensionally. She's had conversations with Sananda, whom we know as Jesus. She complained to Sananda; her father, Zachary; and the prophet Ezekiel that the Starseed are being neglected, that they are suffering too much as they live in mortal bodies. There's too much pain.

I was about to say goodbye when Jan revealed that there was about to be a Transition to a new age, and that it would involve tremendous suffering, natural disasters, death on a massive scale. She'd had a vision of the future, and it had shocked her. "It looked like an atom bomb. I sensed all the devastation and the grief and the pain."

She began to cry. The Transition, she said, could happen in a matter of weeks. She was quite distraught and I reassured her as best I could. Jan Bingham and I lived on different planets, and barely inhabited the same universe, but we both struggled with our fear of bottomless pits. We hugged goodbye, and as I drove away I found myself hoping that, somehow, she'd survive the Transition, and reach the next level of consciousness.

Sagan made his own transition in the later years of his life. He did not grow by adopting new belief systems that were designed to reassure and empower him, or by abandoning his ability to reason out a problem, or by determining that science provides an extremely limited take on the nature of reality. He grew by becoming a nicer, humbler, more decent human being. He learned to take care of the people around him.

As a younger man, he was overbearing, quick to lacerate an opponent for mushy thinking. He had the full complement of arrogance, as a male, a scientist, an acclaimed professor, a TV star. But his brushes with death had clearly changed him, and he made good on his vow, as he entered middle age, to be a better husband and a better father. It was his third try at it, and he finally got it right. Everyone who had known the younger Sagan preferred the older one, even his first wife, Lynn Margulis, who marveled at the transformation.

He was diagnosed with myelodysplasia in the fall of 1994, and for the next two years he struggled to stay alive. He endured two rounds of chemotherapy. His sister, Cari, donated the bone marrow he needed to survive. To prevent his body from rejecting the marrow, he had to take, in one sitting, seventy-two pills labeled "BIOHAZARD." These essentially wiped out his immune system, and would have killed him outright had he not had the bone-marrow transplant immediately. In the meantime, he could have been killed by a single rogue microbe—some humble example of life. From his hospital bed he wrote a moving piece for *Parade* magazine: "There are scientific problems whose outcomes I long to witness—such as the exploration of many of the worlds in our solar system and the search for life elsewhere."

It seemed almost impertinent of nature to confront Sagan with the question of life on such an individual scale. He had scrutinized images of Venus, Mars, the moons of Jupiter and Saturn; had flown, empathetically, with robotic probes reconnoitering the outer limits of the solar system. He'd written the "Life" entry in the *Encyclopaedia Britannica*! Now he was desperate to keep his own heart beating.

He knew it was unlikely that he'd see the great breakthrough in exobiology.

"I'd rather there be extraterrestrial life discovered in my lifetime than not. I'd hate to die and never know," he told me that day I met him in Seattle. But then he quickly qualified his wish: He was never going to accept flimsy evidence for extraterrestrial life. It had to be nailed down.

In fact, Sagan had something of a self-created dilemma on that score. In his last years, he suspected that he himself might have detected a number of alien radio signals. He and Paul Horowitz of Harvard conducted a five-year search called Project META. They were doing an all-sky survey, looking for strong signals near the 1,420-megahertz line of neutral hydrogen and its harmonic at 2,480 megahertz—magic frequencies for SETI people since the Cocconi-and-Morrison paper of 1959. On thirty-seven occasions they detected strong, narrowband radio signals of . . . something. Most could be explained away as terrestrial interference (a passing plane, for example) or "processor errors." Moreover, none of the thirty-seven signals was repeated. But what really caught Sagan's attention was that the five strongest signals came from the direction of the galactic plane. If the strongest signals were from terrestrial interference or random processing errors, then they shouldn't be clustered along the axis of the galaxy.

Sagan said the chance of this preferential location being accidental is "something like half a percent." He strongly suspected that he had detected an artificial signal. But he hastened to add, "That's not strong enough to be sure. It's certainly suggestive. You know, it sends a kind of chill down your spine, your palms get moist, your breathing gets heavy."

Horowitz did not have the same heavy-breathing response. In fact, it was a source of some conflict between the two scientists. They struggled to come up with a mutually acceptable manner of describing this apparent preferential location of the five suspicious signals. How can responsible scientists mention something like that? When they finally published a paper in the *Astrophysical Journal*, they pretty much buried the finding. The abstract had a single line, "However, the strongest signals that survive culling for terrestrial interference lie in or near the Galactic plane." The body of the paper said, "While this may be due to the statistics of small numbers, it is just what would be expected for transmissions from civilizations within the Milky Way, or from a previously unknown astrophysical source of narrow-band radio emission distributed with Galactic stars, gas, and dust."

Later, in his book *Pale Blue Dot*, Sagan was more effervescent about the strange radio "events." He offered up a bit of "extravagant speculation": Let us assume, he wrote, that the five signals are, in fact, beacons from alien civilizations. By factoring in how much time was spent observing the different patches of sky, it should be possible to estimate how many radio transmitters there are in the entire galaxy. "The answer is something approaching a million," he wrote. A million civilizations—precisely the number he'd been favoring for decades! His five inexplicable events suggested, for the great optimist, a galaxy thick with consciousness.

But maybe it meant nothing at all. He had to allow that possibility, too. Maybe the radio events were associated with an astrophysical phenomenon, something unrelated to biology. It was no accident that he did not bring up these signals until page 361 of his book. He could not claim a discovery. In this game, there are no "almost" breakthroughs. This isn't horseshoes or hand grenades—it's not enough to be close.

And so the life of Carl Sagan wound down, day by day, with the greatest mystery of the universe unanswered.

The Sagans and Goldins had dinner one night at the Four Seasons in Washington, and they ran over all the great things happening in the world of exobiology. Europa was thrilling. The new pictures of the icy moon were slowly trickling in from a tiny, auxiliary antenna on the Galileo space probe. There was all the excitement about the Mars rock, and about the extrasolar planets. Occasionally during the dinner Sagan would appear to nod off; he was exceedingly weak. Then he'd perk back up, and continue to spill forth his ideas about life in the universe. Goldin was amazed. Near death, Sagan remained the same—a restless spirit, dreaming, thinking through the possibilities.

I spoke to Sagan only a few times. During the final telephone interview, I heard the doorbell in the background. Sagan put down the phone briefly, and I could hear the conversation that ensued. At the door was an exterminator. He had stopped by to spray for carpenter ants.

"What are you spraying? What chemical?" Sagan asked. "You know its structure? You know its *chemical formula?*"

The exterminator gave the name of a chemical.

"That's just a *name*," Sagan said. "Do you have a structural diagram of the molecule?"

The exterminator eventually produced a diagram. Sagan approved the molecule.

A scientist needs evidence. Faith is not part of the game. The axiom applies in matters both great and small. Sagan's greatest professional achievement may have been his ability to stick to science and resist the incredible allure of sentimental thinking. He had wanted to tap into the life force of the cosmos, to be part of a galactic community, but he didn't want to be Percival Lowell. He won that battle. He was a scientist to the end.

There were no deathbed conversions.

Three separate memorial services paid tribute to Carl Sagan, one at Cornell, one in Pasadena, and a final one in New York City, at the Cathedral of St. John the Divine. The cathedral event was a touch odd in that Sagan was essentially an atheist, even if he refused to wear the label. Stephen Jay Gould showed up; and Neil Tyson of the Hayden Planetarium; and Philip Morrison, who got the SETI thing going way back when; and several theologians, including the Reverend Joan Brown Campbell (she said that of course Sagan was an atheist—which sounded like her way of finally winning an old argument) and the Very Reverend James Parks Morton. Anchoring the lineup was the vice-president of the United States. They spoke of Sagan's achievements, all his scientific and political adventures, his charity, his compassion, his efforts to save the Earth from destruction. None of the luminaries were quite so moving in their comments as Ann Druyan, Sagan's grown son Jeremy, and his teenage daughter, Sasha. But, ultimately, it was Carl Sagan himself who stole the show. Someone had thought to bring a tape. It was Sagan reading from *Pale Blue Dot*. His voice boomed through the cathedral, fully occupying the vast space, a verbal thunder, reverberating from the vault, each word Saganistically enunciated, the italics flying, the consonants threatening to stick as though a stammer might kick in, then suddenly exploding forth—the whole Sagan catalogue of oral moves. He was speaking from beyond the grave. Strikingly, he was not speaking about outer space. He spoke about the Earth, as seen from the Voyager 1 spacecraft as it left the solar system. Sagan had persuaded NASA to take a picture of the Earth from billions of miles away. The Earth was a featureless blue speck.

"That's here. That's home. That's us. On it, everyone you love, everyone you know, everyone you ever heard of, every human being who ever was, lived out their lives. The aggregate of our joy and suffering, thousands of confident religions, ideologies, and economic doctrines, every hunter and forager, every hero and coward, every creator and destroyer of civilization, every king and peasant, every young couple in love, every mother and father, hopeful child, inventor and explorer, every revered teacher of morals, every corrupt politician, every superstar, every supreme leader, every saint and sinner in the history of our species, lived there—on a mote of dust suspended in a sunbeam."

Then he was gone. The cathedral grew quiet. Nobody could compete with that.

The celebration soon ended. The mourners filed out of the cathedral and down the front steps, and dispersed through the streets of Manhattan—special creatures, moving quickly, thinking, searching.

Acknowledgments

Early on, a friend suggested that what I needed to do, more than anything, was to get myself launched into space. I fear I failed miserably on that score, and remained earthbound, and indeed seemed to spend a great deal of time in small dim rooms with low ceilings—not an astronaut but a burrowing, nocturnal creature. On the other hand, anyone who produces a book will know that it feels a bit like being blasted off the planet. You drift in the vacuum. You crash on the far side of the moon, beyond radio contact with your friends and loved ones. The only way to get home again is through the kindness and generosity of dozens of people who serve as the book's Ground Control.

Two institutions directly and lavishly supported this work. John Logsdon, director of the Space Policy Institute at George Washington University, took me aboard for a year as a visiting fellow, which is to say that I was a parasite, feeding off the brilliance of a group of scholars who possess a keen understanding of what humans are doing in space. Their filtered wisdom is throughout these pages.

The Washington Post, my employer, has been generous at every stage of the book. I am indebted in particular to David Von Drehle, Steve Coll, and Mary Hadar, three of the best newspaper editors in the country, all of whom encouraged me, corrected my errors, and gave me space in the paper to tell some of the stories that eventually turned into *Captured by Aliens*.

Among those who read the manuscript and offered expert criticisms, but who are not responsible for any mistakes or misapprehensions, are Martin Harwit, Bob Park, Dwayne Day, John Logsdon, and Richard Berendzen.

Gene Weingarten, my longtime editor, mentor, and friend, read the earliest version and ordered me to change it in several ways that improved it dramatically. Pat Myers, as she has before, labored to elevate the text to the level of standard English. Michael Congdon showed once again that a great agent is also a great critic. Walter Isaacson put in a good word at a crucial moment. My editor Alice Mayhew of Simon & Schuster believed in the premise of the book even when that premise remained unclear; she was, in other words, a visionary.

A number of scientists and thinkers gave special assistance: Lou Friedman, Chas Beichman, Kevin Zahnle, Jack Farmer, Chris Chyba, Chris McKay, Bruce Jakosky, Michael Meyer, John Rummel, Chuck Klein, Bill Schopf, Jerry Soffen, John Kerridge, and Phil Klass. I am grateful to Ann Druyan for providing research material from the archives of Carl Sagan.

Thanks also to: Roger Labrie, Jennifer Thornton, Marc Fisher, Tom Shroder, Roger Launius, Roddy Young, Karl Pflock, Megan Kemble, Trish Mastrobuono, Henry Herzfeld, Patti Cohen, Deb Heard, Glenn Frankel, Len Downie, Layla Hearth, Susan Faludi, Russ Rymer, Naomi Wolf, David Shipley, Tony Horwitz, and Geraldine Brooks; and to Jim and Emily Notestein.

My brother, Kevin, let me sleep on his couch and showed me how to win big money at the track (theoretically).

And finally I benefited immensely from the editing skill, spiritual depth, and patience of Mary Stapp. Thank you Mary and thank you Paris, Isabella, and Shane—lovely examples of life on Earth.

Notes

1. DIALING UP ANDROMEDA

14   In the passenger seat, Sagan ate scraps: Specifically, he ate stale garlic bread. Drake recounts their commute from the beach to Arecibo in his memoir, *Is Anyone Out There*, co-authored by Dava Sobel (Delacorte, 1992).

15   mere "plausibility arguments,": Interview with author.

16   It was such a simple question: A good account of the famous lunch with Fermi is in the 1993 revised version of Walter Sullivan's *We Are Not Alone*, originally published in 1964 (McGraw-Hill). Although his title is melodramatic and rather definitive on the existence of intelligent extraterrestrials, the book is a perfectly responsible and comprehensive look at the evidence, including claims about UFOs.

16   Arthur C. Clarke novel *Childhood's End*: The aliens, creepily, are dead ringers for the ancient image of the devil, complete with horns and tail. What's remarkable is that many of Clarke's themes persist today in one way or another, including the idea that humans are on the brink of a radical change in consciousness, an evolutionary upgrade, signaled by an increase in telepathy.

17   A bunch of astronomers and visionaries: The group included Giuseppe Cocconi, Philip Morrison, Dana Atchley, Barney Oliver, Melvin Calvin, John Lilly, Su Shu Huang, and Otto Struve, the last serving as host and chairman. Sagan was only twenty-seven, working with Calvin, a chemist, on the origin of life. On the second day of the conference, Calvin learned that he'd won the Nobel Prize for his work on photo-

synthesis. This fabled meeting, sponsored by the Space Science Board of the National Academy of Science, has been written up by virtually every author to address the topic of ET. See, among others, David Fisher and Marshall Jon Fisher, *Strangers in the Night*, Counterpoint, 1998, pp. 184–92; and Barry Parker, *Alien Life: The Search for Extraterrestrials and Beyond* (Plenum, 1998), pp. 167–69. For an example of what mainstream scientists talk about at a serious conference about extraterrestrial intelligence, see John Billingham, ed., *Life in the Universe* (The MIT Press, 1981).

22 I wrote up a profile of Sagan: "The Final Frontier? All Carl Sagan Wants to Do Is Understand the Universe. All He Needs Is Time," *Washington Post*, May 30, 1996, p. C1.

2. THE GOLDIN AGE

25 "In the twenty-first century . . .": Goldin said this at the January 1998 meeting of the American Astronomical Society in Washington.

27 "Maybe we will learn": Transcript of Dan Goldin's remarks at Steps to Mars Conference, July 15, 1995, Washington, D.C., reprinted in *Strategies for Mars: A Guide to Human Exploration*, ed. Carol R. Stoker and Carter Emmart (American Astronautical Society, 1996).

29 Yes, there is probably life there: The effusive statement, taken out of context in a wire report, came from John Delaney of the University of Washington.

29 And something even bigger was in the offing: Goldin told me later that he didn't know about the Mars-rock research when we first discussed the great unknowns of science.

3. THE BIGGEST UNKNOWN

34 appeared to buzz the White House: On July 20, 1952, and subsequent nights there were blips on the radar screens at Washington National Airport's control tower. Some radio operators saw, outside, a fiery-orange sphere. The Air Defense Command sent interceptor jets to the scene, but the pilots didn't see any saucers. Witnesses on the ground reported seeing strange lights. In Donald Keyhoe's account, in *Flying Saucers from Outer Space*, there is no question that the blips are saucers: "From their controlled maneuvering, it was plain they were guided—if not manned—by highly intelligent beings. They might be about to land— the capital would be a logical point for contact. Or they might be about to attack." The sightings caused a media sensation. Harvard astronomer Donald Menzel, however, said a temperature inversion over Washington—warm air on top of cold—caused reflections of radar waves.

34   Bill Clinton . . . told . . . Webster Hubbell: This is mentioned in Hubbell's memoir, *Friends in High Places* (Morrow, 1997), p. 282, and elaborated upon by Hubbell in an interview with the author.

34   *Confirmation:* My favorite part of *Confirmation* is the prolonged discussion of a piece of footage shot by the Space Shuttle in 1991. It shows a little white dot moving near the Earth. There's a flash of light, and the dot suddenly changes direction. Something streaks by the dot. The show raises the possibility that this is an enormous spaceship, with a nonhuman pilot (since a human couldn't survive the G-forces involved in the sudden change in direction), and that it was fired upon by a secret American Star Wars base in Australia. But the show, to its credit, gives another possibility: that these are just tiny particles of ice drifting by the shuttle's window. It might be the frozen waste of the shuttle itself. I thought that captured the essential dichotomy of the UFO debate. (Alien spaceship—or frozen pee? You make the call.)

35   just a *little bit invaded*: Phil Patton's fine book *Dreamland* (Villard, 1998) offers the thought that the fascination with alleged flying saucers at Nevada's "Area 51" is a result of decades of Cold War secrecy about experimental aircraft.

37   legitimate exobiology and "kooky" ufology: A good discussion of the gap between mainstream science and ufology can be found in John B. Carlson's introduction to Richard Hall's *Uninvited Guests: A Documented History of UFO Sightings, Alien Encounters & Coverups* (Aurora Press, 1988). Hall, meanwhile, is a careful and thoughtful ufologist whose various writings are excellent references for anyone looking for serious UFO research.

38   cosmologies that blow our minds: Stephen Hawking's *A Brief History of Time* (Bantam, 1988) showed that cosmology and abstruse physics could be the subject of a best-selling book, but I find Timothy Ferris to be the best explicator of modern cosmology for a popular audience. See *The Whole Shebang* (Simon & Schuster, 1997) and, for a somewhat breezier read, *Coming of Age in the Milky Way* (Simon & Schuster, 1986).

40   Let me offer a short list: Which became a story in *The Washington Post*, "Pushing the Boundaries of the Cosmic and the Microscopic," August 11, 1996, page A1. These questions condensed from the dust and fumes of my labors over the years writing the column "Why Things Are." For an engaging account of many of these unanswered questions, see John L. Casti, *Paradigms Lost: Tackling the Unanswered Mysteries of Modern Science* (Avon Books, 1989).

40   the question of "dark matter": Unless there is more matter out there than anyone's found so far, the universe won't have enough mass to slow down its expansion, and will slowly drift apart and grow colder. The final result is the Heat Death of the Universe. If there is enough dark matter to "close" the universe, it will eventually collapse in a Big Crunch. A third option is that the amount of matter is at a critical den-

sity that will allow the universe to expand forever, but ever more slowly, so that it sort of glides along for all eternity. The fate of the universe is, one must acknowledge, not a trivial issue. Nonetheless, it does not make my list, on the grounds that the fate of the universe is determined by the answers to the first two questions. And it lacks urgency.

4. ALIENS IN THE ARCHIVES

43   aliens . . . are a dusty old idea: There are several books that collectively tell the story of humankind's fascination with extraterrestrial life, and to which I have turned repeatedly. They are Steven J. Dick's *The Biological Universe* (Cambridge University Press, 1996), Michael J. Crowe's *The Extraterrestrial Life Debate, 1750–1900* (Cambridge University Press, 1986), and Karl Guthke's *The Last Frontier* (Ithaca, 1990).

44   "infinite worlds both like and unlike . . .": Quoted in Crowe, *Extraterrestrial Life Debate*, p. 3.

44   "It is unnatural in a large field . . .": Quoted in Michael D. Papagiannis's essay in *Extraterrestrials: Where Are They?*, ed. Ben Zuckerman and Michael Hart (Cambridge University Press, 1995), p. 103.

44   ". . . It may be conjectured . . .": Ibid., p. 8.

45   "How can we be masters . . .": Quoted in Paul Davies, *Are We Alone?: Philosophical Implications of the Discovery of Extraterrestrial Life* (Basic Books, 1996), p. 6.

46   aliens must have hands: "Either then the Gentlemen that live there must have Hands, or somewhat equally convenient, which is no easy matter; or else we must say that Nature has been kinder not only to us, but even to Squirrels and Monkeys than them." Christiaan Huygens, *The Celestial Worlds Discover'd: Or, Conjectures Concerning the Inhabitants, Plants, and Productions of the Worlds in the Planets* (Frank Cass & Co., 1968), p. 73.

46   "The excellence of the intelligent creatures . . .": Norman Horowitz, *To Utopia and Back* (W. H. Freeman, 1986), p. 54.

46   "It would be presumptuous . . .": Quoted in Crowe, *Extraterrestrial Life Debate*, p. 199.

47   Lowell saw four hundred canals: Dick, *Biological Universe*, p. 87.

47   "There is strong reason to believe . . .": Quoted in ibid., p. 74.

47   "In our exposition of what we have gleaned about Mars . . .": Percival Lowell, *Mars as the Abode of Life* (The MacMillan Co., 1908), p. 214.

48   "A mind of no mean order": Percival Lowell, *Mars and Its Canals* (1906), quoted in Logsdon et al., *Exploring the Unknown: Selected Documents in the Evolution of the U.S. Civil Space Program*, vol. 1, Organizing for Exploration, NASA SP-4407 (Government Printing Office, 1995), pp. 55–56.

49   "It would imply that man is an animal . . .": Quoted in Dick, *Biological Universe*, p. 48.

49 "Professor Pickering of Harvard Observatory . . .": *The Papers of Robert H. Goddard*, vol. 1, *1898–1924* (McGraw-Hill, 1970), pp. 64–65.

5. THE GATEKEEPER

53 "There was a kind of view . . .": Interview with author.

53 ". . . I feel it in my bones . . .": Recalled by Dyson, interview with author.

55 ". . . dumbest communicative civilization in the galaxy": Richard Berendzen, ed., *Life Beyond Earth and the Mind of Man* (NASA, Scientific and Technical Information Office), p. 63.

55 Many people would say: Although many professors found Sagan somewhat insufferable, others revered him. One night at a scientific conference, when everyone was hoisting beers in a pub to end a long day of lectures, I asked a couple of his colleagues if they were annoyed by Sagan's fame, his money, his love of the camera, his rogue intellectual wanderings into fields where he had little expertise, and so on. They looked at me in stunned silence. Then one of them said, "He was our f—— hero, man."

55 "Which of the stars is my best bet?": Frank Drake and Dava Sobel, *Is Anyone Out There?: The Scientific Search for Extraterrestrial Intelligence* (Delacorte Press, 1992), p. 119.

56 "I have used them in university commencement addresses . . .": Lester Grinspoon, *Marihuana Reconsidered* (Harvard University Press, 1971), p. 114.

56 Sagan was completely outgunned: At this meeting, Sagan submitted a copy of an article he had written for the *Encyclopedia Americana* in which he pointed out that many people may discover that aliens can replace the gods deposed by modern science. Sagan noted the lack of participation of people who "strongly disbelieve in the extraterrestrial origin of UFOs" but said that this view was "not necessarily one I strongly agree with." See transcript, Symposium on Unidentified Flying Objects, Committee on Science and Astronautics, U.S. House of Representatives, July 29, 1968.

57 Richard Hoagland: Hoagland has some bona fides outside of ufology. He worked as a consultant for CBS News during the Apollo program. Arthur C. Clarke credits Hoagland as the first person to explore the possibility that life exists in an ocean under the surface of Europa.

61 "Maybe there is one hiding . . .": Carl Sagan, *Pale Blue Dot: A Vision of the Human Future in Space* (Random House, 1994), p. 57.

62 a Russian scientist: An alternate spelling is Iosif Shklovsky. This was one of the first times Sagan reached out to a Russian counterpart, but not the last. He would, over the years, spend a great deal of time and energy fostering collaboration between American and Soviet scientists. It was

his contribution to world peace, an effort to stave off World War III. I think of it as a sign of his desire for "contact" in a larger sense. The Soviets were isolated. They were an alien civilization. Shklovskii had been, for twenty years, denied permission to travel outside Eastern Europe. Contact with the Soviets prefigured the more difficult job of bringing together humans and extraterrestrials.

62   Sagan arranged for the English translation: In an obituary that Sagan wrote about Shklovskii decades later, he initially said the collaboration was "with his concurrence" but then changed the language to "at his suggestion." To enable readers to know what part of the book was the Russian's and what part the American's, Sagan used little arrows to bracket his contributions. The result is a bit distracting—the reader feels compelled to keep track of who's advancing what theory—but it also proved to be the seminal work on the subject. See I. S. Shklovskii, Carl Sagan, *Intelligent Life in the Universe* (Holden-Day, 1966).

62   A 1976 *New Yorker* profile: Henry S. F. Cooper's two-part piece, "A Resonance with Something Alive," June 21 and June 28, 1976, is essential reading for the Saganologist. It later became the first half of a book, *The Search for Life on Mars: The Evolution of an Idea* (Holt, Rinehart and Winston, 1980).

63   Shklovskii and Sagan hooked up in Armenia: Sagan edited a fine compilation of the discussions, *Communication with Extraterrestrial Intelligence* (MIT Press, 1973).

63   buoys around black holes: Ibid., p. 198.

6. SPACE AGE BLUES

65   The Space Age all but died: For a comprehensive historical treatment of the development of the space program, see Logsdon et al., *Exploring the Unknown: Selected Documents in the Evolution of the U.S. Civil Space Program*, vol. 1, Organizing for Exploration, NASA SP-4407 (Government Printing Office, 1995). Another splendid book is Howard McCurdy's *Space and the American Imagination* (Smithsonian Institution, 1997).

65   "For a being with a lifespan of 3000 years . . .": Ben Zuckerman and Michael Hart, eds., *Extraterrestrials: Where Are They?* (Cambridge University Press, 1995), p. 3.

66   "Since very few astronomers believe . . .": Ibid, p. 7.

66   Shklovskii published an article in a Russian journal: A one-page contemporaneous summary of the article is in the SETI file at the NASA History Office.

67   he assumed that the aliens would be like humans: This is Drake's reading of the defection.

67   the trekkies were so numerous: Stewart Brand, ed., *Space Colonies* (Whole Earth Catalog, 1977), p. 73.

69   ". . . the future in space would be just like *Star Trek*": Piers Bizony, *2001: Filming the Future* (Aurum, 1995), p. 156.

69   forced the cancellation of three moon landings: Essay by Sylvia K. Kraemer, in Logsdon, *Exploring the Unknown*, p. 622.

70   "Enlightenment thinkers believe . . .": E. O. Wilson, "Back to the Enlightenment," *Free Inquiry*, Fall 1998, p. 21.

71   "You'll notice immediately . . .": Gerard O'Neill, *The High Frontier: Human Colonies in Space* (William Morrow, 1977), pp. 14–15.

71   "Industrial operations on Earth . . .": Brand, *Space Colonies*, p. 13.

73   "They may, however, prefer to go the other way . . .": *Space Settlements: A Design Study*, from the 1975 Summer Faculty Fellowship Program in Engineering Systems Design, NASA Ames Research Center, chapter 7. This document can be found at http://science.nas.nasa.gov/Services /Education/SpaceSettlement/.

73   "In the future, the Earth . . .": Ibid., chapter 7.

73   "I regard Space Colonies": Quoted in Brand, p. 34.

74   The hero, Billy Pilgrim: Kurt Vonnegut, Jr., *Slaughterhouse-Five*, 25th anniversary ed. (Delacorte Press/Seymour Lawrence, 1994), p. 111.

7. THE MARS EFFECT

75   What few people know is that in 1976: For this chapter the author made use of extensive documentation about Viking at the NASA History Office. The Viking files include a transcript of the press conference in which Sagan discussed the absence of any sign of life. Jerry Soffen, Chuck Klein, and Gilbert Levin gave multiple interviews, and Klein provided an unpublished manuscript describing the mission.

77   Venus had a surface temperature: Steven J. Dick, *The Biological Universe* (Cambridge University Press, 1996), pp. 128–35.

80   "Martian turtles": Carl Sagan, *The Cosmic Connection* (Anchor Press, 1973), p. 45.

80   Sagan authored half a dozen articles: On file in NASA History Office.

80   At one point Sagan went to Soffen: In an interview, Soffen said, "Carl was a dreamer. True romantic. He always was searching for the shortcut. Something no one else had seen."

81   "The rolling nature of the countryside . . .": Eric Burgess, *To the Red Planet*, p. 90. The book is particularly valuable for its photographs of the participants in the mission.

84   "more probable than not" that his experiment detected life: This was Levin's position when he spoke at the Mars conference commemorating the tenth anniversary of the Viking mission. See Gilbert V. Levin, Pa-

tricia A. Straat, "A Reappraisal of Life on Mars," Proceedings of the NASA Mars Conference July 21–23, 1986 (American Astronautical Society, 1986). By 1997, Levin had reached the conclusion that there were no persuasive nonbiological explanations for the Labeled Release experiment results. The results couldn't be duplicated in the laboratory using mere chemistry, he told the author. In his academic papers, he strengthened his language. A paper presented by Levin at a conference in San Diego stated: "It is concluded that the Viking LR experiment detected living microorganisms in the soil of Mars." See Levin, "The Viking Labeled Release Experiment and Life on Mars," Society of Photo-Optical Instrumentation Engineers, 1997.

84   Jim Hartz asked Sagan: From transcript, NASA History Office.

87   "There are no bushes, trees, cacti, giraffes . . .": Transcript of press conference, NASA History Office.

88   ". . . the irresistible allure of the old Mars": Norman Horowitz, *To Utopia and Back* (W. H. Freeman, 1986), p. xii.

88   ". . . We are alone . . .": Ibid., p. 146.

8. COURSE CORRECTION

90   The author was assisted in the preparation of this chapter by an unpublished account of the Voyager Record by artist Jon Lomberg. Carl Sagan, Frank Drake, Ann Druyan, Tim Ferris, Jon Lomberg, and Linda Salzman Sagan, *Murmurs of Earth: The Voyager Interstellar Record* (Random House, 1978), is the definitive story of the record. Sagan's *The Cosmic Connection* (Anchor Press, 1973) not only details the Pioneer 10 plaque but is a lovely snapshot of a scientist with a groovy streak—as Lomberg's manuscript states, "Carl resonated with the vibes of the 60s, unusually strongly for a scientist of the era. . . . Carl was not stuffy about whom he talked to, and included among his friends artists, poets, science fiction writers and even an alumnus of Barnum and Bailey's Clown School."

92   "Mankind's first serious attempt . . .": Sagan, *Cosmic Connection*, p. 17.

92   "schematic representation": Ibid., p. 18.

94   "Watching Voyager flash . . .": Sagan et al., *Murmurs of Earth*, p. 124.

94   "at least a dozen . . .": Ibid., pp. 34–35.

94   ". . . respect for these intelligent co-residents . . .": Ibid., p. 151.

95   Ann Druyan first met Sagan: Druyan, interview with author.

96   Something didn't feel right: Druyan, interview with author.

99   Then the mail started coming: These letters are from Sagan's personal papers, provided by Ann Druyan.

9. THE NAYSAYERS

104   "It is not enough . . .": Reinhold Breuer, *Contact with the Stars: The Search for Extraterrestrial Life* (W. H. Freeman, 1982), pp. 225–26.

106   "I wonder why you take Tipler so seriously . . .": Personal papers of Dyson.

108   wrote a blistering response: Copy of the letter obtained by the author.

109   "A scientist who devotes his life . . .": Quoted by Jared Diamond in a tribute to Sagan, "Kinship with the Stars," *Discover*, May 1997, p. 48.

110   For Sasha he would try, finally, to be a fully engaged father: In a phone call before he died, Sagan told me, "I've learned a lot from just being alive, from living experience. I think I'm more capable now than I was in my early twenties in fathering. In being a parent. Just from experience, you know better what to do, you're less frantic in illnesses." When he was younger, he said, men didn't spend so much time with their kids. "It wasn't part of the culture."

110   his thirteen-year-old son: Nick Sagan, interview with the author.

111   ". . . the vastness is bearable only through love": Carl Sagan, *Contact* (Pocket Books, 1985), p. 430.

10. THE SAVIOR

115   Hardly anyone wanted the job: Bryan Burrough, *Dragonfly: NASA and the Crisis Aboard Mir* (HarperCollins, 1998), pp. 241–43.

116   Goldin earned the reputation of a bully: The quoted criticisms come from two former NASA officials and one still employed by the agency, in interviews with the author. Since I mention Leonard Fisk in this section, I will note that he is not one of these critics.

116   not . . . "warm and fuzzy": Goldin told me, "Hey, look, if I was warm and fuzzy, I wouldn't be me. I don't think I'm nasty, but there are people who don't get their way, and they get upset. I do not tolerate, once a decision has been made, insubordination. That can be abrasive."

117   insisted that the agency revert to form: Among those whom Goldin credits with inspiring him was Sagan. Sagan urged Goldin to start a comprehensive Mars-exploration program rather than rely on the single, massive, expensive Mars Orbiter that later blew up. See William J. Broad, "Even in Death, Carl Sagan's Influence Is Still Cosmic," *The New York Times*, November 30, 1998.

11. THE PILLARS OF CREATION

121   the astronomer James Jeans: Timothy Ferris, *Coming of Age in the Milky Way* (Anchor Books, 1989), p. 166.

121   When James Keeler: Ibid., p. 162.

122   Edwin Hubble started searching: A great source of information is Gale E. Christianson, *Edwin Hubble: Mariner of the Nebulae* (Farrar, Straus & Giroux, 1995).

123   I . . . stumbled across one of these: *Lost in Space*, episode 2, "The Derelict," 1965, on videotape from Twentieth Century Fox Home Entertainment.

126   And so the "Pillars of Creation" image: Eventually I ran into Paul Scowen, one of the scientists responsible for the Pillars of Creation (the other was his colleague Jeff Hester). "The colors are not terribly real. They're supposed to be representative," Scowen said. Blue, for example, represents hydrogen gas, even though in real life such gas is green, he said. I asked why he needed to change the colors. "The picture looked better," he said.

12. THE WORM

132   The idea that meteorites might contain life: Even as these Martian rocks were being found in Antarctica, scientists were thinking out loud about how Mars might be useful for discovering the origin of life on Earth. A 1990 National Academy of Sciences report noted that Mars, unlike Earth, has no tectonic plates, and thus no recycling of its crust. Mars has an inert surface, and preserves its history perfectly. Therefore, said the academy, Mars is a good place to look for fossils of early life that would, in turn, help solve the mystery of life's origin on Earth. ("Not only does present knowledge of the ancient Martian surface indicate that early environments on Mars were similar to those at the same time on Earth, the environmental record of Mars' first billion years is potentially far better preserved than on Earth, where continuing tectonic activity has destroyed almost all evidence. Samples of Mars can fill the gap in the Earth's geological record." [Space Studies Board, National Research Council, *The Search for Life's Origins: Progress and Future Directions in Planetary Biology and Chemical Evolution* (National Academy Press, 1990), p. 75.]) The assumption of the academy report was that a spacecraft would have to fly to Mars and scoop up some old rocks and look for samples of fossil life. No one countenanced the thought that a piece of Mars containing the great discovery would be found lying on the Antarctic ice sheet. This is in keeping with a rule of anticipated discoveries: We assume we will have to plan our breakthroughs. No one plans a surprise.

134   "You found *what?*": This is Huntress's recollection.

135   Clinton wanted to know: Per Panetta, interview with author.

135   Hubbell said he wasn't satisfied with the answers: Interview with author.

The UFO and JFK enigmas, I should note, are oddly similar cultural mysteries. They have a tendency to consume entire lives. Some of the Kennedy-conspiracy theorists graduated to UFO studies—Jim Marrs, for example, whose book *Crossfire* provided material for Oliver Stone's famously paranoid movie *JFK*. Marrs more recently published *Alien Agenda*, an attempt to explain what, precisely, the beings from outer space are trying to do here on Earth. The assassination and the alien invasion have in common an official orthodoxy that, to many people, seems fatally flawed. For both subjects, there is a never-ending bounty of material to be reviewed and questioned. And both cases show that reality does not always become clearer through closer examination.

137   J. William Schopf: Schopf had been a graduate student at Harvard under Elso Barghoorn, a legendary paleobiologist who was instrumental in the discovery of microfossils in two-billion-year-old rocks. So began the march of life into the very earliest history of the planet, a process that has continued in recent years. There is debate about possible biological markers in 3.85-billion-year-old rocks examined in Greenland. Schopf is a star of this ongoing drama, for he has found the oldest microfossils, roughly 3.5 billion years old, in rocks from Australia. Scientists often refer to "Schopf-like" fossils. With Barghoorn long dead, Schopf has become the dean of ancient life on Earth. In that role, he has zealously defended the field against the encroachment of false evidence.

137   "I'm certain we will find life . . .": Interview with author.

138   "I had been brought in": This piece of writing, provided to the author, later became part of a book. See J. William Schopf, *Cradle of Life: The Discovery of Earth's Earliest Fossils* (Princeton University Press, 1999), p. 305.

13. STAR TREKKING

142   look at Mars for the seasonal waves: Gilbert V. Levin, "Re-Examine Evidence of Life," *Space News*, October 14–20, 1996, p. 13.

144   "Mars is a dartboard . . .": Interview with author.

144   "turd-like shapes": Quoted in Charles Petit, "Pieces of the Rock," *Air & Space*, May 1997, pp. 36–41.

147   The space scientists were going to take the measure: Although the Hubble Space Telescope got all the press and made all the pretty pictures, there are other "Great Observatories" coming on line for NASA. They include the Space Infrared Telescope Facility (SIRTF); the Laser Interferometer Space Antenna (LISA); the Next Generation Space Telescope (NGST); the Far Ultraviolet Spectroscopic Explorer (FUSE); and the Stratospheric Observatory for Infrared Astronomy (SOFIA). This last is a jumbo jet—a Boeing 747—that is entirely dedi-

cated to the task of shuttling an infrared telescope into the thin air high above Earth. NASA has also built an X-ray telescope and a gamma-ray telescope. That's just the American operation. The Europeans and Japanese have their own gadgets, including one, the Italian-operated Beppo-Sax orbiting gamma-ray observatory, that managed one day to detect the most powerful explosion ever seen. Beppo-Sax saw a flash of gamma rays emitted from a point in space about twelve billion light-years away. A little math showed that whatever caused the flash might have been a billion times brighter than an ordinary supernova. The theorists went into a dither trying to understand what could cause such energy. Maybe a neutron star spiraled into a black hole, or maybe it was something else. "We have a real bull market in observations and a real bear market in theories," complained theorist John Bahcall.

All the while that these astronomers were discussing the need for a fantastic Planet Finder telescope, there was a planet, or something very like a planet, sitting in the constellation Taurus, awaiting detection by the Hubble Space Telescope. It would not be seen until an obscure astronomer in Pasadena named Susan Terebey found it. She wasn't looking for planets but, rather, studying young stars, and just happened to see, in one of her images taken by the Hubble's near-infrared camera, an astonishing bit of fireworks. There was an object that appeared to have been hurled from a double-star system. Connecting the object with the binary stars was a long streak of light, like a vapor trail behind a jet. She determined that the trail was 130 billion miles long, more than a thousand times the distance of the Earth from the Sun. This strange, streaking object was too faint to be a star. It looked more like a large, luminous young planet. All signs pointed to that conclusion, including the trail, which apparently was a relatively clear area inside the enveloping dust cloud that is typical of young star systems. The planet had punched through the dust and left behind this tube of light.

Susan Terebey may not have realized it explicitly, but she was confirming a fundamental fact about scientific discoveries. They are not necessarily made by the people who are supposed to make them. No one planned for her to be the first person to obtain an optical image of a planet outside our solar system (indeed, many people challenged her research, saying they didn't think she'd proved that it was a planet). No one at NASA headquarters was likely to have ever heard her name, and she had never given a speech at a scientific conference. But you don't have to be Saganesque to make a great discovery.

147   Huntress realized Goldin had misunderstood the meaning: Huntress told the author, "When he saw it, he saw in his mind something different. He saw going to another star."

14. TO INFINITY AND BEYOND

155   "The public's sense of the size of the universe . . .": Kim Stanley Robinson, speech at a symposium sponsored by the Space Policy Institute at George Washington University, November 1996. A record of the proceedings is available from SPI.

159   Thousands of people in Phoenix: A great account of this is by Tony Ortega, "Starship Stupors," *Phoenix New Times*, March 5–11, 1998. March 13, 1997, was a crazy night in Arizona. Comet Hale-Bopp was in the sky, almost at its closest approach. About 8:30 p.m., a V-pattern of lights could be seen traveling in the night sky. Some people said they were connected—a giant boomeranglike craft, perhaps. But a home video by a resident named Terry Proctor clearly showed that the lights were not connected. Judging from eyewitness reports, they were moving about four hundred miles an hour. They were, in short, planes. An amateur astronomer named Mitch Stanley saw them. "They were planes. There's no way I could have mistaken that," he told the *New Times*. In any case, there was a separate event that night that has gotten even more press: A series of bright lights hovered in the sky southwest of Phoenix, then disappeared. The footage of the lights ran, sensationally, on the late evening news in Phoenix. Was it another enormous UFO? No, the Maryland Air National Guard, of all things, was in Arizona for winter training, and had dropped some flares. The ufologists are not convinced by this explanation, needless to say.

15. A DIFFERENT VIBRATION

164   The frog's head explodes: John E. Mack, *Abduction: Human Encounters with Aliens* (Ballantine, 1994), p. 384.

166   Nor did it matter that the Air Force: Curiously enough, there's a nice summary of the CIA's involvement with UFOs on the CIA's own Web site. The author, Gerald K. Haines, states that in the early years after the Kenneth Arnold sighting the CIA was concerned that the Soviets might have secret weapons that were causing the phenomenon. The 1952 sightings in Washington (later revealed to be radar blips) prompted the CIA to form a special study group to review the situation. One official urged that the CIA keep its interest in the subject secret from the media and the public, "in view of their probable alarmist tendencies." Haines concludes: "This concealment of CIA interest contributed greatly to later charges of a CIA conspiracy and coverup."

169   I visited *The X-Files'* set: My profile of Gillian Anderson appeared in *Allure* magazine: "Planet Hollywood," December 1997, pp. 148–52.

16. SAGAN IS APPALLED

172 "confidence, optimism . . .": E. O. Wilson, in *Free Inquiry*, Fall 1998, p. 21.

172 For his sixtieth birthday: The proceedings, including the question-and-answer transcript, are contained in *Carl Sagan's Universe*, ed. Yervant Terzian and Elizabeth Bilson (Cambridge University Press, 1997).

173 What bothered Sagan: In his book *The Dragons of Eden* (Random House, 1977) he writes about pseudoscience: "These are by and large, if I may use the phrase, limbic and right-hemisphere doctrines, dream protocols, natural—the word is certainly perfectly appropriate—and human responses to the complexity of the environment we inhabit. But they are also mystical and occult doctrines, devised in such a way that they are not subject to disproof and characteristically impervious to rational discussion" (p. 238).

174 He believed they may not come from our universe: "This appears to come from some other dimension, and enters into our physical reality," Mack told me in a brief phone interview.

17. THE ANOMALY PROBLEM

176 The Condon Report had coldly and cruelly declared: The first page of the report includes the much-quoted statement: "Our general conclusion is that nothing has come from the study of UFOs in the past 21 years that has added to scientific knowledge. Careful consideration of the record as it is available to us leads us to conclude that further extensive study of UFOs probably cannot be justified in the expectation that science will be advanced thereby." Ufologists argue that the report does not succeed in debunking a number of sightings and anomalies. I was most struck that the report claims that the term "UFO" should be pronounced "OOFO."

18. THE OUTER LIMITS

186 Aliens are angels with a scientific veneer: The best exploration of this topic is Keith Thompson's *Angels and Aliens: UFOs and the Mythic Imagination* (Addison Wesley, 1991). The book is a lucid, fair interpretation of the cultural phenomenon, which no doubt hampered its retail success. See also Michael Shermer, *Why People Believe Weird Things* (W. H. Freeman, 1997).

186 "What we see in the UFO culture . . .": Paul Davies, *Are We Alone?* (Basic Books, 1996), p. 133.

190 "At first I thought it was a dead child": Col. Philip J. Corso (Ret.) with William J. Birnes, *The Day After Roswell* (Pocket, 1997), p. 32.

191 The description by the Hills was inconsistent: Jacques Vallee, *Dimensions* (Ballantine, 1988), pp. 105–6.

193 Jan told me she occupied a human body: She has written about her experiences under the name of Janis Abriel. A sample:

> Gaia: The true cosmic name of the Earth. Gaia is an entity who attained a state of pure knowledge and integrity, and then chose to become a planet on which other entities could work their way to achieving this same state.
>
> God: The human concept of God was encoded in the human genetic DNA by the architects of the mortal, corporeal body of the human species. These architects were not from the Earth, but were from another clan in the Universe.

19. HYPNOTIZED

200 an alien of the praying-mantis variety: This might be a good moment to discuss the concept of "nutty" ideas. "Nutty" is a relative term. What's crazier—a truck driver believing in flying saucers, or a tenured professor believing in them? They might have slightly different takes on the issue, but society didn't invest tens of thousands of dollars in educational resources to help the truck driver become an arbiter of truth and a teacher of young people. The truck driver doesn't put a fancy title behind or in front of his name. I think it's fair to say that having a Ph.D. doesn't make a person inherently more credible on the question of whether aliens are visiting the Earth, but simply makes it more embarrassing when the person believes something that is obviously not so.

20. HEAVEN'S GATE

208 Applewhite and . . . Nettles: My friend and colleague Marc Fisher of *The Washington Post* did much of the biographical reporting on Marshall Applewhite and Bonnie Nettles, and broke the story of Applewhite's decision to be castrated. Another *Washington Post* reporter, Don Baker, did an extensive interview with Dick Joslyn when the story first broke; I have used his original notes as a reference.

210 a classic work of sociology called *When Prophecy Fails:* Leon Festinger, Henry W. Riecken, and Stanley Schachter (University of Minnesota Press, 1956).

212 The Aetherius Society . . . teaches: Many months later, I revisited

Aetherius to witness its Thursday-night "Operation Prayer Power" session. I learned that George King had departed the physical plane. Alan Moseley was firmly in charge of the church. About three dozen well-dressed people of different ages and ethnicities, male and female, streamed into the building at around 7:45 p.m., many looking as if they'd just come from work. A few wore robes, and many who didn't quickly obtained a robe to throw over their street clothes. Most of the robes were red. As the people filed into the hall of worship, Moseley gave me a briefing.

"What you will be seeing tonight are people conveying, for want of a better word, biomagnetic energy into a battery." The battery would store the energy for later use during worldwide crises.

I couldn't tell what the irreducible element was of the Aetherius Society—whether it was prayer, prophecy, belief in aliens, or this quasi-scientific biomagnetism stuff. Moseley told me, "Everything we've done is physical, it's practical, it's not just an idea." It was important to him that I acknowledge that they weren't just a bunch of wackos who had no understanding of the laws of physics. Other religions were like that, he implied. "If you actually look at the tenets of some of the other religions, you think, do people actually believe this?"

212   Unarius Academy of Science: A fuller account of Unarius, and insight on millennial thought, can be found in Alex Heard's *Apocalypse Pretty Soon* (Norton, 1999).

21. ROSWELL CRASHES

217   Woodstock with an alien theme: The *Albuquerque Journal* reported that promoter John Brower was charging $90 a night for permits to camp at the Corn Ranch "crash site," but never got more than five people to camp there. (René Romo, "Crash Site Flops as UFO Woodstock," July 5, 1997, p. A1.)

219   "The metal was as thin as newsprint": Kevin D. Randle and Donald R. Schmitt, *UFO Crash at Roswell* (Avon, 1991), p. 29.

220   Anderson "likes to tell tall tales . . .": Philip Klass, *The Real Roswell Crashed-Saucer Coverup* (Prometheus, 1997), p. 76.

220   The case never pivots on any single, ridiculous individual: It is impossible to reach the end of the shaggy-dog story of Roswell. This is a problem with the UFO myth in general. Even the people who are obsessed with UFOs and spend their lives in pursuit of the elusive alien reality are forced to admit that they are barely scraping at the outer flange of the mystery. They are gathering flakes, fragments, and crustal extrusions. No one person can claim to have absorbed the entire phenomenon. The only sane choice is to have a subspecialty, like cattle mutilations.

In 1996, a ufologist named Kent Jeffrey decided that, contrary to the orthodoxy within the UFO community, no flying saucer crashed at Roswell. His conclusion startled and confounded many of his fellow researchers. Jeffrey, an airline pilot, had been a leader in an initiative called the Roswell Declaration, which called on Congress to grant immunity to any witness who wanted to step forward with evidence about the crash. But Jeffrey's continued pursuit of the case led him to believe that there simply wasn't anything there. He went so far as to bring an important witness to the crash debris to Washington for a lengthy session of hypnosis. The witness failed to reveal anything that sounded as though it was of extraterrestrial origin. Jeffrey's conclusions were published in an article in the journal of the Mutual UFO Network and then reprinted in the *Journal of Scientific Exploration*. The *JSE* also published two critiques of Jeffrey's article. One, by Michael Swords, raised the expertise problem: "I hold the belief that there are only about three, four, or five persons on the planet who could be designated as ufological experts on Roswell (in its many facets), and that neither Jeffrey nor I are among them." The real story may be that prolonged inquiry into a controversial matter does not dissuade people from reaching a demonstrably absurd conclusion. Undoubtedly there's a simple selection effect at work: Only people who are true believers would spend the time researching the issue with enough depth to become "experts." Anyone else would go on to something more worthwhile, like napping.

225   When reporters demanded to see the fragment: The *Albuquerque Journal* quoted Paul Davids, a movie producer who promoted the alleged discovery, defending the decision not to bring the actual extraterrestrial sample to the press conference: "When you have one sample and one sample only, the risks are just too great. This is as rare as the Shroud of Turin." (Leslie Linthicum, "Alleged UFO Fragment Unveiled," July 5, 1997, p. A1.)

228   "What I saw was a bewildering collection . . .": Michael Hesemann and Philip Mantle, *Beyond Roswell: The Alien Autopsy Film, Area 51, & the U.S. Government Coverup of UFOs*, foreword by Jesse Marcel, M.D. (Marlowe & Company, 1997), p. xi.

22. THE HOLE IN THE STORY

230   Aliens are a *personal* matter: See Elaine Showalter's *Hystories: Hysterical Epidemics and Modern Media* (Columbia University Press, 1997). Showalter's courageous book is essential reading for anyone interested in, and appalled by, the spread of pseudoscience disguised as therapy, and its devastating effect on the very people it would purport to help. Another thoughtful examination of alien abduction can be found in C. D. B. Bryan's *Close Encounters of the Fourth Kind: Alien Abduction, UFOs, and the Conference at MIT* (Alfred A. Knopf, 1995).

231   The aliens are AWOL: The notion that "we'll never know" has been advanced by Jacques Vallée, the French ufologist who inspired François Truffaut's character in Spielberg's *Close Encounters*. Vallée is outside the mainstream of ufology, because he doesn't seem to think that UFOs are spacecraft piloted by alien beings. In fact, he doesn't claim to know what they are. Something may be happening, he thinks, that is forever beyond the comprehension of human beings. He wrote in 1975 that the countless sightings of UFOs "stand as a monument to the limits of our understanding" (*The Invisible College* [Dutton], p. 5). What Vallée seemed to want his readers to understand was that these UFOs were legitimate unknowns, not hoaxes or misinterpretations of trivial phenomena. The Vallée approach might be called the exaltation of the unknown. A perfect example can be found in another book by Vallée, co-written with J. Allen Hynek, called *The Edge of Reality*: "The UFO represents an unknown but real phenomenon. Its implications are far-reaching and take us to the very edge of what we consider the known and real physical environment. Perhaps it signals the existence of a domain of nature as yet totally unexplored, in the same sense that a century ago nuclear energy was a domain of nature not only unexplored but totally unimagined and even, in fact, unimaginable in the framework of the science of that day. All this makes the study of UFO reports a tremendously fascinating subject, combining a sense of adventure into the unknown with the exhilaration of skirting the edge of reality and even the fear of what might be revealed beyond the edge of what we consider to be reality." (J. Allen Hynek and Jacques Vallée, *The Edge of Reality: A Progress Report on Unidentified Flying Objects* [Henry Regnery Company, 1975], p. vii.)

234   "Sometimes, once in a very long while": *The Day After Roswell*, p. 273.

235   It is not the evidence of extraterrestrial creatures but, rather, the idea of the Alien: There's a nifty essay by Larry Niven, the science-fiction writer, called "The Alien in Our Minds," in *Aliens: The Anthropology of Science Fiction*, ed. George E. Slusser and Eric S. Rabkin (Southern Illinois University Press, 1987). Niven writes, "We have dealt with alien intelligences for all of the time that humans have had human brains" (p. 4). Niven draws a parallel between human-alien and man-woman, parent-child, tribesman-stranger. He then goes on to discuss the constraints on the emergence of intelligent life, and favors the notion that water worlds are common. "Intelligent whales and octopi may be waiting for us all across the sky" (p. 11).

25. ON TO MARS

255   Mars was the Where: For the history of Mars exploration plans, the author found helpful an article by Frederick I. Ordway III, "Mars Mission Concepts: The Von Braun Era," in Carol R. Stoker, Carter Emmart,

eds., *Strategies for Mars: A Guide to Human Exploration* (American Astronomical Society, 1996). My account of the Mars Society conference ran
in *The Washington Post*, August 23, 1998, p. F1.

257 "When on Mars, do as the Martians do": Robert Zubrin with Richard
Wagner, *The Case for Mars: The Plan to Settle the Red Planet and Why We
Must* (The Free Press, 1996), p. xvi.

27. THE FACE

269 "Isn't it peculiar what tricks . . .": Richard Hoagland, *The Monuments of
Mars*, 1987), p. 5; also alluded to in Carl Sagan, *The Demon-Haunted
World* (Random House, 1996), p. 53.

271 The builders may have evolved on the parent planet: Tom Van Flandern, interview with author.

272 Goldin felt the taxpayers deserved: Interview with author.

273 "It was probably sculpted": Sagan, *Demon-Haunted World*, pp. 55–56.
It's true that the Face on Mars looks like a face. But there's also a crater
with a smiley-face in it, and a volcano that looks like Ted Kennedy.
There's a nebula in space that looks like Abe Vigoda. And for my
money, the anthropomorphic marvel of them all is the "stick figure," an
arrangement of billions of galaxies covering much of the known universe and looking strikingly like a child's primitive depiction of a human
being.

273 Malin never got the chance to question Sagan directly: Malin always
liked Sagan, for a specific reason. When Malin wrote his thesis, about
Mars, he began with a dedication to his mother, Beatrice, "who gave her
life for mine." Everyone who read the thesis simply blew right past the
dedication without remarking upon it. But not Sagan. "Tell me the
story of your mother," he said to Malin. And Malin told Sagan that his
father had died and his mother had raised him and his sister by giving
up her ambitions and working as a secretary. She never remarried. Only
her kids mattered to her. That Sagan would care about this domestic
drama gave him a permanent sanctity in Malin's universe. (Interview
with author.)

28. SILENCE IN GREEN BANK

278 He read a paper by Philip Morrison and Giuseppe Cocconi: "Searching
for Interstellar Communications," *Nature*, September 1959. An account
of the origin of the Morrison-Cocconi idea can be found in Steven J.
Dick's *The Biological Universe* (Cambridge University Press, 1996), pp.
415–16. The Morrison-Cocconi paper included a statement that SETI
researchers still utter: "The probability of success is difficult to estimate; but if we never search, the chance of success is zero."

281   Lederberg felt the Drake Equation didn't deserve its popularity: Interview with author.

282   It is called "optical SETI": Among those now venturing into optical SETI is Paul Horowitz, Sagan's collaborator on Project META. Horowitz has put a great deal of historical information about the idea on his Harvard Web site. Charles Townes, co-inventor of the laser in 1960, suggested within a year that it could be used for interstellar communication. Laser technology remained primitive compared with radio technology, but advances have been so rapid in recent years that the SETI Institute held a series of workshops to examine once again the various strategies for searching for signals. Out of those workshops came the decision to back optical SETI as well as the traditional radio searches.

29. THE MYSTERY CONSTRAINT

289   The Mystery Constraint could be carbon: *Hubble Space Telescope News*, published on Web site of Space Telescope Science Institute, December 10, 1998.

289   gamma-ray bursts: *Journal of the British Interplanetary Society*, vol. 52, p. 19.

30. THE RULES OF LIFE

297   Over the centuries, knowledge slowly percolated: Among the books that a lay person might turn to for a clear discussion of what life is, exactly, are Christian de Duve's *Vital Dust: Life as a Cosmic Imperative* (Basic Books, 1995) and Robert Shapiro's *Origins: A Skeptic's Guide to the Creation of Life on Earth* (Summit, 1986). A fascinating but somewhat more difficult book that discusses life as a natural outcome of the laws of complexity is Stuart Kauffman's *At Home in the Universe: The Search for the Laws of Self-Organization and Complexity* (Oxford University Press, 1995).

300   "Looking at the single known biology on Earth . . .": Gerald Joyce, "The RNA World: Life Before DNA and Protein," in *Extraterrestrials: Where Are They?*, ed. Ben Zuckerman and Michael Hart (Cambridge University Press, 1995), p. 139.

300   "no more resembles a bacterium . . .": Quoted in Steven J. Dick, *The Biological Universe* (Cambridge University Press, 1996), p. 361.

301   Silicon is sometimes suggested: Shklovskii and Sagan's 1966 book *Intelligent Life in the Universe* (p. 229), citing the chemist George Pimentel of Berkeley, states that silicon-based life should be searched for in low-temperature, nonaqueous environments with high ultraviolet radiation.

302   the universe is "predisposed": Michael D. Papagiannis, in *Extraterrestrials*, ed. Zuckerman and Hart, p. 105.

31. THE BRAIN OF AN ALIEN

307   McMenamin's argument: See Bennett Daviss, "Cast Out of Eden," *New Scientist*, May 16, 1998, p. 30.

308   Woodpecking, wrote Diamond: Jared Diamond in *Natural History*, June 1990, reprinted in *Extraterrestrials: Where Are They?*, ed. Ben Zuckerman and Michael Hart (Cambridge University Press, 1995), p. 163.

312   "a crude system of pantomimes": Steven Pinker, *The Language Instinct*, (HarperPerennial, 1995), p. 337.

312   Sagan and Druyan . . . contended: See Pinker, p. 336. He takes the quote from an excerpt of Carl Sagan and Ann Druyan's *Shadows of Forgotten Ancestors* (Random House, 1992) that ran in *Parade* magazine, September 20, 1992.

313   The Assumption of Mediocrity may need to be retired for a while: Sagan, who eventually preferred the phrase "Principle of Mediocrity," would certainly disagree with my point. In *Pale Blue Dot* (Random House, 1994), Sagan wrote: "If I had to guess—especially considering our long sequence of failed chauvinisms—I would guess that the Universe is filled with beings far more intelligent, far more advanced than we are. But of course I might be wrong" (p. 33). This optimistic assessment of life in the universe is responsibly advertised as nothing more than a plausibility argument. Still, we should ask what, exactly, does Sagan mean by a "far more intelligent, far more advanced" species? This is the language of the water cooler paradigm of life beyond Earth. It's not enough that we make the assertion that there is life out there that has evolved to a state of intelligence. No, these beings are always more intelligent, more advanced. What, we should ask, is being measured here? Do these intelligent creatures write better stories than we do? Tell funnier jokes? Have more flamboyant, more garish professional wrestling matches? Or do we believe that intelligence is measured solely on the math portion of the galactic SAT? The traditional view seems to be that intelligence and advancement is a function of technology, particularly of the ability to engage in space travel. The faster your spaceship, the smarter you are, allegedly. Or perhaps the most advanced species are the ones that have mastered biotechnology. But as humans slowly venture down that path, we realize that some "advances" may be unethical or grotesque. Could it be that by "advanced" we refer to the ability of a civilization to live in peace and harmony with other cultures and with its environment? Even that might be taken too far—you can imagine a really boring alien civilization that sits around feeling groovy all day, humming, napping, painting watercolors, and

never once mustering the energy to engage in a barroom brawl. Clearly this is an issue about which people need to spend more time arguing.

32. DEATH FROM ABOVE

317   "Dolphins can't protect the planet": Interview with author.

320   "Kaboom! And you're out": Interview with author.

323   unimaginably gradual processes: Catastrophism had been popular until the early nineteenth century. Then came Sir Charles Lyell, a godlike figure who argued that the world was shaped by forces so gradual no human could perceive their workings. A canyon could be carved by the merest trickle of water. Mountains crept skyward millimeter by millimeter. Lyell was a uniformitarian, believing that the laws of nature were constant in time and space, on Earth and in the heavens, now and forever. He ridiculed tales of catastrophes, deluges, the freezing of the planet, and so on. Such stories were not scientific. Charles Darwin then extended gradualism into the biological realm—species, too, appeared and disappeared through incremental genetic mutations and the imperceptibly subtle pressures of this vague phenomenon called "natural selection." Nature was the perfect subject for tweedy tenured professors using the same faded yellow lecture notes year after year.

34. STRANGE FUTURES

341   Gott's time-travel mechanism: Time travel always raises the causality problem. How could the universe make sense if you could go back in time and, for example, kill your father before you were born? Wouldn't that mean you never had a chance to grow up and build the time machine? The solution, suggested by Kip Thorne of Caltech (and I'm paraphrasing), is that the past can't be changed. There's only one way that things happened, and the time traveler is part of that narrative. Thus you could go back in time, but even if you wanted to change the course of events you wouldn't. You wouldn't because you didn't!

341   In 1993, Gott published a paper: J. Richard Gott III, "Implications of the Copernican Principle for Our Future Prospects," *Nature*, vol. 363 (May 27, 1993), pp. 315–19.

342   Forecasting is a risky business: An example can be found in Alvin Toffler's *Future Shock* (Random House, 1970). Chapter 11 is titled "The Fractured Family" and offers various hysterical predictions about the end of motherhood and the rise of parents-for-hire and so forth. He quotes a psychoanalyst, William Wolf: "The family is dead except for the first year or two of child raising" (p. 211). "Professional parents"

will raise children rather than the "amateurs" who just happen to be the biological parents. "Given affluence and the existence of specially-equipped and licensed professional parents, many of today's biological parents would not only gladly surrender their children to them, but would look upon it as an act of love, rather than rejection" (p. 216).

343 "unraveling the secrets of Nature": Michio Kaku, *Visions: How Science Will Revolutionize the 21st Century* (Anchor Books, 1997), p. 10.

343 "Human intention is entering matter . . .": Interview with author.

343 "Life spans are not inherently finite": Interview with author.

344 "How would you like a child by Abraham Lincoln?": Interview with author.

35. THERE MUST BE MORE

348 But they reserve the right to reject: The political objection to science is explored in Jodi Dean's *Aliens in America* (Cornell University Press, 1998), an academic's rummaging of the UFO movement for deeper social and political truths. Her premise is that a belief in aliens is fundamentally a political act. It challenges the authority of the scientific elite. "Given the political and politicized position of science today, funded by corporations and by the military, itself discriminatory and elitist, this attitude toward scientific authority makes sense. Its impact, moreover, is potentially democratic. It prevents science from functioning as a trump card having the last word in what is ultimately a political debate: how people will live and work together" (p. 8).

350 "a triumphant human progress toward real knowledge": Brian Appleyard, *Understanding the Present: Science and the Soul of Modern Man* (Doubleday, 1992), pp. 75–76.

351 "science bore the seeds of its own destruction": John Broomfield, *Other Ways of Knowing* (Inner Traditions, 1997), p. 229.

351 It has an implicit oath to seek the truth: In the influential book *The Marriage of Sense and Soul* (Random House, 1997), Ken Wilber, a prolific writer with an interest in both Eastern and Western belief systems, argues that science is value-free: "It tells us what *is*, not what *should be* or *ought to be*. An electron isn't good or bad, it just is; the cell's nucleus is not good or bad, it just is; a solar system isn't good or bad, it just is." (Introduction, p. x.) And so people turn to religion for their meaning, their guidelines about what is good and bad.

352 *Newsweek* ran a cover story declaring "Science Finds God": July 20, 1998. The article was inspired by a conference in Berkeley, "Science and the Spiritual Quest," organized by a physicist turned theologian, Robert John Russell. The *Newsweek* piece liberally quotes scientists of faith as well as theologians who wish to be viewed as scientifically

aware. The article states: "Now the very science that 'killed' God is, in the eyes of believers, restoring faith. Physicists have stumbled on signs that the cosmos is custom-made for life and consciousness." The article doesn't mention the "many-universes" idea that some have offered as an alternative to an intelligent designer.

A delicious literary treatment of this topic is John Updike's *Roger's Version* (Alfred A. Knopf, 1986), in which a curmudgeonly professor of religion is confronted by a young computer whiz who thinks he's found, in the constants of nature, the proof of God's existence. Although the novel is clotted with other literary agendas—it's a perverse reconfiguring of *The Scarlet Letter*—Updike describes the debate brilliantly. His professor, Roger Lambert, complains, "Even Aquinas, I think, didn't postulate a God Who could be hauled kicking and screaming out from some laboratory closet, over behind the blackboard" (p. 21). Lambert says later, "Whenever theology touches science, it gets burned. . . . Only by placing God totally on the other side of the humanly understandable can any final safety for Him be secured" (p. 32).

See also Victor J. Stenger, "Has Science Found God?," *Free Inquiry*, Winter 1998–99. The article reports that a National Academy of Science survey showed that only 7 percent of its members believe in a personal Creator, down from 29 percent in 1914.

352   Brandon Carter's "Anthropic Principle": The principle, in "weak" and "strong" versions, can be used by partisans on either side of the "science-finds-God" debate. The weak version is the one that states that only in universes compatible with the rise of intelligent life will there be physicists trying to find out why the universe is designed the way it is. The strong version (or one of the various strong versions) states that the universe has been designed, intentionally, for the emergence of human beings. See John D. Barrow and Frank J. Tipler, *The Anthropic Cosmological Principle* (Oxford University Press, 1986).

36. PERSONAL JOURNEYS

355   David McKay and Everett Gibson haven't given up; nor have their critics: See "Requiem for Life on Mars? Support for Microbes Fades," *Science*, vol. 282 (November 20, 1998). McKay is quoted as conceding that many of the "worms" found by the Houston team are too small to be bacteria, and are merely the jutting edges of mineral crystals, altered by the coating process used when samples are examined with a scanning electron microscope.

357   James Annis, an astrophysicist, recently argued: Press release from *New Scientist*, posted on EurekAlert! Web site, January 20, 1999. Paul Davies is quoted as dissenting, saying that although gamma-ray bursts are po-

tentially lethal they happen too quickly to sterilize an entire planet: "If the drama is all over in seconds, you only zap half a planet. The planet's mass shields the shadowed side."

361 Hoagland entertained the possibility: Posted December 8, 1998, on Hoagland's Web site, enterprisemission.com. In a later posting he said, "It is possible that no habitable world exists at EQ Pegasi and what was picked up was a robotic space probe from another star system."

366 "something like half a percent . . .": Interview with author.

366 When they finally published a paper: "Five Years of Project META: An All-Sky Narrow-Band Radio Search for Extraterrestrial Signals," *Astrophysical Journal*, vol. 415 (September 20, 1993).

367 "extravagant speculation": Carl Sagan, *Pale Blue Dot: A Vision of the Human Future in Space* (Random House, 1994), p. 361.

INDEX